# Southern Rockies Wildlands Network VISION

*A Science-Based Approach to Rewilding the Southern Rockies*

July 2003

*Lead Authors*

Brian Miller • Dave Foreman • Michelle Fink • Doug Shinneman

Jean Smith • Margaret DeMarco • Michael Soulé • Robert Howard

*For additional information contact:*

Southern Rockies Ecosystem Project
Margaret DeMarco, Executive Director
4990 Pearl East Circle, Suite 301
Boulder, CO 80301
303.258.0433

www.RestoreTheRockies.org
Vision@RestoreTheRockies.org

## ABOUT THE CONSERVATION GROUPS:

**The Southern Rockies Ecosystem Project (SREP)** is a non-profit conservation biology organization working to protect and restore large, continuous networks of land in the Southern Rockies ecoregion of Colorado, Wyoming and New Mexico. SREP realizes our vision for a healthy ecoregion by connecting networks of people in order to connect networks of land. By connecting people and landscapes, the rich biological diversity of the Southern Rockies will be maintained and restored.

**The Denver Zoo** provides a wildlife conservancy which offers high-quality experiences in an urban recreational setting; provides environmental education which inspires public awareness of global conservation; and engages in scientific programs which make meaningful contributions to the conservation of animals and their ecosystems.

**The Wildlands Project** protects and restores the natural heritage of North America through the establishment of a connected system of wildlands.

Published in the United States by The Colorado Mountain Club Press
710 10th St. #200, Golden, CO 80401
(303) 996-2743; 1 (800) 633-4417
CMC Press Staff: Gretchen Hanisch and Terry Root
email: cmcpress@cmc.org
website: http://www.cmc.org



Manufactured in Canada

All illustrations by: Evan Cantor, cover illustration - Lost Creek Wilderness
Maps by: Southern Rockies Ecosystem Project
Typography and interior design by Todd Cummings and Margaret DeMarco

Southern Rockies Wildlands Network Vision/ Brian Miller, Dave Foreman, Michelle Fink, Doug Shinneman, Jean Smith, Margaret DeMarco, Michael Soulé, Robert Howard

ISBN 0-9724413-6-0
Library of Congress Control Number: 2003112973

# CONTENTS

*Range after range of mountains*
*Year after year after year*
*I am still in love.*
-Gary Snyder


## ACKNOWLEDGEMENTS

We would like to acknowledge the many people who helped formulate this Vision for the Southern Rockies. Their support is greatly appreciated:

**Science Committee:** Brian Miller, (Denver Zoo and Southern Rockies Ecosystem Project), Michael Soulé, (Southern Rockies Ecosystem Project), Michelle Fink (Southern Rockies Ecosystem Project), Doug Shinneman (University of Wyoming, Southern Rockies Ecosystem Project), and Tim Hogan (University of Colorado Herbarium, Southern Rockies Ecosystem Project).

**Southern Rockies Ecosystem Project:** Jean Smith, Michelle Fink, and Margaret DeMarco.

**Wildlands Project:** Dave Foreman, Reed Noss, Kathy Daly, Jen Clanahan, Todd Cummings, Leanne Klyza Linck, Bob Howard and John Davis.

**Expert Reviewers:** Kevin Crooks (University of Wisconsin-Madison), Barbara L. Dugelby, and Tina Arapkiles (Sierra Club, Southern Rockies Wolf Restoration Project).

**Additional authors contributed greatly to the Vision:** Mark Pearson (San Juan Citizens Alliance), Kathy Daly (Wildlands Project), Jen Clanahan (Wildlands Project), Hannah Gosnell, Mary Wisz (University of California at Berkeley), Kurt Menke (Bird's Eye View), Rich Reading (Denver Zoo), Matt Clark (New Mexico Wilderness Alliance), Dave Parsons, Jeff Kessler, Kyran Kunkle, Paul Paquet, Mike Phillips, and Mike Seidman.

**Supporters of the Southern Rockies Wildlands Network Vision include:** Aveda, Conservation Technology Support Program, Denver Zoological Foundation, Environmental Systems Research Institute, Ettinger Foundation, Foundation for Deep Ecology, Fund for Wild Nature, Harder Foundation, Maki Foundation, Mennen Environmental Foundation, New-Land Foundation, Norcross Wildlife Foundation, Oak Lodge Foundation, Patagonia, Peradam Foundation, Recreational Equipment, Inc., Southern Rockies Conservation Alliance, Town Creek Foundation, Turner Endangered Species Fund, Turner Foundation, Wild Oats, Winslow Foundation, and Working Assets.

**Thank you to the following individuals and organizations who provided technical expertise, attended workshops or provided feedback for the Southern Rockies Wildlands Network Vision:** John Amoroso (National Wildlife Federation), Jorge Andromidas (Grizzly Project), Greg Aplet (The Wilderness Society), O'Connor Bailey (Southern Rockies Ecosystem Project), Tom Beck, Nannette Bragin (Denver Zoo), Steve Cain (U. S. National Park Service), Carlos Carroll (Klamath Center for Conservation Research), Christine Canally (San Luis Valley Ecosystem Council), Amanda Clements (Bureau of Land Management), Richard Compton (White River Conservation Project), Lance Craighead (Craighead Environmental Research Institute), Lisa Dale (Citizens for Arapaho-Roosevelt), Karin Decker (Colorado Natural Heritage Program), Melissa Decker (Land and Water Fund for the Rockies), Rob Edward (Sinapu), Denis Hall (Gunnison Basin Biodiversity Project), Allison Jones (Wild Utah Project), Suzanne Jones (The Wilderness Society), Jeff Kessler (Biodiversity Conservation Alliance), Kurt Kunkle (Colorado Environmental Coalition), Bill Martin (Colorado State Land Board), Marianne Martin (University of Colorado Environmental Center), Dave Mattson (U.S. Geological Survey), Roz McClellan (Rocky Mountain Recreation Initiative), Kurt Menke (Bird's Eye View), Craig Miller (Defenders of Wildlife), Erik Molvar (Biodiversity Conservation Alliance), Betsy Neely (The Nature Conservancy), Chris Pague (The Nature Conservancy), Dave Parsons (U. S. National Park Service), Mike Phillips (Turner Endangered Species Fund), Jonathan Proctor (Predator Conservation Alliance c/o Denver Zoo), Jeremy Puckett (Western Slope Environmental Resource Council), Richard P. Reading (Denver Zoological Foundation), Polly Reetz (Audubon Society of Greater Denver), Kimberly Riggs (Sinapu), Erin Robertson (Center for Native Ecosystems), Michael Robinson (Center for Biodiversity), Nancy Ryke (Pike and San Isabel Forest), Sandy Shea (High Country Citizens Alliance), Sloan Shoemaker (Aspen Wilderness Workshop), Doug Smith (U.S. National Park Service), Jacob Smith (Center for Native Ecosystems), Rocky Smith (Colorado Wild, Inc.), Steve Smith (Sierra Club, Southwest Region), Vera Smith (Colorado Mountain Club), John Stansfield (Pikes Peak Group, Sierra Club), David Stern (Denver Zoo), Daniel B. Tinker (University of Wyoming), Steve Torbit (National Wildlife Federation), and George Wuerthner.

Artwork provided graciously by Evan Cantor, and poetry generously contributed by Gary Snyder.

# INTRODUCTION

Brian Miller, Margaret DeMarco

*THIS VISION IS BOLD, YET PRACTICAL. AMBITIOUS, YET ACHIEVABLE -- SCIENTIFICALLY CREDIBLE, AND HOPEFUL. TOGETHER, THESE CHARACTERISTICS CREATE AN OUTSTANDING OPPORTUNITY TO PRODUCE REAL CHANGE IN THE SOUTHERN ROCKIES ECOREGION.*

*WE ENVISION A SOUTHERN ROCKIES ECOREGION THAT IS WHOLE -- A VAST, CONNECTED LANDSCAPE WHERE NATIVE SPECIES THRIVE AND NATURAL ECOLOGICAL PROCESSES MAINTAIN A HEALTHY BALANCE. SUCH A HOLISTIC VISION TRANSCENDS POLITICAL AND HUMAN-MADE BOUNDARIES AND ADDRESSES LARGE LANDSCAPES ON THE ECOSYSTEM AND ECOREGIONAL LEVEL.*

World-renowned for their striking beauty and high mountain topography, the Southern Rockies are one of North America's gems. The Southern Rockies ecoregion contains a diversity of life. From alpine tundra to ponderosa pine forests and sagebrush grasslands, over 500 vertebrate species find their home in the Southern Rockies and a rich variety of plants and invertebrate species can also be found within its borders. Over 270 species of butterflies and 5,200 species of moths make the Southern Rockies the second leading hotspot in North America for the insect order Lepidoptera (Shinneman et al. 2000). The Southern Rockies maintains this abundance partially because of its continuous stretches of wild, remote and undeveloped lands.

And yet, this biodiversity and abundance is threatened, as are many wild places in North America, due to human expansion and development: native species have been extirpated, old-growth forests logged, wild and powerful rivers dammed and polluted, and land degraded and developed. *The State of the Southern Rockies Ecoregion*, a recent report by the Southern Rockies Ecosystem Project (SREP) (Shinneman et al. 2000), provides a detailed ecological assessment of the Southern Rockies. Chapters 2 and 3 of this Vision condense a background of natural history, conservation status, and human demographics from *The State of the Southern Rockies Ecoregion* and a complementary report on reintroducing wolves to the Southern Rockies (Phillips et al. in press). For more discussion on wounds and threats to the Southern Rockies, please see Chapter 5 of this volume.

In order to address the threats to the Southern Rockies, SREP, in conjunction with the Wildlands Project and the Denver Zoo, produced the Southern Rockies Wildlands Network Vision. This Vision calls for ecological restoration that is based on healing ecological wounds: the Vision identifies wounds to the land and then considers the anthropogenic causes for each, addressing not only the symptoms and the disease, but also the root cause(s) of the illness.

A comprehensive approach to healing the wounds of the Southern Rockies requires a full analysis of the current conditions in the ecoregion, as well as an assessment of our goals and approach. Chapter 1 of this document outlines this vision for a Wildlands Network Design and the elements that create it. It is a prescription for the future. We recognize that national parks, wilderness areas, and wildlife refuges have accomplished a great deal for nature. But over time, protected areas have been surrounded by roads and degraded landscapes. Now the protected areas are too isolated to sustain viable populations of large animals, let alone many ecological and evolutionary processes. To overcome this we must address very large landscapes (continental), and heal areas that have been wounded (Soulé and Terborgh 1999). This logic has led to the Southern Rockies Wildlands Network Vision you see before you. This Vision is both a prescription for the ecoregion itself, and an important piece of a larger picture creating contiguous wildlands, known as MegaLinkages, across North America.

Under the guidance of the Wildlands Project, other regional conservation plans are striving toward a north to south MegaLinkage through the Rocky Mountain chain (Figure 1.1). The southern part of the Southern Rockies Wildlands Network overlaps the northern part of the New Mexico Highlands Wildlands Network. The New Mexico Highlands Wildlands Network, in turn, overlaps the Sky Islands Wildlands Network. In the northwestern section of the Southern Rockies Wildlands Network, we overlap slightly with the Heart of the West Wildlands Network. In turn, the Heart of the West overlaps with the Yellowstone to Yukon link. As a visual model, MegaLinkages would allow grizzly bears and wolves to move safely between Mexico and the Arctic.

The Southern Rockies Wildlands Network is a vital piece of the puzzle that connects these reserve designs together. This larger landscape vision is the future of conservation planning and will ultimately protect and restore the Southern Rockies ecoregion within a living, dynamic wildlands network throughout North America.

This Vision is only the first step toward a working, living and changing plan for the Southern Rockies. In short, what you see before you is a hypothesis to test. As knowledge accumulates, methods improve or change, and conservation opportunities arise and fall, successive iterations will modify the conservation plan, and therefore, the Network Vision, of the Southern Rockies. If this Vision stimulates thought and activity toward better methods for conservation, then it has achieved an important outcome.

Throughout this document we refer to the Network Design and the Network Vision for the Southern Rockies. They are different terms. A Wildlands Network Design is a pro-active, landscape-based conservation *map* that designates areas as core protected, wildlife movement and riparian linkages, or compatible-use areas. The overall Southern Rockies Wildlands Network Vision is an integrated and realistic *approach* to maintaining and restoring viable populations of native species within a healthy ecoregion. We will refer to our conservation map as the Network Design and this document in its entirety, with emphasis on implementation measures to be taken in the Southern Rockies, as our Vision.

The mission of our Vision is to protect and rewild the regional landscape. "Rewilding" recognizes the importance of top-down regulation to healthy ecosystems. It emphasizes large core wild areas, functional connectivity across the landscape, and the vital role of keystone species and processes, especially large carnivores (Noss and Soulé 1998, Soulé and Terborgh 1999). This does not mean that we wish to ignore isolated populations of less-charismatic species facing extinction. The needs of such species are covered by The Nature Conservancy (TNC) in an excellent plan titled *Southern Rocky Mountains: An Ecoregional Assessment and Conservation Blueprint* (Neely et al. 2001). We endorse that plan, and we view our plan (emphasizing large carnivores) as complementary to TNC's. We look forward to further cooperation integrating both plans on behalf of Nature.

Because of the landscape-scale approach to our Vision, we cannot delve into the same level of detail as a local plan. We hope, however, that this regional plan for the Southern Rockies can complement and aid local groups' efforts. For example, the Upper Arkansas and South Platte Project applied the regional data produced from SITES, the reserve design optimization computer program used in the Network Design, to a local plan—The Upper Arkansas and South Platte Inventory. The SITES output matched very well with the local inventory that was based on field work and expert opinion. The SITES analysis covered the federally protected Wilderness Areas, the areas proposed for wilderness protection, and potential connections for animal movement. In addition, it also identified several areas of biological importance that were not initially included in the Inventory. Thus, just as local groups can inform the regional plan, so too can the regional plan add to the local effort. The results of these models are found in Chapter 8 and the methods used are described in Chapter 7.

Chapter 4 offers a general philosophical and historical background for conservation strategies and reserve design. In this chapter, we lay the groundwork for our scientific approach. In Chapter 9 our conservation Vision for the Southern Rockies is revealed, with specific areas of importance outlined in a Network Unit Analysis. Chapter 9 is the centerpiece of the document, and it unites the scientific models with local expert opinion to produce a plan of action. And, finally, Chapter 10 introduces ideas for on-the-ground implementation of this Southern Rockies Wildlands Network Vision – the next steps of action and movement on behalf of this plan.

In addition to the main chapters are appendices on our chosen focal species: American marten, beaver, bighorn sheep, black bear, cutthroat trout, grizzly bear, lynx, pronghorn, and wolf. These chapters discuss the species' demographics and their importance in the Southern Rockies. In many cases, these species act as excellent habitat quality indicators and wilderness quality indicators. Management recommendations for each species are then presented.

Current reintroduction efforts of the Canada lynx into Colorado are an important step toward rewilding. The lynx is thus a focal species in the flagship species category (see Chapter 4, Appendix 1, Miller et al. 1998). This category recognizes the value of lynx for public education and conservation campaigns. For this iteration, we did not use lynx movements to highlight areas of ecological importance because of the tendency for reintroduced animals to explore widely (one lynx released in Colorado wandered to Nebraska). As the reintroduced lynx settle into an ecological routine, future drafts can incorporate their ecological needs.

We propose that the next important step for the Southern Rockies is reintroduction of the gray wolf. We outline some issues of wolf recovery in the focal species account found in the appendix. For more detail on the potential for gray wolves in the Southern Rockies, see a report titled *Feasibility of Reintroducing the Gray Wolf to the Southern Rockies* by Phillips et al. (in press).

Our Vision is being edited at the same time as this report. Because of the common interests of both documents, parts of the report are included within our Vision. You will find this noted at the beginning of each chapter if that is the case. Duplication is done with author's permission.

As a final introductory word, in this document we use the metric method, a global standard of measurement, instead of the American method. We offer a quick summary table, with conversions rounded off, for reference:

- 1 hectare (ha) = 2.5 acres—in other words, 0.4 ha is the same as an acre. For example, when the Wilderness Act refers to areas of land exceeding 1,000 acres, the metric equivalent is 400 ha.

- 1 meter (m) = 39 inches—thus a 13,000 ft. mountain is 4,000 m.

- 1 kilometer (km) = 0.6 miles—in other words, 1.6 km is the same distance as a mile.

- A change of 1 degree Celsius ($C^{O}$) = a change of 1.9 degrees Fahrenheit—the freezing point of 32 degrees Fahrenheit equals 0 $C^{O}$.

# SECTION I:

# BACKGROUND FOR THE SOUTHERN ROCKIES WILDLANDS NETWORK VISION

*In my native place*
*there's this plant:*
*as plain as grass*
*but blooms like heaven.*

-Issa

# 1 A VISION FOR THE WILDLANDS NETWORK DESIGN

Dave Foreman

*(This chapter was originally written for the New Mexico Highlands Wildlands Network Vision and has been modified, with the author's permission, for the Southern Rockies Wildlands Network Vision.)*

## 1. Introduction

Earth is now clearly in a mass extinction event—the 6th Great Extinction in the last 500 million years (Diamond 1992, Leakey and Lewin 1995, Pimm 2001, Wilson 2002). Although this mass extinction began 40,000 years ago when behaviorally modern humans spread out from Africa (Martin and Klein 1984, Diamond 1992, Ward 1997, Klein and Edgar 2002), it has reached monumental proportions at the beginning of the 21st century. Unlike previous mass extinctions, which were caused by physical forces (asteroid strikes, geological events), this 6th Extinction is caused solely by the impact of *Homo sapiens* (Soulé 1983, Mayr 2001). It is widely recognized that direct killing by humans, habitat destruction and fragmentation, invasion and competition by alien species, disease, and pollution are the general causes of current extinctions (Wilcove et al. 1998). Stemming this alarming tide of extinction will require conservation vision and action at local, regional, and continental scales.

Both the traditional conservation and modern conservation biology movements have long recognized that protected areas are the best way to safeguard species and habitat. In 1980, Soulé and Wilcox wrote that protected areas were "the most valuable weapon in our conservation arsenal" (Soulé and Wilcox 1980: 4). Protected areas (national parks, wilderness areas, wildlife refuges, etc.) have been central to the strategy of conservationists in North America and throughout the world (Hendee et al. 1990, Foreman and Wolke 1992, Foreman 1999, Nash 2001). In the early 20th century, however, ecologists recognized that traditional protected areas were proving inadequate; species and ecosystems were still being lost (Shelford 1926, Shelford 1933, Leopold 1937). Although the goals of protected areas have included the preservation of an enduring resource of wilderness (The Wilderness Act 1964, Scott 2001) and of self-regulating ecosystems (Soulé pers. comm.), protected areas and protected area systems have fallen short because of:

- Poor ecosystem representation in protected areas and degradation of ecosystems both within and outside protected areas;
- Isolation of protected areas and fragmentation of habitat between protected areas;
- Extirpation or extinction of native species, especially keystone and foundation species[1];
- Loss or degradation of ecological processes, especially fire, natural hydrology, and predation;
- Invasion by disruptive exotic species;
- Pollution and consequent ecological problems, including global climate change and atmospheric ozone depletion.

However, it is important to understand that National Parks, Wilderness Areas, and Wildlife Refuges have accomplished much good conservation. Without existing protected areas in North America, the state of Nature would be far grimmer. The problem is twofold: there have not been enough protected areas, and the areas that have been protected generally were not selected using biological or eco-

[1] **Keystone species** affect the structure, function or composition of an ecosystem significantly through their activities, with the effect disproportionate to their numerical abundance (Power et al. 1996). **Foundation species**, like keystone species, enrich ecosystem function in a unique and significant manner, but occur at much higher densities.

logical criteria. Hence, many kinds of ecosystems—especially the most productive—are not well represented. To heal these wounds, conservation must now 1) address very large landscapes, ultimately continental in scope, and 2) attempt ecological restoration based on rewilding (Soulé and Terborgh 1999). Instead of mere island-like protected areas, a continental wildlands network (of core wild areas, wildlife linkages, and compatible-use lands) is needed to meet the habitat requirements of focal species and to support natural disturbance regimes. Moreover, this network must be based on the scientific approach of rewilding (Soulé and Noss 1998), which recognizes the fundamental role of top-down regulation of ecosystems by large carnivores (Terborgh et al. 2001, Miller et al. 2001), and large carnivores' need for secure core habitats, largely roadless, and for habitat connectivity between core wild areas (Soulé and Noss 1998, Soulé and Terborgh 1999). Fully protected cores such as wilderness areas are central to this approach.

While such a continental vision is bold and ambitious, it follows in the footsteps of early conservation visionaries. In the 1920s and 1930s, eminent ecologist Victor Shelford and the Ecological Society of America called for a careful inventory and planning for a United States system of natural areas protecting all ecosystem types (Shelford 1926, Shelford 1933). Wilderness Society co-founder Benton MacKaye based his vision for the Appalachian Trail on regional planning (Sutter 2002). In drafting the Wilderness Act, Howard Zahniser planned for a national system of Wilderness Areas (Scott 2001). The world-class system of national parks, wildlife refuges, wild and scenic rivers, and wilderness areas in Alaska was the result of decades of careful planning by government and citizens to protect entire ecosystems and represent all habitats in Alaska (Nash 2001). More recently, conservation groups have undertaken huge, detailed, statewide inventories of potential Wilderness Areas in Western states (Fish 1987, Arizona Wilderness Coalition 1987, Utah Wilderness Coalition 1990, Price et al. 1998, California Wild Heritage Campaign 2002, Colorado Wilderness Network 2002, Southern Rockies Conservation Alliance ongoing, and Biodiversity Conservation Alliance in Wyoming ongoing).

During the 20th century, most conservation work was defensive. Citizen conservationists fought to protect wildlands and wildlife against dams, logging, mining, road building, development, and bad grazing practices. The 1964 Wilderness Act clearly states that its purpose is to protect natural areas from threats of development:

> *In order to assure that an increasing population, accompanied by expanding settlement and growing mechanization, does not occupy and modify all areas within the United States and its possessions, leaving no lands designated for preservation and protection in their natural condition, it is hereby declared to be the policy of the Congress to secure for the American people of present and future generations the benefits of an enduring resource of wilderness.*[2]

Samuel Hays (1996), one of today's great conservation historians and a citizen conservationist who campaigned for the 1975 Eastern Wilderness Areas Act, writes "wilderness proposals are usually thought of not in terms of perpetrating some 'original' or 'pristine' condition but as efforts to 'save' wilderness areas from development."

Without the dedicated effort of citizen conservationists to stop the exploitation of wild places, dams would flood the Grand Canyon, Gila Wilderness Area, Dinosaur National Monument, and other "protected" areas; oil and gas wells and pipelines would crisscross the Arctic National Wildlife Refuge and Bob Marshall Wilderness Area; hardly any ancient forest would exist in the National Forests and National Parks; DDT would likely have brought the bald eagle and peregrine falcon to extinction; paved scenic highways would wind along ridge tops in most National Parks; and there would be no Wilderness Act and National Wilderness Preservation System. Without citizen suits to protect endangered species, dozens of species of native fishes, birds, amphibians, mammals, and invertebrates would be no more. Without books, articles, scientific papers, and public educational materials, hardly anyone would be aware of the threats to Nature. Without the kind of detailed citizen conservationist work that developed statewide wilderness area proposals beginning in the 1960s, the current 42,400,000 ha (~100,000,000 acres) National Wilderness Preservation System would be far smaller and less ecologically representative.

## 2. Vision

Some criticize conservationists for defensive actions, for fighting "brush fires," and for "doom-and-gloom" prophesizing. But without them, Earth would fast become unlivable for many species—including humans. Nonetheless, it is not enough for conservationists to only react to urgent threats against Nature. Too much talk of the looming dark of mass extinction, global climate change, and a thoroughly domesticated Earth leaves people depressed, distraught, and

[2] The Wilderness Act. 1964. Public Law 88-577 (16 U.S. C. 1131-1136) 88th Congress, Second Session, September 3, 1964 in Watson, Jay, ed. 1998. The Wilderness Act Handbook third edition (revised). The Wilderness Society, Washington, D.C.

without hope—and thus without the energy to take action.

Throughout the 20th century, citizens, scientists, and land managers have also worked positively for a future where humans and wild Nature can coexist. In addition to the reaction against unbridled development, a positive vision for the future is inherent in the statement of purpose for the Wilderness Act. While conservation has a long heritage of envisioning a positive future, this aspect of conservation did not come fully into its own until the spring of 1991 when Michael Soulé suggested a meeting of a few leaders to develop a *hopeful* hundred-year vision of what North America should be. They formed the Wildlands Project to continue visionary conservation planning, but with a view encompassing all of North America and grounded in recent ecological research and theory (Wildlands Project/Wild Earth 1992, Wildlands Project 2001). The Wildlands Project and its cooperators, including the Southern Rockies Ecosystem Project, believe that producing science-based wildlands network designs and clear conservation visions not only gives hope, but also leads to more effective on-the-ground efforts by conservation organizations, individuals, and agencies.

Thus, the strategy of the Wildlands Project is to design a visionary continental network of regional wildlands networks, identify the steps necessary for conservation of species and processes within those networks, communicate the designs and visions, and work to catalyze and coordinate their creation. Such continental conservation must be transboundary, with cooperation between nations in design and management of wildlands networks. Many groups throughout the world have been inspired by the Wildlands Project's vision and are developing similar wildlands networks (Johns 2001).

Of highest priority to the Wildlands Project is reconnecting, restoring, and rewilding four MegaLinkages that will tie North American ecosystems together for wide-ranging species and ecological processes. These MegaLinkages are 1) Pacific, from Baja California to Alaska; 2) Spine of the Continent, from Central America to Alaska through the Rocky Mountains and other ranges; 3) Appalachian, from Florida to New Brunswick; and 4) Boreal, from Alaska to the Canadian Maritimes (Figure 1.1). The Southern Rockies Wildlands Network Vision is a key part of this continental conservation in the Spine of the Continent MegaLinkage.

Designing this North American MegaLinkage Vision and helping conservationists across the continent to make it a reality will 1) strengthen conservation groups and others in the urgent and local defense of places and species; 2) strengthen longer-term wilderness and endangered species campaigns, and ecological restoration; 3) heal fragmentation in both the ecological landscape and in the conservation movement; and 4) inspire conservationists and the public with hope for the future. The Southern Rockies Ecosystem Project is pleased to be part of this effort.

## 3. Elements of a Wildlands Network Vision

A Wildlands Network Vision is distinguished from all other conservation area design approaches, whether traditional or science-based, by the combination of several specific elements. These elements together make wildlands network visions bold, hopeful, scientifically credible, and practical. The integration of all of these elements in the Wildlands Network Vision offers a comprehensive and effective approach to large-scale conservation planning and implementation. By sharing these elements, wildlands networks across the continent are consistent and connect with one another.

Each of these elements is discussed elsewhere in this document; here they are briefly identified:

- *Rewilding*
  Wildlands networks are explicitly based on the scientific approach of rewilding, which emphasizes large core wild areas, functional connectivity across the landscape, and the vital role of keystone species and processes, especially large carnivores.
- *Healing the Wounds Goal-Setting*
  Ecological restoration is now recognized as essential in conservation. Wildlands networks approach restoration through "healing-the-wounds" goal setting.
- *Expert Design*
  The initial Wildlands Network Design is mapped based on expert opinion.
- *Three-track Approach*
  Wildlands networks are designed by a three-track approach of ecosystem representation, special elements, and focal species planning. A wildlands network strives to represent all ecosystems, and to identify and protect rare species occurrences and other sites of high biodiversity values in core wild areas. Wildlands networks are also based on the habitat needs of focal species—organisms used in planning and managing protected areas because their requirements for survival represent factors important to maintaining ecologically healthy conditions.
- *Focal Species Modeling*
  The initial Wildlands Network Design is tested and revised by computer modeling techniques that variously include SITES, PATCH, and Least Cost Path Analysis.

- ***Fieldwork***
  On the ground fieldwork is necessary for inventorying road systems, potential wilderness boundaries, ecological condition, focal species presence, barriers to wildlife movement, and special elements to improve the accuracy of the wildlands network design.
- ***Wilderness Areas as Cores***
  Wildlands networks are anchored by a core system of Wilderness Areas on public lands.
- ***Conservation on Private Lands***
  Private lands, voluntarily managed for conservation by landowners, play a key role in wildlands networks.
- ***Compatible-Use Lands***
  In addition to core wild areas and wildlife linkages, public and private lands managed for compatible resource and recreational use are a key part of wildlands networks and provide habitat and dispersal connectivity for a wide variety of species.
- ***Linkages to other Wildlands Networks***
  Connectivity within a wildlands network is a fundamental part of Wildlands Network Design, but connectivity to adjacent wildlands networks is also important for wide-ranging species and ecological processes.
- ***Specific Units***
  Wildlands networks are built from many individual units of land, including federal, state, county, tribal, and private, that are specifically proposed or recognized as cores, linkages, and compatible-use lands.
- ***Unit Classification and Management Guidelines***
  Wildlands networks have consistent, detailed management recommendations and guidelines for the different land unit classifications proposed.
- ***Focal Species Management Recommendations***
  Management recommendations for focal species are a key part of Wildlands Network Visions.
- ***Conservation Action***
  A Wildlands Network Vision is an abstract exercise unless an implementation plan is conceptualized simultaneously. Wildlands Network Design and planning for implementation must proceed on parallel tracks at the same time and with constant feedback.
- ***Cataloging Compatible Conservation Initiatives***
  In any region where a Wildlands Network Design effort is underway, there are many other complementary conservation efforts going on as well, which are incorporated in the Wildlands Network Vision.
- ***Economic Incentives***
  Wildlands Network Visions propose economic incentives that promote human interaction with the land that conserves, rather than destroys, wild Nature.
- ***Expert Review***
  Critical, ongoing review of Wildlands Network Visions by scientists and conservation groups is an important way of ensuring accuracy and effectiveness.
- ***Continental Vision***
  The Southern Rockies Wildlands Network Vision is part of a continental vision for a North American wildlands network based on four megalinkages proposed by the Wildlands Project.

The Wildlands Project's vision for rewilding North America does not replace or diminish the need for vigilant defense against schemes to domesticate the whole landscape, but it does add a positive blueprint for all conservation work, a context for wildlands and wildlife defense. This vision is distinctive because it is *bold, hopeful, scientifically credible, and practically achievable.*

The combination of these four characteristics is unprecedented in conservation. Of particular note are the combination of boldness and hopefulness with scientific credibility and a practical blueprint of how to achieve the vision. Human beings—conservationists are no exception—need hope to carry on. They also need facts and rational arguments. The Southern Rockies Wildlands Network Vision is one of hopefulness and "do-ability." By bringing together networks of people to work toward a positive future where networks of wildlands fit in with a civilized human community, we may at last achieve harmony between humans and Nature.

*We paddle forward, backstroke, turn,*
*Spinning through the eddies and waves*
*Stairsteps of churning whitewater.*
*above the roar*
*hear the song of a Canyon Wren.*

*A smooth stretch, drifting and resting.*
*Hear it again, delicate downward song*

*ti ti*          *ti ti tee tee tee*

-Gary Snyder

# 2 NATURAL LANDSCAPES OF THE SOUTHERN ROCKIES ECOREGION

Doug Shinneman, Brian Miller

*(For more information on the condition of the Southern Rockies ecoregion, please see "The State of the Southern Rockies Ecoregion", a report by the Southern Rockies Ecosystem Project).*

## 1. Introduction

The Southern Rockies natural landscape is not only ruggedly beautiful; it is also diverse and complex, with a rich pattern of landforms that support an equally rich array of natural communities and species. While the Southern Rockies have lost several native species as a result of human settlement, persecution, and overuse of natural resources, and many species and ecosystems are at risk, a significant portion of the region's natural landscape remains relatively intact. These remaining natural areas are important to regional and global conservation goals, as they are capable of supporting biological elements both unique to and representative of the Southern Rockies. Moreover, they offer increasingly rare opportunities to rewild Nature and native species. Although humans have altered the natural landscape, conservation opportunities still abound in the Southern Rockies.

To assess the potential of a Wildlands Network Vision for the Southern Rockies, we must present the appropriate ecological context. We thus describe the present natural environment and discuss the condition of the region's many natural communities and species. The scientific names of plant species given follow the standards developed by Kartesz (1994), as reported and updated in the PLANTS database (USDA NRCS 2002).

## 2.Geological and Landform Background

The specific focus of this Wildlands Network Vision is the Southern Rockies ecoregion (Figure 2.1). The Southern Rockies are the highest ecoregion on the North American continent, with 20% of land area resting above the elevation of 3,000 meters (roughly 10,000 feet, Shinneman et al. 2000). As stated in *The State of the Southern Rockies Ecoregion* (Shinneman et al. 2000: 2),

> *An ecoregion is a large landscape area that has relatively consistent patterns of topography, geology, soils, vegetation, natural processes and climate. These natural components, patterns, and processes also help to create and influence the myriad of smaller ecosystems within an ecoregion.*

The unique characteristics of an ecoregion, and its component ecosystems, are forged over the eons. Powerful geologic forces carved the landscape we see in and around the Southern Rockies.

The land currently occupied by the Southern Rockies was covered by the sea for billions of years, and the Ancestral Rocky Mountains arose roughly 300,000,000 years before present (ybp) in a tropical climate near the Earth's equator, when the land was part of the supercontinent Pangaea (Knight 1994). Erosion leveled these ancient mountains about 260,000,000 years ago, and the remaining sediment left the red sandstones that can still be seen (Elias 2002). Some examples include the rock that has tilted to nearly vertical position along the Front Range of Colorado (Elias 2002).

About 70,000,000 ybp, colliding tectonic plates raised

the Rocky Mountain chain in the "mountain building" known as the Laramide Orogeny (Elias 2002). Fault lines developed from the collision, and the earth began to rise on one side of a given fault and drop on the other. For example, the top of Longs Peak is 6.5 km above similarly aged granite that is buried below the plains on the other side of the fault line (Wuerthner 2001). The uplift also fractured the crust from the San Juan Mountains near what is now Golden, Colorado in the Front Range, and that allowed mineralized solutions to rise in the vents. When these solutions solidified, they created the veins of gold, silver, copper, lead, zinc, and other ores that fed the mining economy of Colorado over the last century and a half (Elias 2000, Wuerthner 2001). The Laramide Orogeny took 30,000,000 years to complete. The resulting mountains generally run in a north and south direction (Benedict 1991, Knight 1994, Elias 2002).

After the Laramide Orogeny, volcanism, erosion, deposition and uplift continued to modify the Southern Rockies' mountainous landscape. The San Juan Mountains of southern Colorado, the Never Summer Range west of Rocky Mountain National Park, and the Jemez-Naciemento Mountains of New Mexico are volcanic. Ancient volcanic activity in these areas has left calderas, ancient lava flows, volcanic dikes, and extinct, eroded volcanic domes (Ellingson 1996). The Mosquito Range and the Sangre de Cristo Range are mainly upturned sedimentary rock (Elias 2002). Open areas surrounded by mountains (Laramie Basin, North Park, Middle Park, South Park, and the San Luis Valley) were formed as blocks in the earth's crust moved down via folding and faulting (Elias 2002).

Over the last 2,000,000 years the main forces were localized volcanism and extensive glaciation, with as many as 17 major glacial episodes during the Pleistocene epoch (Benedict 1991, Blair 1996, Flannery 2001). Glaciation in the Southern Rockies, however, was probably less spectacular than in the Central and Northern Rockies (Elias 2002). Still, the ranges of the Southern Rockies show classic high-mountain topographical features, such as alpine cirques and tarns, glacial moraines, broad U-shaped valleys, and glacial-outwash plains at lower elevations. Today glaciers are small in extent and limited to high elevation cirques, but periglacial activity, water flow, and wind continue to erode and shape the region's mountainous landscape (Benedict 1991, Blair 1996). It is within this latter time frame that humans entered North America and the present extinction event began.

On today's east slope, the Southern Rockies descend into a complex assortment of mesas, foothills, hogbacks, parallel ridges, and rocky outcroppings. The topography then unfolds into the short-grass prairie, a drought-driven system in the rain-shadow of the mountains (Flores 1996). On the west slope, the mountains subside into rugged canyons and mesas, including the massive White River and Uncompahgre Plateaus, contradicting the region's popular image of a land of jagged high-peaks.

## 3. Climate

While weather reflects day-to-day changes, climate is the long-term atmospheric condition of an area (Wuerthner 2001). Climate has enormous impact on the fauna, flora, and ecological processes of a region. The climate of the Southern Rockies is a temperate, semi-arid steppe regime (McNab and Avers 1994), with generally sunny weather, warm summers and moderately cold winters.

The north-south trend in mountain ranges and the inland location are major factors affecting climate in the Southern Rockies, leading to wetter western slopes, drier eastern slopes, and magnification of global temperature changes (Flannery 2001, Wuerthner 2001).

Other factors affecting climate in the Southern Rockies include latitude, major weather patterns (such as winter storm tracks and jet stream locations), topographic aspect, and elevation (Benedict 1991, Knight 1994). Lower elevations tend to have hot summers and cool winters with semi-desert levels of moisture, while higher elevations are cooler and wetter with short growing seasons. The highest elevations experience long, harsh, and snowy winters.

Climate differences can be quite pronounced over short distances. Portions of the San Luis Valley in Colorado average about 18 cm of precipitation a year, while some locations in the nearby San Juan Mountains receive more than 140 cm, mainly in the form of snow (*e.g.*, Wolf Creek pass averages about 11 m of snow per year). Temperatures can also vary greatly over relatively short distances. Boulder, Colorado (at 1,631 m) has an average July high temperature of 30.5 C°, while at nearby Berthoud Pass (3,480 m) the average July high temperature is only 16.5 C° (Western Regional Climate Center Database 2001).

The mountain chains can be significant barriers to movement of air masses. Due to prevailing westerly weather patterns, the western mountains tend to be wetter than eastern slopes, with most precipitation coming in the form of snow (Neely et al. 2001). In fact, while roughly 60% of the Southern Rockies' surface area drains eastward, more than 75% of the precipitation falls on the west slope (Benedict 1991). This influences the spatial distribution of aquatic and riparian ecosystems in the Southern Rockies, as well as human uses and diversions of water. The latter have had dramatic impacts on aquatic ecosystems.

Because there are large barometric differences between the tops of the high peaks and the plains below, and the surface distance between the peaks and plains is short, the Southern Rockies can experience high winds—sometimes greater than hurricane force. For example, in 1997 winds in excess of 120 mph blew down an estimated four million trees in the Routt National Forest.

The biggest storms occur in winter. The jet stream generally flows across the northern United States and Canada, but in winter it deflects far enough south to influence northern Colorado (Wuerthner 2001). However, the driest time of year is early winter, since much of the snowfall occurs in March and April. Summer rainfall comes as late-afternoon showers that last an hour or two. But the summer evening cools rapidly because of low humidity. This cooling breaks up the clouds, leaving a clear night and clear sunrise.

## 4. Drainage Basins and Aquatic Systems

As moisture-laden weather systems pass over the Southern Rockies, the mountains squeeze out rain and snow, creating an ecoregion that is generally wetter than surrounding areas. Much of this moisture derives from two sources. The Pacific waters off Mexico benefits the San Juan Mountains (Elias 2002), and the Gulf of Mexico largely deposits moisture on the eastern Front Range (Elias 2002). This high-elevation moisture forms the headwaters of some of the continent's major river systems. West of the Continental Divide, water flows to the Pacific Ocean via the Colorado River to the Gulf of California. On the eastern slope, water travels to the Atlantic through the Gulf of Mexico by two main routes: the Rio Grande drains directly into the Gulf; and the North/South Platte Rivers and Arkansas River empty their aquatic loads indirectly via the greater Mississippi/Missouri River system.

The Southern Rockies have nearly 48,000 km of perennial creeks, streams, and rivers scattered throughout the region (Shinneman et al. 2000), ranging from clear, cold, fast high-mountain creeks to relatively slow moving, wide, lower-elevation rivers. Natural deepwater lakes are numerous but roughly 90% are found above 2,700 meters in elevation (Colorado Water Resources Research Institute 2001). Wetlands of various types are found throughout the region, from willow (*Salix* spp.) carrs scattered throughout the high country to large playa lakes in the San Luis Valley. Groundwater and aquifers occur throughout the region. The largest is the San Luis Valley Aquifer, which supports numerous shallow wetlands and springs (Pearl 1974).

## 5. Natural Processes and Landscape Patterns

Natural processes play important roles in maintaining ecological integrity, and they include energy flows, nutrient cycles, hydrologic cycles, succession of natural community types, pollination, predator-prey relationships, and disturbance regimes (Noss and Cooperrider 1994). These processes make ecosystems diverse, dynamic, resilient, and naturally evolving. Fires, floods, wind storms, landslides, insect infestations, and diseases help to create varied natural landscape mosaics over space and time by influencing the composition, physical structure, and function of ecosystems and landscapes.

Spatial and temporal characteristics of natural disturbances within an ecosystem type define a disturbance regime (Pickett and White 1985). In the Southern Rockies, fire is a particularly important disturbance agent. In general, the dense continuous crown-cover in upper montane and subalpine forests supports occasional but extensive stand-replacing fires, while lower montane and foothill forests experienced low-intensity surface fires carried by fine surface fuels like grasses (Veblen 2000). The region's grassland and shrubland fire regimes are less well understood (Knight 1994). However, within many community types, fires vary in intensity and size over space and time, creating a shifting mosaic of patch age structures and patch types (Pickett and White 1985). Disturbed patches typically go through various "successional stages" over time, until a relatively stable stage, such as an old-growth forest, eventually returns (Knight and Wallace 1989). In other cases, such as old-growth ponderosa pine forests, a regime of low-intensity surface fires may actually maintain relatively steady-state conditions over long periods by thinning forest stands and maintaining large, old trees and grassy understories (Covington and Moore 1994). Yet even these forests can experience rare, severe stand-replacing disturbances, and for some ecosystems, these less predictable and more variable disturbance regimes may even be the "norm" (Reice 1994). Because natural disturbances support dynamic and healthy ecosystems and provide habitat for native species, human alteration and disruption of natural disturbance regimes in the Southern Rockies is of concern

for many ecologists (*e.g.*, Veblen and Lorenz 1991, Kipfmueller and Baker 2000, Romme et al. 2000).

One way to assess the condition of Nature is to examine how natural communities are spatially distributed in a mosaic across a landscape, such as an ecoregion or a watershed; important indices include patch size, patch configuration, boundaries between patches, and connectivity (Forman and Godron 1986). However, depending on the ecological element or process of interest, the appropriate scale, detail, and resolution at which to measure landscape structure may vary (Wiens 1997). For instance, in the Southern Rockies subalpine forests often cover hundreds of thousands of contiguous hectares, and can be viewed as one large patch or matrix community that dominates a given landscape area. Yet, within the forest matrix, smaller patches of different forest ages (e.g., old-growth stands, dense pole-sized stands) and cover types (aspen forest, montane riparian shrublands) will exist, due to disturbance histories and environmental gradients.

Recognizing that these different landscape patterns exist at different scales has great relevance for species conservation. While a habitat generalist such as elk (*Cervus elaphus*) or wolf (*Canis lupus*) may easily move through the subalpine forest landscape matrix, American marten (*Martes americana*) may be sensitive to the natural or human-induced patchiness within the forest matrix. A marten's dispersal success is limited by the amount of connected forest habitat with dense stands of old trees and downed snags (Buskirk and Ruggiero 1994). Although much of the landscape before European colonization was patchy, other areas consisted of extensive expanses of forest that were continuous for many interior dependent species (Knight and Reiners 2000). The loss and fragmentation of such large interior habitat, due to logging, road-building, and residential development, is of increasing concern to scientists, land managers, and conservationists in the Southern Rockies (e.g., Knight et al. 2000).

## 6. Ecosystems and Natural Communities

An ecosystem is an area where plants, animals and other organisms interact with each other and the non-living physical environment, including soil, rock, dead organic matter, air, and water. The distribution of ecosystem types, as well as the plants and animals associated with each, is dependent on numerous environmental variables including climate, water availability, topography, geology, soils, and elevation. Within the Southern Rockies, elevation and aspect are particularly influential, as these environmental factors strongly affect temperature, moisture availability, wind, and solar radiation levels (Knight 1994). A latitudinal gradient is also relevant to ecosystem distribution, as illustrated by the elevation of tree line in the Southern Rockies varying from 3,200 meters in the north to 3,800 meters in the south (Benedict 1991).

Ecosystems have long been classified within broad elevation life zones (*e.g.*, Merriam 1890). In the Southern Rockies, the major life zones and their general elevations include the alpine (>3,200 m), subalpine (2,800 m – 3,200 m), upper montane (2,300 m – 2,800 m), and lower montane-foothills (<2,300 m, Neely et al. 2001). A more specific classification identifies natural communities under the levels of the U.S. National Vegetation Classification system (Grossman et al. 1998) with further refinement into "associations." The Nature Conservancy indicates that there are 411 plant associations (Neely et al. 2001) in the Southern Rockies. These, in turn, can be grouped under the alpine, subalpine, montane, and lower montane-foothills ecological zones. For example, the *subalpine fir/whortleberry* community association is nested within the *spruce-fir forest* ecological system, which in turn is nested within the subalpine zone (Figure 2.2).

In the following discussion, we depart somewhat from these standardized classification systems to describe the Southern Rockies' biological communities within 13 coarsely grouped major natural ecosystem types (following Shinneman et al. 2000).

### *Semi-Desert and Sagebrush Shrublands*

#### *Description*

These shrubland types collectively comprise about 16% of the Southern Rockies land area (Shinneman et al. 2000). Semi-desert shrublands are generally found below 2,300 meters in elevation, where summer temperatures are hot and precipitation sparse, and typically on poorly drained saline soils. These ecosystems cover an extensive area in the San Luis Valley, Upper Rio Grande Basin, and portions of the lower Gunnison Valley. Dominant species include greasewood (*Sarcobatus vermiculatus*), four-winged saltbush (*Atriplex canescens*), shadscale (*Atriplex confertifolia*), and winterfat (*Krascheninnikovia lanata*, Dick-Peddie 1993, Knight 1994). Sagebrush shrublands at lower-elevations are often dominated by big sagebrush (*Artemisia tridentata*), while mountain sage-

Big sagebrush (*Artemisia tridentata*)

brush (*Artemisia tridentata vaseyana*) communities occur in cooler, more mesic, mid-elevations up to 3,000 m (Knight 1994). Sagebrush shrublands can form extensive matrices covering millions of hectares, especially in the major valleys and intermountain basins, such as the Gunnison Basin, North Park, Middle Park, and San Luis Valleys in Colorado (Neely et al. 2001).

Although the disturbance history is not well known, fires historically played a role in sagebrush shrubland ecology. Evidence suggests that mean fire return interval of sagebrush shrublands varies from 20 to 100 years, depending on site conditions (Knight 1994). Historically sagebrush steppe was probably a mosaic of productive grasses, other shrub cover, and sagebrush patches of varying ages, due to site conditions that encouraged a variable mix of species and because sagebrush is easily killed by fire (Knight 1994). Other leading disturbance agents include heavy herbivory by grasshoppers, bison, and other ungulates, as well as severe drought. Shrublands are important winter range for native deer (*Odocoileus* spp.) and elk, and native predators such as coyote (*Canis latrans*), mountain lion (*Puma concolor*), grizzly bear (*Ursus arctos*), and wolves hunted these shrublands (Fitzgerald et al. 1994, Bennett 1994).

These shrubland ecosystems contain significantly different plant and animal communities due to their broad ranges in environmental conditions. They have been recorded with 157 vertebrate species (Shinneman et al. 2000). Common plant species include rabbitbrush (*Chrysothamnus* spp.), snakeweed (*Gutierrezia sarothrae*), and bitterbrush (*Purshia tridentata*), and sagebrush shrublands may have well-developed grass and forb cover. Representative animal species include western rattlesnake (*Crotalus viridis*), collared lizard (*Crotaphytus collaris*), ferruginous hawk (*Buteo regalis*), golden eagle (*Aquila chrysaetos*), sage grouse (*Centrocercus urophasianus*), sage sparrow (*Amphispiza belli*), green-tailed towhee (*Pipilo chlorurus*), loggerhead shrike (*Lanius ludovicianus*), coyote, pronghorn (*Antilocapra americana*), elk, mule deer (*Odocoileus hemionus*), desert cottontail (*Sylvilagus audubonii*), black-tailed jackrabbit (*Lepus californicus*), Wyoming ground squirrel (*Spermophilus elegans*), deer mice (*Peromyscus maniculatus*), white-tailed prairie dog (*Cynomys leucurus*), and Gunnison's prairie dog (*Cynomys gunnisoni*).

Western rattlesnake (*Crotalus viridis*)

***Conservation Concerns***

In many shrubland areas, livestock overgrazing and fire suppression have decreased palatable forbs and grasses favored by wildlife and increased unpalatable woody plants (Knight 1994). Other human disturbances include removal of sagebrush to increase forage for livestock, conversion to cropland, oil and gas exploration, and invasion of exotic species, especially cheatgrass (*Bromus tectorum*). Shrubs catch the blowing snow, and the deep root system of sage passes water down to the water table. Loss of sage from these systems has reduced stream flow and lowered soil moisture (Knight 1994).

These perturbations have collectively replaced, fragmented and altered the species composition and disturbance regimes of sagebrush communities. This habitat destruction has caused population declines of shrubland dependent species, including sage grouse, sage sparrow, Brewer's sparrow (*Spizella breweri*), white-tailed prairie dog, and Gunnison's prairie dog (Braun 1995, Knick and Rotenberry 1995, Johnson and Braun 1999, Neely et al. 2001). Bison (*Bison bison*), grizzly bear, and wolves have been extirpated by direct human persecution (Fitzgerald et al. 1994). Sagebrush shrublands are poorly represented in Southern Rockies' protected areas, with less than 3% of their total area in National Parks, Research Natural Areas, and Wilderness Areas.

## Montane Shrubland

***Description***

Montane shrublands make up slightly more than 8% of the region. These shrubland ecosystems are typically found below montane forests and above grasslands, semi-desert shrublands, or piñon-juniper woodlands. They are generally in semi-arid sites between 1,700-2,600 meters, although they may occur over 3,000 meters on south-facing slopes (Benedict 1991). Gambel oak (*Quercus gambelii*) shrublands (and sometimes the woodland form) dominate the western and southern portions the Southern Rockies, while mountain mahogany (*Cercocarpus montanus*) is dominant in the semi-arid foothills in the northeastern portion of the ecoregion.

Other common montane shrubland species include serviceberry (*Amelanchier alnifolia*), snowberry (*Symphoricarpos oreophilus*), skunkbrush (*Rhus trilobata*), bitterbrush, ninebark (*Physocarpus* spp.), and several currant species (*Ribes* spp.). These shrublands often have well-developed herbaceous understories of bunchgrasses and forbs. Characteristic animals include eastern fence lizard (*Sceloporus undulatus*), western rattlesnake, Virginia's warbler (*Vermivora virginiae*), scrub jay (*Aphelocoma californica*), spotted towhee (*Pipilo maculatus*), rock wren (*Salpinctes obsoletus*), sharp-shinned hawk (*Accipiter striatus*), red fox (*Vulpes vulpes*), ringtail (*Bassariscus astutus*), bobcat (*Lynx rufus*), western small-footed myotis (*Myotis ciliolabrum*), mule deer, bushy-tailed woodrat (*Neotoma cinerea*), brush mouse (*Peromyscus boylii*), mountain cottontail (*Sylvilagus nuttallii*), and rock squirrel (*Spermophilus variegatus*). Due to their mid-elevation position, these shrublands often contain a mix of species from different elevation life zones, and several species, such as elk and deer, winter in these shrublands. One hundred and sixty-one vertebrate species have been observed in mountain shrubland (Shinneman et al. 2000).

Fire encourages the establishment and spread of mountain shrubland ecosystems as an early seral stage (Floyd-Hannah et al. 1996), although there is evidence they may also exist as stable communities where dry climate and soil conditions permit (Benedict 1991). Mountain mahogany leaves are heavily grazed by native herbivores, and oak acorns are an important food source for deer, elk, wild turkey (*Meleagris gallopavo*), squirrel (*Sciurus* spp.), black bear (*Ursus americanus*), and many other wildlife species (Knight 1994, Fitzgerald et al. 1994). Many shrublands also support a rich array of insects, which attracts high numbers of insectivorous birds and reptiles (Floyd-Hannah et al. 1996).

#### *Conservation Concerns*

Development in montane shrubland ecosystems is reducing and fragmenting thousands of hectares of valuable wildlife habitat, especially along Colorado's eastern slope and portions of the San Juan Mountains (Shinneman et al. 2000). Fire suppression may eventually result in altered community composition and the displacement of native shrub communities by trees. Also, if not regenerated by fire, Gambel oak woodlands will become senescent and acorn production will decline, diminishing a vital food source for wildlife. Montane shrublands are not well represented in protected areas, especially oak woodlands. Less than 3% of the total area of oak woodlands in the Southern Rockies is in nature preserves.

### *Piñon-Juniper Woodland*

#### *Description*

Piñon-juniper woodlands cover extensive areas in the southern and western foothills and mesa tops, and comprise about 13% of the Southern Rockies. They generally occur between 1,700-2,400 meters in elevation, but occasionally reach higher than 2,700 meters on south-facing slopes. Along their lower elevation range, in hot and dry conditions, piñon-juniper woodlands occur as sparsely wooded savannas (Dick-Peddie 1993). At these elevations regularly occurring fire may have thinned these woodlands, supporting grassy understories and preventing the encroachment of these trees into neighboring grasslands and shrublands (Dick-Peddie 1993). In contrast, at the upper elevational range, these woodlands occur in relatively dense stand conditions and are often interspersed with ponderosa pine (*Pinus ponderosa*) and Gambel oak. Although these woodlands are often associated with periodic surface fires, a recent study in Mesa Verde National Park in southwestern Colorado suggested that infrequent large crown fires characterize both the modern and historical fire regime of piñon-juniper woodlands in that area, based on analysis of stand age and fire-scar data (Floyd et al. 2000).

Piñon-juniper woodlands have variable plant species compositions depending on site conditions and regional location, but dominant species include piñon pine (*Pinus edulis*), one-seed juniper (*Sabina monosperma*), Utah juniper (*Sabina osteosperma*), and Rocky Mountain juniper (*Sabina scopulorum*). Piñon pine reaches its northern terminus of the ecoregion in an isolated patch along the Colorado-Wyoming border. Alligator juniper (*Juniperus deppeana*) occurs in the very southern portions of the Southern Rockies. Grassy understories include Junegrass (*Koeleria macrantha*) and Indian rice grass (*Achnatherum hymenoides*); and numerous cacti and shrub species can also be found. The 181 vertebrate species in piñon-juniper woodlands include eastern fence lizard, tree lizard (*Urosaurus ornatus*), collared lizard, golden eagle, piñon jay (*Gymnorhinus cyanocephalus*), bushtit (*Psaltriparus minimus*), blue-gray gnatcatcher (*Polioptila caerulea*), coyote, gray fox (*Urocyon cinereoargenteus*), mountain lion, ringtail, Townsend's big-eared bat (*Plecotus townsendii*), pallid bat (*Antrozous pallidus*), mule deer, Mexican woodrat (*Neotoma mexicana*), rock squirrel, Colorado chipmunk (*Tamias quadrivittatus*), and piñon mouse (*Peromyscus truei*). Piñon nuts are an important food source for many species of wildlife, and were equally important for native peoples as well.

### *Conservation Concerns*

Overgrazing by livestock in these arid ecosystems has reduced forage for wildlife and exposed soils that were easily eroded (Flores 1996). "Chaining" and "roller-chopping" (practices in which large tracts of piñon-juniper woodlands are mechanically removed to improve land for domestic livestock) have eliminated areas of old piñon-juniper woodlands. Collectively, livestock grazing, mechanical removal, logging trees for firewood, and housing development have greatly altered and fragmented piñon-juniper woodlands in some areas. Fire suppression and grazing may encourage piñon and juniper trees to invade adjacent shrublands and grasslands (*e.g.*, Miller and Rose 1999). Piñon-juniper woodlands/savannah have less than 6% of their total area within protected lands (Shinneman et al. 2000).

## Ponderosa Pine Forest and Woodland

### *Description*

Ponderosa pine forests cover about 12% of the region and are generally found throughout the Southern Rockies in the foothill and montane zones between 1,500-2,700 meters in elevation. Ponderosa pine forests exist in variable patch sizes, but form large matrix communities along the eastern slope of the Front Range and the foothills of southern Colorado and northern New Mexico. Ponderosa pine forests may extend above 2,700 meters in elevation where thin, dry soils occur on south-facing slopes. These forests are typically dry and warm, and snowfall does not appreciably accumulate for long periods during the winter. At lower elevations ponderosa pine trees are often interspersed with piñon, juniper, and oak; in other cases, open ponderosa pine woodlands dominate the landscape at the grassland-foothill ecotone (Dick-Peddie 1993, Knight 1994). At higher elevations and in mesic sites, stands can be fairly dense, often mixing with other tree species such as lodgepole pine (*Pinus contorta*), Douglas-fir (*Pseudotsuga menziesii*), or quaking aspen (*Populus tremuloides*, Veblen et al. 2000).

Frequent, low-intensity, surface fires historically thinned many ponderosa pine forests, maintaining large, fire-adapted, old trees in open, park-like conditions with grassy understories (Veblen and Lorenz 1991, Covington and Moore 1994). However, in some locations in the Southern Rockies, especially on cooler, more mesic sites, denser stands of ponderosa pine also experienced a mixed-fire regime with occasional stand-replacing fires (Veblen et al. 2000, Brown et al. 1999). Ponderosa pine is also susceptible to mountain pine beetle (*Dendroctonus ponderosae*) outbreaks which are capable of killing trees over large areas (Schmid and Mata 1996).

Understory shrubs include bearberry or kinnikinnick (*Arctostaphylos uva-ursi*), common juniper (*Juniperus communis*), mountain mahogany, wax currant (*Ribes cereum*), and numerous herbaceous plants such as mountain muhly (*Muhlenbergia montana*), Arizona fescue (*Festuca arizonica*) Oregon grape (*Mahonia repens*), wild geranium (*Geranium* spp.), pasque flower (*Pulsatilla patens*), and mountain ball cactus (*Pediocactus simpsonii*). Ponderosa pine forests also support a rich diversity of animals, including 129 vertebrate species (Shinneman et al. 2000). Characteristic species include the bullsnake (*Pituophis melanoleucus*), eastern fence lizard, flammulated owl (*Otus flammeolus*), Mexican spotted owl (*Strix occidentalis lucida*), Lewis' woodpecker (*Melanerpes lewis*), Williamson's sapsucker (*Sphyrapicus thyroideus*), Steller's jay (*Cyanocitta stelleri*), mountain bluebird (*Sialia currucoides*), western tanager (*Piranga ludoviciana*), red crossbill (*Loxia curvirostra*), pygmy nuthatch (*Sitta pygmaea*), mountain chickadee (*Poecile gambeli*), Cassin's vireo (*Vireo cassinii*), black bear, mountain lion, long-eared myotis (*Myotis evotis*), mule deer, porcupine (*Erethizon dorsatum*), Abert's squirrel (*Sciurus aberti*), least chipmunk (*Tamias minimus*), and golden mantled ground squirrel (*Spermophilus lateralis*).

### *Conservation Concerns*

Historical and current logging has resulted in a substantial loss of ponderosa pine old-growth forest habitat; probably less than 5% remains in the Southern Rockies (Shinneman et al. 2000), and most remaining old trees exist in small, isolated patches (Romme et al. 2000). Old forest conditions support species such as Abert's squirrel and Mexican spotted owl. Logging, fire suppression, and livestock grazing have created dense stands of younger trees that are susceptible to unnaturally large, catastrophic fires and insect outbreaks (Harrington and Sackett 1992, Romme et al. 2000). Thinning (logging) and prescribed fire are methods used to try to return these altered forests to their pre-settlement structures and composition (Covington and Moore 1994), especially around residential development (US Forest Service 1997a, 2000, City of Boulder 1999). However, this approach may be misguided in areas where these forests were naturally dense and experienced stand-replacing fires, especially in mesic sites and upper elevations (Shinneman and Baker 1997, Veblen et al. 2000). Any restoration effort must actually retain remaining old trees; it must not increase road densities or edge habitat, destroy interior habitat and roadless areas, alter landscape structure, and aid the spread of weedy species (Shinneman and Baker 2000, Romme et al. 2000). Other significant human impacts in ponderosa pine forests include heavy recreation, extensive road-networks, and exploding residential development. Such effects have significantly fragmented ponderosa pine forest habitats in places such as Colorado's Front Range foothills (Shinneman et al. 2000, Theobald 2000). Only about 4% of ponderosa pine forests in the region fall under protected land status (Shinneman et al. 2000).

## Douglas-Fir Forest

### *Description*

Douglas-fir forests cover just over 2% of the region and are generally found between 1,700-2,700 meters in elevation, usually occurring on cooler and less xeric sites than ponderosa pine, although the two trees often occur together (Goldblum and Veblen 1992). At the higher end of its elevation range, and on north-facing slopes at lower elevations, Douglas-fir may form pure stands, but it can also be found with blue spruce (*Picea pungens*), aspen, Rocky Mountain juniper, and lodgepole pine (Benedict 1991, Dick-Peddie 1993). In the southern part of the ecoregion Douglas-fir is often codominant with white fir (*Abies concolor*, Benedict 1991, Dick-Peddie 1993). On exposed ridge tops and dry south-facing slopes, Douglas-fir can exist in open, park-like stands along with limber pine (*Pinus flexilis*) and bristlecone pine (*Pinus aristata*, Benedict 1991, Knight 1994).

Historically, low intensity, surface fires occurred fairly regularly and maintained stands of large, old, fire-tolerant trees on drier sites (Goldblum and Veblen 1992), while occasional large, stand-replacing fires occurred when there was enough moisture to support dense forests stands (Veblen et al. 2000). Other disturbance agents include outbreaks of western spruce budworm (*Choristoneura occidentalis*) and Douglas-fir bark beetles (*Dendroctonus pseudotsugae*) which are capable of defoliating or killing Douglas-fir trees over large areas (Schmid and Mata 1996).

Several shrub species are commonly found in Douglas-fir forests, including common juniper, ninebark, snowberry, Rocky Mountain maple (*Acer glabrum*), Oregon boxleaf (*Paxistima myrsinites*), thimbleberry (*Rubus parviflorus*), Oregon grape, and Woods' rose (*rosa woodsii*). Herbaceous understories often include heartleaf arnica (*Arnica cordifolia*) and Arizona fescue, while in moister locations, especially north-facing slopes and narrow ravines, numerous species of mosses and lichens, such as old man's beard (*Usnea hirta*), are found. The 81 recorded vertebrate species are similar to those of other montane coniferous forests, and characteristic species include northern goshawk (*Accipiter gentilis*), hairy woodpecker (*Picoides villosus*), hermit thrush (*Catharus guttatus*), dark-eyed junco (*Junco hyemalis*), ruby-crowned kinglet (*Regulus calendula*), elk, pine squirrel (*Tamiasciurus hudsonicus*), and red-backed vole (*Clethrionomys gapperi*).

### *Conservation Concerns*

Similar to ponderosa pine, these forests suffer from the effects of historical fire suppression, logging, loss of old-growth habitat, heavy recreation, and residential development. Only about 4% of Douglas-fir forests in the region are protected (Shinneman et al. 2000).

## Montane Mixed-Conifer Forest

### *Description*

Under a broad definition, "mixed conifer" forests occur throughout the Southern Rockies, but ecologists apply the term to the region's middle elevation (2,270-2,880 m) conifer stands in southern Colorado and northern New Mexico. These mixed-conifer stands can be dominated by a variety of tree species, including ponderosa pine, white fir, Douglas-fir, southwestern white pine (*Pinus strobiformis*), bristlecone pine, corkbark fir (*Abies lasiocarpa arizonica*), Engelmann spruce (*Picea engelmannii*), and blue spruce. The forests generally grow under more mesic and cool conditions than foothill forests, but depending on site conditions, various combinations and abundances of conifer species may be present. Due to their relatively mesic conditions and dense stand structures, these forests were historically subjected to less frequent fires than foothill forests and experienced a mixed fire regime with occasional stand-replacing fires (Jamieson et al. 1996, Dick-Peddie 1993). Western spruce-budworm and Douglas-fir bark beetle infections are also significant disturbance agents (Lynch and Swetnam 1992). Mixed-conifer forests make up about 4% of the Southern Rockies' land area.

Because of the diversity of conifers, understory plant species and animal composition are also diverse and somewhat characteristic of other conifer forests. Animal species include Abert's squirrel, pine squirrel, white-breasted nuthatch (*Sitta carolinensis*), pygmy nuthatch, Mexican spotted owl, black bear, mule deer, and porcupine. This ecosystem also harbors one of the highest numbers of Lepidopteran species (butterflies and moths) on the continent.

### *Conservation Concerns*

Similar to ponderosa pine and Douglas-fir forests, these forests suffer from the effects of fire suppression, logging, road building, loss of old-growth habitat, heavy recreation, and residential development. Roughly 11% of these forests is protected.

### *Lodgepole Pine Forest*

#### *Description*

Lodgepole pine forests cover almost 7% of the Southern Rockies, mainly in the northern half of the ecoregion, where they are extensive. Lodgepole pine forests are found only in scattered patches in southern Colorado and are absent in northern New Mexico (except where planted). They are generally found between 2,600 and 3,000 meters above sea level. Lodgepole pine grows under a variety of conditions, but is usually found on cool, dry sites. Spruce and fir are more dominant on mesic sites (Knight 1994). Snowfall is often heavy, while summers can be quite warm with intermittent periods of drought (Knight 1994).

Lodgepole pine forests are often considered a "pioneer" to other forests (an early successional stage). However, infrequent, stand-replacing fires often perpetuate lodgepole pine over other tree species growing in the understory of mature stands, including Douglas-fir, Engelmann spruce, and subalpine fir (*Abies lasiocarpa*, Knight 1994). Fires can vary in size and intensity in these forests, creating spatially complex mosaics of mature forest and relatively open, uneven-aged stand conditions (Knight 1994). Infrequent, stand-replacing fires can burn tens of thousands of hectares, and subsequent regeneration, especially when coupled with phenotypes that have serotinous cones (cones that remain closed until opened by intense heat), often results in large patches of even-aged forest with dense stands of sapling/pole-sized trees (Kipfmueller and Baker 2000). Such forest stands are commonly referred to as "doghair" stands. Other disturbance agents capable of destroying stands over large areas include windthrow, mountain pine beetle, and disease (Knight 1994, Schmid and Mata 1996). Dwarf mistletoe (*Arceuthobium americanum*) is a parasitic plant that can deform and reduce vigor in individual trees over large areas, but also provides an important source of food, cover, and nesting sites for many species of wildlife (Kipfmueller and Baker 1998).

Because these forests are often in dense stands with closed-canopy conditions, lodgepole understory may be sparse; however, forbs and shrubs that are common include heartleaf arnica, pine drops (*Pterospora andromedea*), grouse whortleberry (*Vaccinium scoparium*), Woods' rose, kinnikinnick, common juniper, and buffaloberry (*Shepherdia canadensis*). Typical of the 83 recorded vertebrate species include sharp-shinned hawk, Steller's jay, gray jay (*Perisoreus canadensis*), white-breasted nuthatch, Clark's nutcracker (*Nucifraga columbiana*), downy woodpecker (*Picoides pubescens*), brown creeper (*Certhia americana*), pine siskin (*Carduelis pinus*), ruby-crowned kinglet, elk, black bear, American marten, mule deer, pine squirrel, red-backed vole, and porcupine.

#### *Conservation Concerns*

Clear-cutting and shelterwood logging combined with extensive road building have severely fragmented many of the region's lodgepole pine forests (Reed et al. 1996). (See spruce-fir forests, below, for associated ecological affects.) In addition, the ability of land managers to allow large fires to burn unimpeded is increasingly restricted as people continue to build homes in and near these mountain forest habitats. The presence of these homes leads to fire suppression, controlled burns, and thinning, all of which may alter forest composition, structure, and natural function. Roughly 14% of lodgepole forests exist on protected lands, but it is unlikely that the full diversity of the lodgepole pine community is represented within these protected areas. Moreover, many of these protected forests are not large enough or not sufficiently connected to provide for unimpeded natural disturbance regimes or movement of native species.

### *Aspen Forests*

#### *Description*

Aspen forests comprise 8% of the Southern Rockies and are found throughout the region. These deciduous forests are most common between 2,400-3,000 meters in elevation and can occur under a wide range of site conditions, but forests of large aspen trees generally occur in moist, cool sites. Aspen forests often occur in small groves, but can form extensive stands, especially in the southern and western portions of the Southern Rockies. Aspen often become established after a disturbance, such as fire, destroys other forest types. They typically precede (in successional stage) conifer species such as Douglas-fir and spruce which grow in the understories of shady aspen groves (Romme et al. 2001). However, under certain conditions, aspen may also form stable, pure stands and sometimes maintain old forest conditions for long periods (Knight 1994, Romme et al. 2001). Aspen are susceptible to fungal diseases and numerous leaf-eating insects, as well as herbivory by deer and elk, which eat aspen bark and sprouts.

Columbine (*Aquilegia* spp.)

Species composition varies widely depending on site conditions, but in drier areas stands often have grassy understories, while under moist conditions there is a thick forb cover, including lupine (*Lupinus* spp.), columbine (*Aquilegia* spp.), wild geranium, heartleaf arnica, and cow parsnip (*Heracleum maximum*). Aspen forests support 81 recorded vertebrates (DeByle 1985), including black bear, silver-haired bat (*Lasionycteris noctivagans*), elk, mule deer, deer mouse, western jumping mouse (*Zapus princeps*), northern pocket gopher (*Thomomys talpoides*), long-tailed vole (*Microtus longicaudus*), and masked shrew (*Sorex cinereus*). Beaver (*Castor canadensis*) depend on aspen for food and dam-building material. Dozens of songbird species prefer to nest in old aspen forests, including many cavity-nesting birds such as the red-naped sapsucker (*Sphyrapicus nuchalis*), purple martin (*Progne subis*), mountain bluebird, violet green swallow (*Tachycineta thalassina*), white-breasted nuthatch, and house wren (*Troglodytes aedon*). Aspen forests also support a rich diversity of insects (Jones et al. 1985).

#### *Conservation Concerns*

Increased interest in logging old aspen forests, especially along Colorado's western slope, may eliminate large stands of mature trees that many cavity nesting songbirds, hawks, and owls use (Finch and Ruggiero 1993). Livestock grazing, over-grazing by elk, and fire suppression may negatively alter stand structure and tree species composition (Knight 1994). For instance, over-grazing on winter range by overabundant elk herds in portions of Rocky Mountain National Park in Colorado is contributing to the mortality of established aspen trees and preventing the regeneration of new aspen stands (Baker et al. 1997). Ripple and Larson (2000) reported that aspen overstory recruitment ceased when wolves disappeared from Yellowstone National Park. Wolves are a significant predator of elk, and wolves may positively influence aspen overstory through a trophic cascade caused by reducing elk numbers, modifying elk movement, and changing elk browsing patterns on aspen (Ripple and Larson 2000). Increased levels of residential development fragment and replace these forests (Theobald 2000). Roughly 9% of aspen forests are protected in the region.

### Montane and Intermontane Grasslands

#### *Description*

These grasslands make up about 8% of the region. Montane grasslands are generally small to medium sized patches of meadow among forest ecosystems. Fires and other disturbances may have created some meadows, but most are likely the result of dry, cold growing conditions with nutrient-poor soils that won't support trees (Knight 1994). Intermontane grasslands typically occur in large mountain valleys and mountain "parks". These grasslands can cover hundreds of square kilometers, such as South Park, North Park, the Wet Mountain Valley, and along the fringes of the San Luis Valley in Colorado.

Plant species include bunch grasses such as Idaho fescue (*Festuca idahoensis*) and Thurber's fescue (*Festuca thurberi*) in the north and Arizona fescue to the south. Other common grasses include Junegrass, needle-and-thread grass (*Hesperostipa comata*), oatgrass (*Danthonia* spp.), and mountain muhly. Many shrubs are common, especially big sagebrush. These grasslands also often contain numerous wildflowers, such as lupine, yarrow (*Achillea lanulosa*), mountain golden banner (*Thermopsis montana*), purple locoweed (*Oxytropis lambertii*), paintbrush (*Castilleja* spp.), bellflower (*Campanula* spp.), Colorado false hellebore (*Veratrum tenuipetalum*), penstemon (*Penstemon* spp.) and Columbian monkshood (*Aconitum columbianum*). Bird species include red-tailed hawk (*Buteo jamaicensis*), mountain bluebird, and broad-tailed hummingbird (*Selasphorus platycercus*). South Park's large grassland patches support breeding populations of the federally threatened mountain plover. These grassland ecosystems include 98 vertebrate species, and they often provide important forage for mammal species such as elk, mule deer, and pronghorn. Predator species include coyotes, badgers (*Taxidea taxus*), and historically a heavy presence of wolves (Fitzgerald et al. 1994, Bennett 1994). Other species include mountain cottontail, white-tailed jackrabbit (*Lepus townsendii*), northern pocket gopher, long-tailed vole, masked shrew, Gunnsion's prairie dog, white-tailed prairie dog, and Wyoming ground squirrel. Fire and drought were major disturbances in these ecosystems, and many grasslands evolved with herbivory by bison, elk, deer, and pronghorn (Knight 1994, Neely et al. 2001).

#### *Conservation Concerns*

While herbivory was a factor in all grasslands, it is likely that intermontane grasslands did not evolve with the same degree of bison herbivory as the Great Plains. Thus, heavy livestock grazing has increased big sagebrush cover and reduce forage cover and production in many areas. It has also spread non-native weeds such as Kentucky bluegrass (*Poa pratensis*), Russian thistle (*Salsola tragus*), and cheatgrass (Fleischner 1994, Weddel 1996). Off-road vehicles, housing development, and fire suppression have also degraded and fragmented intermountain grasslands. These grasslands are poorly represented in protected areas -- only about 4% of intermontane and 1% of montane grasslands are protected.

### Limber Pine and Bristlecone Pine Forests

#### *Description*

The area of these unique forests amounts to less than 1%

of the region. They are usually found from 2,300 meters up to tree-line and grow under harsh conditions, typically on upper south-facing slopes, exposed ridges, and rocky outcrops with windy, dry, sunny exposures and generally short growing seasons (DeVelice et al. 1986, Benedict 1991, Dick-Peddie 1993). Limber pine and bristlecone pine occur individually in relatively pure stands, together as co-dominants, or with other conifer species, such as Douglas-fir, lodgepole pine, Engelmann spruce, and subalpine fir. Bristlecone pine is found mainly in the southern two-thirds of the ecoregion. While these forests do not typically form extensive stands, they do occasionally occur in large patches in places such as the Sangre de Cristo Mountains. Due to severe growing conditions, these forests are often sparse with open canopy conditions. Bristlecone pine trees can live longer than 2,000 years in the Southern Rockies (Benedict 1991).

Understory is generally sparse, with herbaceous species such as Arizona fescue, Junegrass, mountain muhly, spearleaf stonecrop (*Sedum lanceolatum*), and littleflower alumroot (*Heuchera parviflora*), while characteristic shrubs include serviceberry, common juniper, and sticky laurel (*Ceanothus velutinus*). Limber pine seeds are an important food source for Clark's nutcrackers and gray jays, and Clark's nutcrackers are important to the reproduction and dispersal of 5-needle pines. Mammals from rodents to bears benefit from the nuts produced by these pines. In general, the 63 vertebrate species found in these forests are similar to those of other mountain coniferous forests.

#### *Conservation Concerns*

Many of these forests remain relatively unaltered, as the trees are generally undesirable for timber values due to twisted wood grain, and the rugged, inhospitable sites are not prime for development and road building. However, in some areas over-grazing has denuded fragile soils and the already sparse understory plant communities. Blister rust (*Cronartium* spp.), a fungus that infects 5-needled pines, has now arrived from Eurasia, and it is anticipated to eliminate 98% of the 5-needled pines before it runs its course (Mitten pers. comm.). The mutualistic relationship between jays and 5-needled pines may be broken during this decline, costing the pines their method of dispersal and the jays and nutcrackers a critical food source.

### *Engelmann Spruce - Subalpine Fir Forests*

#### *Description*

Spruce-fir forests often form vast high-elevation matrix communities, and make up nearly 13% of the Southern Rockies. They generally occur from 2,700 meters to tree line on cool, moist sites where most precipitation falls as snow. Found throughout the ecoregion, Engelmann spruce and subalpine fir sometimes grow in pure stands of either species, but are typically co-dominant tree species (Knight 1994). In other cases, spruce-fir forests are interspersed with lodgepole pine, limber pine, or aspen. At tree line, stunted, windswept versions of these forests (called krummholz) are found interspersed with alpine tundra.

Spruce-fir forests experience stand-replacing crown fires every few hundred years (on average) that are capable of burning thousands of hectares. In addition, spruce beetle (*Dendroctonus rufipennis*) outbreaks can kill most mature trees over hundreds of thousands of hectares (Baker and Veblen 1990, Veblen et al. 1994, Schmid and Mata 1996, Kipfmueller and Baker 2000). Windthrow and wood-rotting fungi are other notable disturbance agents. Due to the variability in disturbances, pre-settlement spruce-fir forest landscapes probably contained a complex mosaic of various stand ages, including old-growth (Rebertus et al. 1992). Old-growth forest stands have complex forest structures with various sized standing trees and numerous downed dead trees, with many large canopy trees that are 300-500 years old (Veblen et al. 1994, Mehl 1992). Because stand replacing disturbances generally have long rotations and cover extensive space, large patches of old-growth forests probably existed in many locations prior to EuroAmerican settlement, especially in topographically sheltered locations less susceptible to windthrow and fire (Knight 1994).

Depending on stand conditions, understory vegetation ranges from dense to open and patchy, and includes blueberry (*Vaccinium* spp.), common juniper, Woods' rose, and numerous herbs, such as heartleaf arnica, wood nymph (*Moneses uniflora*), and clustered lady's slipper (*Cypripedium fasciculatum*). Representative of the 89 vertebrate species recorded in spruce-fir are boreal owl (*Aegolius funereus*), northern goshawk, mountain chickadee, red crossbill, blue grouse (*Dendragapus obscurus*), Townsend's solitaire (*Myadestes townsendi*), olive-sided flycatcher (*Contopus cooperi*), golden-crowned kinglet (*Regulus satrapa*), hermit thrush, elk, black bear, American marten, red-backed vole, pine squirrel, and snowshoe hare (*Lepus americanus*). The region's historic populations of lynx (*Lynx canadensis*) and wolverine (*Gulo gulo*) also inhabited these forests (Seidel et al. 1998).

#### *Conservation Concerns*

Logging and associated road building have fragmented formerly extensive, old-growth forest patches and altered forest structure and composition (Reed et al. 1996). Much of the Medicine Bow National Forest in Wyoming and portions of the Rio Grande National Forest in southern Colorado offer sad examples of logging levels that led to extensive habitat fragmentation (Reed et al. 1996, US Forest Service 1998, Shinneman et al. 2000). In addition, clear-cutting has inhibited stand regeneration (Reed et al. 1996, US Forest Service 1998, Shinneman et al. 2000). Recreation is a concern in many areas, including backcountry travel, summer and winter off-road vehicle use, and ski-area development and expansion (Knight 2000). Roughly 30% of these forests fall within protected land management categories, although mainly for scenic values. More consideration should be given to including the diversity of spruce-fir associated species and natural communities, maintaining natural processes, and connecting forest habitats across the regional landscape. In some areas, the above-mentioned problems have severely fragmented forests, eliminated old growth, and caused subsequent declines in populations of old-growth and forest-interior dependent species, including the northern goshawk, boreal owl, and American marten (Reynolds et al. 1992, Hayward 1994, Buskirk and Ruggiero 1994).

## *Alpine Tundra*

### *Description*

Alpine tundra is found throughout the region above tree line, about 3,300 meters in elevation. These cold, wind-swept ecosystems receive substantial precipitation mostly in the form of snow. In many cases, however, moisture availability is limited as snow is swept clear by persistent high winds. Alpine conditions are typically quite patchy with localized topographic diversity and different plant communities occurring under different site conditions, resulting in a rich mosaic of alpine wetlands, dry meadows, snowfields, fellfields, talus and scree slopes, rock faces, and krummholz forests (Benedict 1991, Knight 1994). About 6% of the region is within alpine habitats.

In general alpine ecosystems represent relatively stable conditions, but natural disturbance includes soil movement, spring snowmelt, expanding snowfields, and burrowing impacts from small mammals such as pocket gophers (Knight 1994). Grizzly bears also historically foraged the tundra, digging for roots, rodents, and other food sources (Fitzgerald et al. 1994). Due to the harsh and brief growing season, succession after disturbance often takes hundreds of years.

Plant communities vary depending on factors such as moisture availability, snow cover duration, solar radiation, and wind exposure. Plants are dominated by low growing shrubs and perennial herbs, including cushion plants, forbs, sedges, and grasses. Representative species include short-fruit willow (*Salix brachycarpa*), arctic willow (*Salix arctica*), diamondleaf willow (*Salix planifolia*), tufted hairgrass (*Deschampsia caespitosa*), alpine bluegrass (*Poa alpina*), bog sedge (*Kobresia myosuroides*), Rocky Mountain sedge (*Carex scopulorum*), moss campion (*Silene acaulis*), Ross' avens (*Geum rossii*), Parry's clover (*Trifolium parryi*), cushion phlox (*Phlox pulvinata*), alpine mountainsorrel (*Oxyria digyna*), Rocky Mountain snowlover (*Chionophila jamesii*), American bistort (*Polygonum bistortoides*), whitish gentian (*Gentiana algida*), sudetic lousewort (*Pedicularis sudetica*), marsh marigold (*Caltha leptosepala*), and redpod stonecrop (*Rhodiola rhodantha*). Lichens and mosses are common. Characteristic species among the 51 vertebrates recorded include white-tailed ptarmigan (*Lagopus leucurus*), brown-capped rosy finch (*Leucosticte australis*), white-crowned sparrow (*Zonotrichia leucophrys*), American pipit (*Anthus rubescens*), horned lark (*Eremophila alpestris*), common raven (*Corvus corax*), golden eagle, short-tailed weasel (*Mustela erminea*), bighorn sheep (*Ovis canadensis*), elk, yellow-bellied marmot (*Marmota flaviventris*), pika (*Ochotona princeps*), northern pocket gopher, and montane shrew (*Sorex monticolus*).

(Fielder 1997)

Yellow-bellied marmot (*Marmota flaviventris*)

### *Conservation Concerns*

In many areas, especially in the southern portion of the ecoregion, grazing by domestic sheep has damaged fragile, native alpine vegetation. Dramatically increasing recreation, especially off-road vehicle use and peak-bagging by hikers, can also trample and destroy alpine vegetation and cause severe erosion. Wolverines roamed and denned in alpine habitats and the grizzly bear once played a major ecological role in alpine areas; both are considered extirpated from the region (Fitzgerald et al. 1994). In addition, the alpine zone is more susceptible to increased ultra-violet radiation and deposition of nitrogen, PCBs, and other organochlorines because it is more exposed to atmospheric conditions and less able to buffer atmospheric inputs (Baron et al. 2002).

## *Aquatic Ecosystems: Wetland and Riparian*

### *Description*

The Southern Rockies contain a diverse range of aquatic, wetland, and riparian ecosystem types. These terms are sometimes used interchangeably, as these ecosystems are often ephemeral, overlapping, and transitional in nature. For instance, playa lakes in the San Luis Valley are shallow and ephemeral and are typically classified as wetlands. Aquatic ecosystems tend to be small patches or linear features on the landscape, and collectively constitute a small portion the region's surface area. However, despite their small size, these ecosystems are among the most valuable to native species in the Southern Rockies.

### *Riparian*

The Southern Rockies have thousands of miles of streams and rivers dispersed throughout the region, from fast, clear, high-mountain streams to slower moving low-elevation rivers. Natural lakes in the Southern Rockies occur above 2,700 meters in elevation, and were formed behind terminal moraines or in depressions left by past glacial activity (Benedict 1991). Ponds are also abundant at higher elevations. There are hundreds of human-made reservoirs covering thousands of hectares, many in lower elevations that were historically devoid of deep-water lakes (Shinneman et al. 2000).

Riparian ecosystems can be thought of as a special type of aquatic ecosystem that occurs along the upland margins of streams, rivers, and lakes, and represents the meeting place of aquatic and terrestrial ecosystems. Riparian ecosystems are distributed throughout the region at all elevations, ranging from narrow linear communities in deep canyons to more extensive forests in broad floodplains. Roughly 3-8% of the Southern Rockies occur within streamside riparian habitat (Shinneman et al. 2000).

### *Wetland*

Wetlands are another type of aquatic ecosystem and occur throughout the region. A wetland can be defined as an area that is covered by water for at least part of the year, and where plants and animals are adapted to life in water or in saturated soils (Cowardin et al. 1979). Wetlands in the Southern Rockies include forested wetlands, willow carrs, fens, marshes, bogs, alpine snow glades, wet meadows, salt meadows, bottomland shrublands, shallow ponds, and playa lakes. Many of the wetlands in the Southern Rockies are seasonal, resulting from spring snowmelt and high water tables. Wet meadows account for the largest acreage of wetlands in Colorado (Jones and Cooper 1993).

### *Interconnection*

Riparian areas and wetlands are actually tightly interconnected hydrologic systems, additionally influenced by the region's geology, soils, topography, weather, plant communities and even animals. For instance, during spring peak-flow period, streams overflow their channels and inundate adjacent floodplains, providing water to wetlands. Water from wetlands and streams seeps below ground, supplying groundwater. In a reciprocal fashion, during times of low precipitation, riparian areas and wetlands may serve as sources of water to recharge creeks and streams, and groundwater may supply water to streams and wetlands via seeps, springs, and direct stream-water recharge (Maxwell et al. 1995).

The beaver, an aquatic "keystone" species (ecosystem shaper), creates pond habitats that benefit and support diverse assemblages of species and natural communities (Naiman et al. 1988). These ponds also trap and store organic material, nutrients, and sediment, which over time build up and transform into riparian communities, marshes, wet meadows, and eventually dry meadows. Streams with more beaver dams also have higher late-summer stream flows, which benefit fish and other wildlife and land owners in the Southern Rockies (Knight 1994).

Aquatic-dependent species and communities have adapted to, and depend on, dynamic and interconnected hydrologic systems. Groundwater recharge to streams and wetlands is often crucial to the survival of aquatic plants and animals during dry periods in the Southern Rockies (Cooper 1993, Power et al. 1997). Processes such as flooding, meandering stream channels, and water recharge from groundwater sustain stream and nearby riparian ecosystems and create dynamic and structurally complex aquatic and riparian ecosystems.

For instance, meandering stream flows and periodic high river flows (especially floods) alter riparian habitats by both washing away and rebuilding streamside landforms such as oxbows, cut banks, point bars, islands, and terraces. These dynamic processes provide diverse and complex mosaics of riparian and flood plain habitat in various successional stages that many aquatic and riparian-associated species require (Gutzwiller and Anderson 1987, Baker and Walford 1995). In the Southern Rockies, this process establishes cottonwood (*Populus angustifolia, P. deltoides*) riparian forests by building new point bars with nutrient-rich sediment layers that are beneficial to seedling development, as well as distributing cottonwood seeds onto these new land forms (Knight 1994).

Riparian vegetation, in turn, provides shade for rivers and streams, creating cooler air and water temperatures that native fish require. Riparian vegetation that falls into streams provides sources of food and nutrients for fish, insects, and other organisms. Riparian root systems deter stream bank erosion by decreasing the velocity of water flow, and by trapping nutrients and sediments, which build stream banks and form nutrient-rich wet meadows and fertile floodplains (Cheney et al. 1990).

Native species compositions vary considerably among the diversity of the region's aquatic ecosystems. Lower elevation shallow lakes and ponds in the Southern Rockies, which tend to be richer in oxygen, organic matter, and other nutrients, support species such as yellow pond-lily (*Nuphar lutea*), arrowhead (*Sagittaria* spp.), duckweed (*Lemna* spp.), northern leopard frog (*Rana pipiens*), wood frog (*R. sylvatica*), and tiger salamander (*Ambystoma tigrinum*), along with numerous "macroinvertebrates," such as crayfish, insects, snails, clams, and leeches, all of which occupy important places in the food web. Plant and animal plankton also populate these waters. Deep, cold, mountain lakes typically provide habitat for fewer species, including the region's native cutthroat trout (*Oncorhynchus clarki*).

Tiger salamander (*Ambystoma tigrinum*)

Streams have different species composition depending on factors such as channel width and depth, oxygen availability, velocity, turbidity, volume, or temperature. For instance, native greenback (*O. clarki stomias*), Colorado River (*O. clarki pleuriticus*), and Rio Grande cutthroat trout (*O. clarki virginalis*) are highly dependent on the clear, cold, well-oxygenated streams with riffles and pools that occur in higher elevations. In contrast, slower, more turbid, less oxygen-rich, and warmer lower elevation rivers contain native fishes such as the roundtail chub (*Gila robusta*), razorback sucker (*Xyrauchen texanus*), and Colorado squawfish (*Ptychocheilus lucius*). The river otter (*Lontra canadensis*) depends on larger streams and rivers, and beaver play important roles by affecting stream habitats and hydrology. Species such as the American dipper (*Cinclus mexicanus*), muskrat (*Ondatra zibethicus*), and water shrew (*Sorex palustris*) are also found in streams. Insect species, such as caddisfly, mayfly, and stonefly nymphs, play important roles in the aquatic food web, as do numerous microinvertebrate species and algae.

The many types of wetlands in the region offer important and varied habitat for numerous plants and animals. Among the wide variety of plant species are bog birch (*Betula pumila*), heartleaf bittercress (*Cardamine cordifolia*), crowfoot (*Ranunculus* spp.), reedgrass (*Calamagrostis* spp.), horsetail (*Equisetum* spp.), bog orchid (*Platanthera* spp.), and yellow pond-lily, as well as numerous species of willows, sedges (*Carex* spp.), rushes (*Juncus* spp.), and pondweeds (*Potamogeton* spp.). Here are the greatest concentrations of amphibian species including northern leopard frog, tiger salamander, and boreal toad (*Bufo boreas*). Many lower elevation wetlands, such as the San Luis Valley's marshes and playa lakes, provide important stopover sites for thousands of migratory waterfowl and shorebirds, such as greater sandhill crane (*Grus canadensis tabida*), American avocet (*Recurvirostra americana*), snowy egret (*Egretta thula*), green-winged teal (*Anas crecca*), and northern pintail (*Anas acuta*). Numerous insect species and other macroinvertebrates also inhabit the ecoregion's wetlands, such as clams, fairy shrimp, flatworms, water striders, and mosquito larvae. Microscopic plant and animal plankton are also abundant, especially in nutrient-rich waters.

As with other aquatic ecosystems, the species composition of riparian communities varies greatly with soils, landforms, and elevation. Whether riparian areas are narrow willow communities in steep mountain canyons or extensive cottonwood forests in broad, low-elevation valleys, they tend to be extremely rich in species diversity relative to surrounding upland communities. More deciduous tree and shrub species occur in riparian ecosystems than in any other ecosystem in the Southern Rockies. They include narrow leaf cottonwood (*Populus angustifolia*), plains cottonwood (*Populus deltoides*), box elder (*Acer negundo*), Rocky Mountain maple, redosier dogwood (*Cornus sericea*), thinleaf alder (*Alnus incana tenuifolia*) gooseberry (*Ribes inerme*), water birch (*Betula occidentalis*), bog birch, New Mexico locust (*Robinia neomexicana*), shrubby cinquefoil (*Dasiphora floribunda*), as well as dozens of species of willows. Conifer species such as blue spruce and white fir are often found. Numerous species of forbs, grasses, sedges, rushes, mosses, lichens, fungi, and liverworts also typify many riparian areas.

Riparian communities support up to 80% of all animal species in the region, which depend on these habitats for food, water, and shelter, or for important life history needs such as breeding or nesting sites (Olson and Gerhart 1982, Floyd-Hannah et al. 1996). Characteristic herpetofauna include tiger salamander, northern leopard frog, wood frog, bull snake, and smooth green snake (*Liochlorophis vernalis*). Characteristic birds include red-tailed hawk, northern harrier (*Circus cyaneus*), belted kingfisher (*Ceryle alcyon*), great blue heron (*Ardea herodias*), western grebe (*Aechmophorus occidentalis*), green-winged teal, American dipper, Lincoln's sparrow (*Melospiza lincolnii*), Bullocks's oriole (*Icterus bullockii*), yellow warbler (*Dendroica petechia*), Wilson's warbler (*Wilsonia pusilla*), and tree swallow (*Tachycineta bicolor*). More songbird species nest in riparian habitat than in any other mountain ecosystem (Mutel and Emerick 1992, Jones and Cooper 1993). Mammals are also abundant, and include black bear, river otter, mink (*Mustela vison*), mule deer, beaver, meadow jumping mouse (*Zapus hudsonius*), water shrew, and montane vole (*Microtus montanus*). Riparian areas also support a rich assortment of insects, including numerous butterflies and dragonflies.

#### *Conservation Concerns*

Human development and water use have destroyed or dramatically altered most species-rich aquatic and riparian ecosystems in the Southern Rockies. Hence, many species at risk of extinction or extirpation in the region are aquatic-dependent or riparian species. In the Southwest, although riparian areas comprise only 5% of the lands managed by the U.S. Forest Service, 70% of the federally threatened and endangered species are dependent upon riparian and aquatic ecosystems (US Forest Service 1997b). In the Southern Rockies, representative riparian and aquatic-dependent species at risk include the boreal toad, Preble's meadow jumping mouse (*Zapus hudsonius preblei*), Rio Grande cutthroat trout, and Ute ladies' tresses (*Spiranthes diluvialis*). There are also hundreds of rare and imperiled wetland and riparian plant communities in the ecoregion (Neely et al. 2001).

Some streams, mountain lakes, and ponds in the Southern Rockies are distant from human-dominated landscapes and remain relatively unpolluted. Others are polluted by acid deposition from nearby power plants, pesticides and herbicides from agriculture, acid and heavy-metal mine drainage, excess nutrients from sources such as septic systems or livestock waste, and increased sedimentation from land uses such as logging, road-building, and recreation. Over-grazing in many areas has destroyed riparian habitat and caused stream bank erosion, leading to warmer, more ephemeral, and more sediment-filled waters which are harmful to native aquatic species (Belsky et al. 1999).

Stream hydrology has been significantly altered as a result of thousands of water storage and diversion projects in the Southern Rockies, which often limit or eliminate floodwaters. Dams and diversions also alter sediment loads, oxygen levels, and water temperatures. These changes have altered stream ecosystems and harmed native aquatic species, such as downstream warm-water fishes (Osmundson et al. 1995). In some cases in the Southern Rockies, streams have been channeled, drastically altering stream ecosystems and destroying riparian habitat.

Dams in the Southern Rockies have damaged many older cottonwood riparian forest communities by impeding flooding and thus preventing regeneration. Flooding is a process essential to the distribution and establishment of cottonwood seedlings and the creation of dynamic riparian ecosystem mosaics (Knight 1994, Busch and Scott 1995). Under the relatively static conditions created by dams, shrubby and exotic species, such as tamarisk (*Tamarix ramosissima*), can more easily invade and out-compete cottonwood seedlings that do manage to become established. Thus, in many places in and near the Southern Rockies, senescent cottonwood forests are being replaced by less biologically diverse, weedy species (Somers and Floyd-Hannah 1996).

Roads are common along the streams and rivers of the Southern Rockies. Poorly designed roads can increase sedimentation to levels that can destroy aquatic habitat and aquatic life (Shinneman et al. 2000). They are sources of gasoline, oil, and other pollution run-off and provide easy access for people and exotic species to sensitive riparian areas. These ecological effects can extend up to 154 m (500 ft.) from the road edge (Forman et al. 1995). The U.S. Forest Service requires that roads be built at least 31 m (100 ft.) from a stream on National Forest lands. In the Southern Rockies many roads do not adhere to this standard.

Before EuroAmerican settlement, wetlands covered about 3% of the state of Colorado, but roughly one-third to one-half of that original wetland acreage has been lost to human development and conversion to croplands (Dahl 1990, Wilen 1995). Lowered groundwater levels may lead to the destruction of groundwater-fed springs and seeps that provide water for wildlife and sustain wetland communities. For instance, thousands of hectares of marsh wetlands in the San Luis Valley may be destroyed as a result of current and proposed groundwater pumping which will likely draw-down water tables a meter or more (Cooper 1993). Loss of groundwater water supply may also threaten some riparian forests, which depend on groundwater when stream levels have dropped (Cooper 1993, Power et al. 1997).

In addition, several exotic species threaten native aquatic species in the Southern Rockies. For instance, many streams and lakes have been stocked with non-native fish species (usually "sport" fishes), which have altered the food chain and prey upon, out-compete, or hybridize with native fishes. These exotic fish species threaten all three subspecies of native cutthroat trout (Young 1995). Other disruptive exotic species of concern include the eastern bullfrog (*Rana catesbeiana*), which threatens native frogs, and purple loosestrife (*Lythrum salicaria*), which displaces native wetland plant species (Rosen and Schwalbe 1995, Rutledge and McLendon no year).

### Plains Steppe and Great Basin Grasslands

#### *Description*

These grassland ecosystems, although peripheral to the Southern Rockies, have many close ecological ties to the Southern Rockies due to animal migration, water and nutrient flows, and other natural processes. They generally occur below 1,800 meters in elevation, with short-grass prairie and occasional mixed-grass and tall-grass prairie communities along the eastern and northern edges of the Southern Rockies (Benedict 1991, Knight 1994), while Great Basin semi-desert grasslands occur at lower elevations along the southwestern fringes (Dick-Peddie 1993). The eastern short-grass prairie is characterized by blue grama grass (*Bouteloua gracilis*), buffalo-grass (*Buchloe dactyloides*), western wheat-grass (*Pascopyrum smithii*), needle-and-thread grass, James' galleta (*Pleuraphis jamesii*), Indian rice grass, fringed sage (*Artemisia frigida*), yucca (*Yucca glauca*), prickly pear cactus (*Opuntia* spp.) and tree cholla (*Opuntia imbricata*, Benedict 1991, Knight 1994). Great basin grasslands are dominated by black grama (*Bouteloua eriopoda*) and many shrub species (Dick-Peddie 1993).

Elk from the mountains historically wintered in these lower elevations (Fitzgerald et al. 1994), and wolves and grizzly bears regulated ungulate and other prey species populations. Black-tailed prairie dog (*Cynomys ludovicianus*) burrows improved nutrient cycling and increased habitat diversity (Whicker and Detling 1988). Large prairie dog towns and heavy grazing by bison provided habitat for numerous other species, such as western rattlesnake, burrowing owl (*Athene cunicularia*), black-footed ferret (*Mustela nigripes*), badger, ferruginous hawk, and mountain plover (*Charadrius montanus*, Miller at al. 1996). Fire also regulated these ecosystems (Knight 1994).

Black-tailed prairie dog (*Cynomys ludovicianus*)

*Conservation Concerns*

Grasslands on the eastern edge of the ecoregion have largely been converted to farmland and urban landscapes (Dick-Peddie 1993, Shinneman et al. 2000, Theobald 2000). Where large grassland areas still exist, they are subjected to heavy livestock grazing and fire suppression (The Nature Conservancy 1998), which alters plant composition and may lead to invasion by woody plant species (Wright and Bailey 1982). Non-indigenous plants comprise between 13 and 30% of the prairie species (Sampson et al. 1998). Wild bison, grizzly bear, wolves, and black-footed ferrets have been extirpated (Fitzgerald et al. 1994, Shinneman et al. 2000), black-tailed prairie dogs have been eliminated from over 98% of their original geographic range (Mac et al. 1998), and elk and pronghorn are scarce or absent from many areas (Fitzgerald et al. 1994).

Bison grazing and wallowing had a large impact on grassland diversity (Detling 1998). Prairie dogs are a keystone species, and grassland inhabited by prairie dogs provides a greater mosaic of vegetation structure, an abundance of prey for predators, burrow systems used by many species, and altered ecological processes (increased nitrogen content, succulence, and productivity of plants, and macroporosity of soil), as compared to uninhabited grasslands. The matrix of ecological boundaries created by prairie dog colonies improves overall diversity of life across a landscape (Paine 1966). After extirpation of prairie dogs and bison, the prairie is ecologically poorer.

### Other Ecosystems of Note

Numerous other "large patch" natural communities exist alone or are nested within the major ecosystems described above. For example, sand dune complexes occur in the San Luis Valley and North Park. Those in North Park are comprised of both active sand dunes (some over 200 meters high) and a stabilized sand dune and swale complex that covers thousands of hectares and supports numerous unique natural communities and species, including several endemic species of plants and insects (Neely et al. 2001). Other important natural features are the many cliffs, canyons, and barren rock faces scattered throughout the region. Many birds nest in cliff habitats, such as the peregrine falcon (*Falco peregrinus*) and black swift (*Cypseloides niger*), and unique species of plants are found in these rocky habitats.

## 7. Diversity and Distribution of Plants and Other Wildlife

The Southern Rockies are a biological meeting place, where species representative of high-elevation Rocky Mountain ecosystems converge with species from adjacent lowland prairie and semi-desert ecosystems, often forming unique natural communities. The region is well known for species such as elk, mountain lion, and ponderosa pine, but thousands of lesser-known species also call the Southern Rockies home. Some of these species are abundant and widespread, found in almost every major habitat type in the region. For example, Colorado has the West's largest elk herd, with over 305,000 individuals (Meyers 2002), and the state also supports roughly 550,000 deer (Colorado Division of Wildlife 2001). In contrast, other species in the region are narrowly restricted to particular habitats and locations, like the dependence of Abert's squirrels on dense, old ponderosa pine stands (Fitzgerald et al. 1994).

Shinneman et al. (2000) estimated that there were at least 328 extant native vertebrate species closely associated with the Southern Rockies' mountain habitats, including 203 birds, 90 mammals, 18 fishes, 10 reptiles, and 7 amphibians. When "peripheral" vertebrate species are added to the list (those species more closely associated with the neighboring short-grass prairie or Colorado Plateau regions but occupying the margins of the Southern Rockies) well over 500 vertebrate species are thought to inhabit the region (Shinneman et al. 2000). Colorado alone has over 2,596 native vascular plants, and lesser-known taxonomic groups have perhaps thousands more species awaiting recognition or discovery (Weber and Whitman 1992, Stucky-Everson 1997, Opler, pers. comm.).

Compared to other ecoregions, the Southern Rockies are rich in bird species, and only two other ecoregions in the U.S. and Canada have a higher total number of mammal species (Ricketts et al. 1999, Shinneman et al. 2000, Neely et al. 2001). The region has the second highest number of Lepidopteran species north of Mexico. There are over 270 species of butterflies and an estimated 5,200 species of moths; this represents over 40% of North America's known moth and butterfly species (Kocher et al. 2000, Opler pers. comm.). In addition, hundreds of these species are globally rare and many are limited to the Southern Rockies. Examples of species endemic to the Southern Rockies include the San Luis Dunes tiger beetle (*Cicindela theatina*), the Jemez Mountains salamander (*Plethodon neomexicanus*), the Weber monkey-flower (*Mimulus gemmiparus*), Uncompahgre fritillary (*Boloria improba acrocnema*), greenback cutthroat trout, brown-capped rosy finch, and banded physa (*Physa utahensis*).

In general, species richness in the Southern Rockies is greatest with lower elevation ecosystems, culminating in low-elevation riparian and aquatic ecosystems. For comparison, over 450 vertebrate species are associated with Colorado's wetland and riparian communities, 129 within foothill ponderosa pine forests, and 51 with alpine habitat (Shinneman et al. 2000). However, another important consideration is that 40% of the region is within private lands, and these lands typically contain biologically rich lower elevation landscapes. Theobald et al. (1998) determined that in Colorado species richness is 46% higher on private versus public lands.

## 8. Species and Communities at Risk

Persecution, intensive and extensive levels of natural resource use, pollution, and erupting human population levels have led to the extinction of at least four species or subspecies that once inhabited the region: yellowfin cutthroat trout (*Salmo clarki macdonaldi*), Eskimo curlew (*Numenius borealis*), Carolina Parakeet (*Conuropsis carolinensis*, which visited the southern foothills and plains), and New Mexico sharp-tailed grouse (*Tympanuchus phasianellus hueyi*, Shinneman et al. 2000, Neely et al. 2001). The grizzly bear, gray wolf, wild populations of bison, black-footed ferret, Canada lynx, and wolverine are considered extirpated from the Southern Rockies (Shinneman et al. 2000). The Rio Grande bluntnose shiner (*Notropis simus simus*), Rio Grande silvery minnow (*Hybognathus amarus*), American eel (*Anguilla rostrata*), freshwater drum (*Aplodinotus grunniens*), sturgeon (*Scaphirhynchus platorynchus*), blue sucker (*Cycleptus elongatus*), Rio Grande shiner (*Notropis jemezanus*), and speckled chub (*Macrhybopsis aestivalis aestivalis*) are also extirpated (Neely et al. 2001). At least four species of birds that historically bred in the region no longer do so: marbled godwit (*Limosa fedoa*), harlequin duck (*Histrionicus histrionicus*), merlin (*Falco columbarius*), and ring-billed gull (*Larus delawarensis*, Andrews and Righter 1992, Neely et al. 2001).

At least 8 plants, 2 invertebrates, and 10 vertebrate species from the region are currently listed as Threatened or Endangered under the U.S. Endangered Species Act (Shinneman et al. 2000, U.S. Fish and Wildlife Service 2001a). Black-tailed prairie dogs are listed as warranting Threatened status, but precluded by other priorities. State Natural Heritage Programs track 101 species native to the Southern Rockies that are globally imperiled (ranked G1-G2), and nearly 300 other species are considered of special concern due to restricted ranges, population declines, and other vulnerability factors (Neely et al. 2001). Shinneman et al. (2000) reported that of the 328 species closely linked to the Southern Rockies' mountain habitats, the state Natural Heritage Program lists about one-half of the amphibians, one-half of all fishes, one-quarter of all birds, and one-fifth of all mammals as vulnerable, imperiled, or critically imperiled.

Finally, as discussed, many natural communities are greatly reduced in extent, natural composition, and function due to activities such as logging, fire suppression, livestock grazing, housing development, agricultural conversion, water use/dams, and pollution. Habitats particularly at risk include old-growth ponderosa pine forest, old aspen forest, low-elevation riparian communities, sagebrush shrublands, montane grasslands, and most wetlands and aquatic systems

(Noss et al. 1995, Shinneman et al. 2000, Neely et al. 2001).

## 9. Conclusion

In the Southern Rockies, few if any ecosystems are unaltered by humans. We all understand history in a given time frame, largely gleaned from our personal experiences. But to really understand the processes that exist in the present, it is necessary to return to a time when there were no people in North America. The footprint of people cannot be separated from the track of ecological history in the Southern Rocky Mountains.

In a short period of 13,000 years, humans from Eurasia have profoundly changed and remolded the ecology of the region in two major waves of colonization. The impact of humans on ecological processes has been accelerating at a rapid rate, particularly since the European wave swept ashore 500 years ago. Where does that leave us when we try to evaluate what is a natural state? The National Park Service has grappled with the question of using an historical benchmark as a goal for present management policies. The debate over what time period would provide the best benchmark, and even what existed at a given benchmark, has sometimes been acrid. And, most benchmarks simply represent points along a declining slope of biological diversity because of human impacts.

It could thus be argued that almost all land has been altered, at least to some extent, making the rewilding idea a gradation along a cline. Rather than arguing whether or not rewilding is necessary, it would be better to analyze where a particular region lands on the regression slope and where the region will fall if present trends continue. Where there has been less damage (*e.g.* polar areas) the term rewilding may not resonate as well as in areas with more damage. In areas with less human impact, the concept of holding the line may be more motivational. Nevertheless, the term rewilding is highly relevant to the Southern Rockies region.

It is obviously not feasible to return to a time when there were no humans in the western hemisphere. Rather than trying to resurrect past conditions at some point over the last 13,000 years, we suggest that it would be much more useful to carefully evaluate present conditions—both ecological conditions and those of human attitudes toward nature. With that knowledge, goals could be based on "where can we go from here?" What can we provide to restore land to a state where it is "self-willed" (where ecological and evolutionary processes, not the economic policies of humans, can dictate the direction and pace)? What philosophies, attitudes, and beliefs of humans need to change? The Southern Rockies Wildlands Network Vision and restoration of large carnivores are such rewilding steps.

Yet, compared to other areas in the U.S., the Southern Rockies still contain many opportunities to protect and restore a vast biological wealth and diversity. We can act. Large matrix communities, such as subalpine forests, remain relatively intact throughout the region and in similar patterns of distribution as when EuroAmericans first settled the region. In addition, many subalpine natural communities with long-rotation disturbance regimes have probably not been significantly altered by fire suppression. Fifty years of effective fire suppression activity has not significantly affected ecosystems that experience stand-replacing events on the order of hundreds of years (Romme and Despain 1989). This may not be true, however, for some subalpine forests (Kipfmueller and Baker 2000). Many areas that have experienced significant anthropogenic disturbance in the Southern Rockies, such as logging, road building, overgrazing, and even damming and fire suppression, can be restored, especially where they exist on publicly owned lands.

As Gosnell and Shinneman point out in the next chapter, roughly 60% of the region is within public ownership, nearly 12% is within strictly protected reserves, and another 13% remains in an unprotected roadless condition (Table 3.1). Thus, collectively, nearly a quarter of the region (~4 million ha) remains in a wild condition and roughly two-thirds (~10 million ha) receives at least some minimal level of protection as publicly owned lands. Protection for the unprotected roadless lands would mean that 22 of the 27 ecological systems would be protected above a 10% level and 12 of 27 would be protected above a 20% level. Moreover, these public, protected, and roadless lands are configured in a pattern of relative connectivity across the landscape. Thus, on an ecoregional scale, some of the key strategies to protecting and restoring wide-ranging species and large landscape natural processes in the Southern Rockies will be to protect ecologically strategic public land that remain at risk, such as roadless lands and wetlands. Also, working with willing landowners could conserve biologically important habitat on private lands. Finally, addressing the issue of barriers that sever the region's connectedness, such as dams and highways, will enhance wildlife movement.

Burrowing owl (*Athene cunicularia*)

*It's just one world, this spine of rock and streams*
*And snow, and the wash of gravel, silts*
*Sands, bunchgrasses, saltbrush, bee-fields,*
*Twenty million human people, downstream, here below...*

-Gary Snyder

# 3 THE HUMAN LANDSCAPE

Doug Shinneman, Hannah Gosnell

*(This chapter is a shortened version of Gosnell and Shinneman in press "The Feasibility of Wolf Reintroduction to the Southern Rockies", edited by Phillips et al. 2003.)*

## 1. Introduction

A successful Wildlands Network Vision requires careful consideration of habitat and ecological conditions, but equally important are the human dimensions of wildlife conservation. A solid understanding of a region's culture, politics, and history can provide conservationists with insight into locals' attitudes regarding wildlife issues, help them identify key stakeholders, and prepare them for potential conflicts and allies.

In this chapter we describe the human landscape of the Southern Rockies, which includes concrete "things" on the land, like people, buildings, roads, dams, and mines, as well as less tangible anthropogenic influences, like political boundaries, land management regimes, and local economies.

Section 2 of this chapter describes the ways in which humans have organized and "divvied up" the land, in terms of ownership, management, protection, and roads. Next, we look at historical and current uses of the land, and associated ecological impacts. Section 4 describes the cultural landscape, past and present. The sections on population, land development, and economic trends use recent data to describe what is currently happening in the region as a whole, as well as in the subregions of interest. The conclusion considers what the findings in this chapter might mean for a Wildlands Network Vision.

## 2. The Lay of the Land

### *Land Ownership*

A complex pattern of public and private land ownership exists in the Southern Rockies (Figure 3.1 and Table 3.1).

**Table 3.1 Land ownership in the Southern Rockies ecoregion.**

| Owner | Hectares | % |
|---|---|---|
| U.S. Forest Service | 7,070,000 | 42.4 |
| U.S. Bureau of Land Management | 1,940,000 | 11.6 |
| Other federal, state, county and city | 870,000 | 5.2 |
| Native American tribal land | 500,000 | 3.0 |
| Private | 6,300,000 | 37.8 |
| Total | 16,680,000 | 100 |

see Figure 3.1 for data sources.

This pattern owes its existence to a history of land acquisition by the U.S. government and transfer of public lands to private interests through mining claims, farmland acquisition under the Homestead Act of 1862, and other means (Wilkinson 1992). The U.S. government also retained large portions of the land and eventually allocated these lands to newly established public agencies such as the U.S. Forest Service, National Park Service, and Bureau of Land Management. Western states were granted "school trust" lands from the federal government, inheriting millions of acres (usually in separate square mile sections) that were intended to raise money for public education.

The federal government owns roughly 55% of the land in the Southern Rockies. When state lands are added to that total, about 58% of the Southern Rockies are within public ownership. The U. S. Forest Service is the single largest landowner.

The Southern Rockies' ecosystem types do not fall evenly across this complex pattern of land ownership. Many lower elevation ecosystems, such as grasslands and shrub lands, are mainly on private, Bureau of Land Management, or state lands, while most land area within high-elevation ecosystems, such as subalpine forests and alpine tundra, occurs on National Forest lands. Moreover, single functioning ecosystems, defined at almost any scale, regularly fall across multiple land ownership types. Because of vastly different management emphases among these entities, comprehensive protection of ecosystems can be difficult.

For example, this complex public-private ownership pattern has concentrated development in lower elevations and valleys that are predominately in private hands, often fragmenting undeveloped, natural habitat that remains on adjacent public lands. Yet, the region also has one large mostly contiguous area of public land that connects the Medicine Bow Mountains in Wyoming to the southern end of the Sangre de Cristos in New Mexico, and the Front Range foothills outside Denver to the Colorado Plateau in the west (Figure 3.1). This intact public land pattern provides hope for maintaining habitat connectivity across the region.

Another human-created landscape pattern important to native ecosystems is the distribution of cities and counties in and around the ecoregion. These cultural and political entities influence land use and development patterns, based on factors such as population growth, local economies, and land use planning policies. For instance, certain counties and cities, such as the City of Boulder, Boulder County, and Jefferson County in Colorado, have ambitious open space land acquisition programs, as well as growth control measures, while most other Southern Rockies cities and counties have no such land conservation initiatives and often very little in the way of effective land use planning (Shinneman et al. 2000).

### *Land Management and Protection*

Since the establishment of the U.S. Forest Reserves in the 1890s, the Southern Rockies have witnessed numerous conservation milestones. Rocky Mountain National Park was established in 1915, and in 1919 the Trapper's Lake area on the White River Plateau was the first National Forest area in the nation managed for wilderness values. Since those early conservation accomplishments, the Southern Rockies have gained 49 federally designated Wilderness Areas and 6 National Parks and Monuments. When other strictly protected lands, such as U.S. Forest Service Research Natural Areas (RNAs), are added to that total, there are at least 1,740,000 ha protected, or roughly 10.5% of the ecoregions land-base. More than 200,000 additional hectares are within areas afforded slightly lesser levels of protection, such as state wildlife areas, National Wildlife Refuges, and Bureau of Land Management Areas of Critical Environmental Concern (ACECs). In addition, the region contains other protected lands, at various levels of stewardship, that are not mapped and thus not included above (*e.g.*, county and city open space lands, private nature reserves, and many private land conservation easements).

As a result, the Southern Rockies contain a fairly high percentage of protected land compared to most other regions in the U.S. However, most of the strictly protected federal lands, such as National Parks and Wilderness Areas, were not selected to represent the full diversity of ecosystems and species in the region. Rather, these areas were typically chosen for protection because of scenic and recreational values, because they contained charismatic wildlife such as elk, or because they had little economic value.

So, many of the Southern Rockies' ecosystems and species are not well represented in the existing system of publicly owned nature reserves, such as National Parks and Wilderness Areas. Biologically rich landscapes, such as low elevation riparian areas and shrub lands, are not covered well in the current system of nature reserves. Shinneman et al. (2000) determined that only 3 of 12 major terrestrial ecosystem types in the Southern Rockies had more than 10% of their total area within strictly protected lands such as National Parks, Wilderness Areas, and Research Natural Areas.

Numerous areas may yet become protected in the Southern Rockies to remedy the above gaps. Recently the U.S. Forest Service identified just over 2,000,000 ha of roadless lands on National Forests that were to be given some protection under the Roadless Area Rule (U.S. Forest Service 2000). At the time of this writing, that finding is being reassessed by the G.W. Bush administration. However, the potential to protect additional wildlands in the region is a significant opportunity, as these lands would bring the total protected area to roughly 4,000,000 ha. Moreover, these lands contain significant amounts of lower elevation ecosystems now poorly protected. Shinneman et al. (2000) predict that if these roadless areas were given protection, nearly 10 of the 12 major ecosystem types they analyzed would have at least 10% of their total area protected, and 5 of the 12 would have protection levels above 25%. In addition, local conservation groups have inventoried roadless areas that the U.S. Forest Service does not yet officially recognize (Shinneman et al. 2000).

Local and regional land trusts have become quite

numerous in the region, buying private properties and establishing conservation easements with willing landowners. For instance, the Colorado Coalition of Land Trusts now has 6 regional and national land trusts and 33 local land trust member organizations working throughout the Colorado portion of the Southern Rockies. Millions of dollars of federal, state, and local government money have been directed toward local open space and state park and wildlife refuge acquisitions, and to acquire significant new federal lands such as the U.S. Forest Service's recent acquisition of New Mexico's 36,000 ha Baca Ranch. Great Outdoors Colorado donates proceeds from the Colorado lottery to protect land, usually in cooperation with other governmental and non-profit entities. The Nature Conservancy as well as the Wildlands Project, have undertaken landscape level conservation planning efforts. In addition, Great Sand Dunes National Monument and Black Canyon of the Gunnison National Park received additional protection through expanded boundaries and Wilderness designation, and wilderness legislation has been introduced in Congress to protect 520,000 ha of primarily Colorado Bureau of Land Management roadless areas.

### *Roads/Infrastructure*

The ecological value of protected land in the Southern Rockies is compromised by ubiquitous roads, which fragment the landscape. Starting with the mining boom of the late 1800s, people wanted more transportation in the Southern Rockies, and wagon roads, stagecoach routes, and railroads soon criss-crossed the region. During this early settlement period, railroads in particular were instrumental not only in getting valuable minerals out of the region, but in bringing resources, new industries, new residents, and tourists into the Southern Rockies. Cities such as Greeley, Colorado and Laramie, Wyoming, sprang up and prospered along railroad lines, and railroads helped to transform the Southern Rockies from a frontier region to a modern industrial economy. Railroads also helped facilitate migration to the region; between 1860 and 1900, Colorado's population grew from 34,277 to 539,700 (Noel et al. 1994).

During the early to mid 1900s the construction of paved automobile roads changed the region even more. In Colorado, paved roads grew from roughly 800 to 6,400 km between 1930 and 1940 (Noel et al. 1994). By the 1950s, with the advent of the interstate highway system and the mobility provided by the modern automobile, easy access was possible into formerly remote mountain locations, promoting further population growth, tourism, and new development industries, including the region's famed downhill ski resorts.

Today there are over 121,000 km of primary and secondary roads in the Southern Rockies (Figure 3.2), not including most residential streets and the thousands of miles of poorly mapped primitive roads (Shinneman et al. 2000). There are over 27,742 km of inventoried roads on National Forest lands in Colorado alone (Finley 1999). Local road densities vary greatly within the region, but are often much higher than expected, even in relatively undeveloped areas. For instance, one study in New Mexico found that road density in Bandelier National Monument and surrounding area averaged over 6.25 km/km$^2$ (Allen 1994). In contrast, there are also several large areas in the Southern Rockies that are relatively devoid of roads, especially those centered in large Wilderness Areas, such as in portions of the San Juan Mountains and on the White River Plateau. However, except for these large wildlands, few areas in the Southern Rockies are more than 6.5 km from the nearest road (Shinneman et al. 2000). Alpine and subalpine habitats have fewer roads than have most other habitats; lower elevation and more biologically diverse habitats are usually the most heavily roaded in the Southern Rockies (Shinneman et al. 2000).

Roads are a concern for many land managers, biologists, and conservationists due to their impacts on native species and ecosystem function (Schoenwald-Cox and Buechner 1992, Trombulak and Frissell 2000). Ecologically deleterious impacts of roads include: 1) increased species mortality due to automobile collisions (Bangs et al. 1989) reduced species mobility for both small and large animals, due to the barrier effect (Fahrig et al. 1995, Foster and Humphrey 1995) increased dispersal of edge-adapted, weedy, aggressive, opportunistic, and parasitic species due to the travel corridor effect (Tysor and Worley 1992, Parendes and Jones 2000); 4) greater human access to habitat interiors and activities such as fuel-wood gathering, hunting, poaching, plant gathering, and recreation in those areas (Lyon 1983, Trombulak and Frissell 2000); 5) increased sediment and pollution runoff into nearby streams and wetlands (Bauer 1985, Forman and Deblinger 2000); and 6) increased likelihood of severe erosion when roads are built on steep slopes (Trombulak and Frissell 2000). Combined, these factors fragment and isolate natural habitat by subdividing formerly intact vegetation patches and creating a "road effect zone" that changes the habitat conditions and species compositions well into the interiors of adjacent natural habitat (Reed et al. 1996, Shinneman and Baker 2000, Forman 2000).

Various factors influence the relative impact that roads have on the environment and species. For instance, a lightly used, primitive dirt road may not restrict some species from moving across, while a busy, paved, four-lane highway may be an impermeable barrier to many wildlife species.

The U. S. Fish and Wildlife Service has identified Interstate 80 in Wyoming and Interstate 70 in Colorado as serious threats to future lynx (*Lynx canadensis*) populations, and state routes are likely to be expanded as traffic increases. On the other hand, primitive roads may have lower collision rates, but they allow access to wildlife areas and raise the probability of poaching. Options such as closing some roads (seasonally or permanently) and constructing wildlife underpasses or land bridges may lessen the negative ecological effects of some road networks. When road densities are high, species such as elk (*Cervus elaphus*), mountain lion (*Puma concolor*), wolves (*Canis lupus*), black bear (*Ursus americanus*), and grizzly bear (*Ursus arctos*) may not persist due to an aversion to roads or negative impacts from increased human hunting, poaching, and harassing (Lyon 1983, Van Dyke et al. 1986, McClellan and Shackleton 1988). For instance, studies suggest that if road densities exceed 0.4 km/km$^2$, wolf populations may not persist (Thiel 1985, Mech 1989).

## 3. Historical and Current Land Uses

Mining, livestock grazing, logging, and water use have had significant ecological impacts on the natural ecosystems of the Southern Rockies. While these activities no longer form the basis of the Southern Rockies' economy, they still play major roles in the shape and condition of the current physical landscape. Mining put the Southern Rockies on the map for many Americans during the late 1800s, and helped to promote westward migration in the United States and the settlement and industrial development of the Southern Rockies. Livestock grazing started in the 1600s in northern New Mexico and now occurs in nearly all grazable locations and ecosystem types throughout the region. Although the Southern Rockies have never been an economically prosperous lumbering region as a whole, 150 years of Euro-American settlement and localized logging booms have had significant impacts on the region's forests. Nearly every drop of water originating in the Southern Rockies has been allocated toward agricultural, industrial and urban uses, manifested in the region's myriad dams, ditches, and tunnels that store and redirect millions of acre-feet of water every year. In this section we discuss in more detail the history and ecological consequences of these land and resource uses.

### *Logging*

Logging in the Southern Rockies has never been as big an industry as in the Pacific Northwest. Nonetheless, most of the 9,500,000 ha of forest and woodlands in the Southern Rockies (59% of the land cover) have been affected by timber cutting, road building, and fire suppression (Shinneman et al. 2000). Ponderosa pine and Douglas-fir forests – both central to the region's foothill forest ecosystems – have been the hardest hit by logging activities, which have often focused on cutting the largest, oldest trees. Of the remaining ponderosa pine (*Pinus ponderosa*) stands in Southern Rockies National Forests, less than 5% are considered to be in an old-growth structural stage, a problematic situation for species that depend on old-growth habitat (Shinneman et al. 2000, Chapter 2—The Natural Landscape).

In addition to eliminating old-growth habitat, logging practices have fragmented habitat with clearcuts and roads. A 1996 study of the spruce-fir and lodgepole pine forests on the Medicine Bow National Forest in Wyoming found that clearcutting there had resulted in a landscape more fragmented than any in the Pacific Northwest (Reed et al. 1996). Forest fragmentation may lead to declines in populations of forest-interior-dependent species, including the boreal owl (*Aegolius funereus*), the northern goshawk (*Accipiter gentilis*), and the American marten (*Martes americana*).

### *Mining*

Mining has had a significant effect on the landscape, as well. There are currently hundreds of active mines in the region. In addition, there are at least 9,700 abandoned mines in the Southern Rockies, many of which continue to pollute terrestrial and aquatic ecosystems in the region (Shinneman et al. 2000). One of the worst offenders is the Summitville Mine in the San Juan Mountains, which was declared bankrupt in 1992. This open-pit, cyanide heap-leach gold mine leaked significant quantities of acidic, metal-rich drainage and cyanide solutions into the Wightman Fork of the Alamosa River, destroying all aquatic life for 27 km downstream (Hinchman and Noreen 1993, Shinneman et al. 2000).

### *Water Use*

Water use in the Southern Rockies has enabled the region to flourish, but contrary to popular belief, as much as 90% of the water diverted from streams goes not to the sprawling subdivisions along the Front Range, but rather to crops like hay and alfalfa, grains, vegetables, and fruit (Riebsame 1997). There are approximately 800,000 irrigated hectares in the Colorado. This is not to say that urban water use is inconsequential. In Denver, for example, over half the water consumed is attributable to outdoor landscaping (Riebsame 1997). The biggest impacts related to water use in the Southern Rockies come from dams and reservoirs, *e.g.*, loss and degradation of stream habitat and riparian habitat, loss of groundwater function, and alterations in stream hydrology (Shinneman et al. 2000). Water issues are exacerbated by drought cycles, and the last few years of drought have heightened concern about conservation, additional storage, and techniques to squeeze more water out of our arid region.

### Agriculture

Agricultural practices have played a major role in the transformation of the Southern Rockies landscape through the conversion of native vegetation and natural communities to croplands and rangelands. Roughly 800,000 ha, or 5% of the Southern Rockies, are currently classified as either dryland or irrigated cropland, and this 5% tends to be some of the most biologically important land in the ecoregion (*e.g.* valley bottoms, riparian areas, and wetlands). The Southern Rockies Ecosystem Project found that approximately 82% of all croplands in the ecoregion are below 2,461 m in elevation, and 10% are within 150 m of a river or perennial stream (Shinneman et al. 2000).

Agriculture in the region, however, appears to be on the decline. In Colorado, New Mexico, and Wyoming, the number of full-time farms decreased between 1992 and 1997, as did the average size of farms in all three states. Land in farms decreased by 4% in Colorado and 2% in New Mexico, but increased by 4% in Wyoming (US Department of Agriculture 1997). Some of the decrease in cropland is absorbed by suburban and exurban development, which are also inhospitable to wildlife.

### Livestock Grazing

Ranching's contribution to the region's economy has been declining, but its effects on the land are still extensive and significant. Most ranchers in the Southern Rockies depend at least partially on public lands for grazing their animals. Typically, cattle and sheep spend summers on high elevation meadows in National Forests, and then in fall are moved to lower elevation rangelands, often Bureau of Land Management grasslands. Figure 3.3 shows active grazing allotments on federal lands by county throughout the region.

Nearly 70% of the 7,160,000 ha of U.S. Forest Service lands in the ecoregion have active grazing allotments on them, and 93% of the 3,320,000 ha of Bureau of Land Management lands in Colorado are actively grazed by livestock. Of the 1,200,000 ha of state school trust lands in Colorado, roughly 80% are actively grazed by livestock. These figures are similar for Bureau of Land Management and state lands in the Wyoming and New Mexico portions of the ecoregion. In addition, livestock grazing is allowed on portions of all three National Wildlife Refuges in the region, as well as in many Wilderness Areas. Given these numbers, Shinneman et al. (2000) estimate that roughly 70%-80% of state and federal public lands in the ecoregion are actively grazed by livestock, and 80%-90% are available to livestock grazing.

The Bureau of Land Management is required to monitor the condition of its rangelands. In 1998, the Colorado Bureau of Land Management estimated that 72% of the 1,240,000 ha where forage condition was rated and classified was in "fair" or "poor" condition. Given that "fair" condition means the land supports less than one-half its historical carrying capacity, these numbers should cause some concern.

Wilcove et al. (1998) estimated that livestock grazing has been a factor in the imperilment status of 33% of federally listed threatened species and 14% of endangered species. Since cattle tend to congregate along stream banks, water quality and stream hydrology can suffer serious negative impacts (Schultz and Leininger 1990). Heavy grazing causes changes in plant species structure and composition (*e.g.*, the proliferation of weeds like cheatgrass (*Bromus tectorum*), and some of those changes can lead to increased soil erosion (D'Antonio and Vitousek 1992). Indeed, rangeland managers often intentionally introduce non-native grasses, such as crested wheatgrass (*Agropyron cristatum*, Noss and Cooperrider 1994).

Large ungulates, predators, and other native animals are negatively affected by grazing as well. Fences for controlling roaming livestock interfere with animal movement, especially pronghorn (*Antilocapra americana*), but also deer (*Odocoileus* spp.) and elk (Noss and Cooperrider 1994). Livestock compete with native herbivores for forage, water, and space, and livestock managers make a practice of eliminating "pests" and predators like prairie dogs (*Cynomys* spp.) and coyotes (*Canis latrans*, Peek and Dalke 1982). Top predators such as wolves and grizzly bears were eliminated in the Southern Rockies in the early 1900s, largely to accommodate the livestock industry (Fitzgerald et al. 1994). The absence of these large carnivores has resulted in unnaturally large elk populations throughout much of the Southern Rockies, which has led to over-browsing of native vegetation, like aspen (*Populus tremuloides*), in some places (Stohlgren 1998, Baker et al. 1997).

The Forest Service and Bureau of Land Management are working with livestock operators to improve the condition of public rangelands, and there has been some progress. However, of the 1,920,000 ha where the Bureau has determined rangeland trends, only 26% showed improvement in 1998.

### *Recreational Uses*

Every year, millions of people visit the public lands of the Southern Rockies for recreation, bringing significant tourist dollars to the region, many of whom come at least in part to see wildlife. Their presence is positive for the economy, but challenging for wildlife managers.

Most people who come to the Southern Rockies for recreation target one of the six National Parks and Monuments in the region or one of the eight National Forests. Recreation on Bureau of Land Management land is on the rise, especially with the growth of off-road vehicle recreation, but it does not rival use of the parks and forests.

The most popular National Forests in the region, measured in "recreation visitor days" (RVDs), are the White River, Pike/San Isabel, and Santa Fe (Shinneman et al. 2000). The White River National Forest ranked fifth in the nation in 1995 in terms of visitor days and is recognized throughout the world for its exceptional outdoor recreation opportunities. Its 900,000 hectares surround major resort areas like Aspen, Vail, and Breckenridge, and provide 13% of all ski visits in the nation. Being only 2-4 hours west of Denver on I-70, it is the primary target of Front Range "weekend warriors." Although the White River National Forest contains only 16% of Forest Service lands in Colorado, it hosts about 30% of the state's National Forest Recreation (U.S. Forest Service 1999). In addition, the Pike/San Isabel area is popular with backpackers, campers, and all-terrain vehicle enthusiasts; and the Santa Fe National Forest is heavily visited because of its close proximity to the urban areas of Santa Fe and Albuquerque.

One of the big impacts associated with recreation in the Southern Rockies is the extensive land development associated with ski areas (*e.g.* parking lots, second homes, condos, resorts, golf courses, and shopping centers). The ski areas themselves fragment high elevation forests with ski runs, chair lifts, and high mountain lodges, and they sometimes dewater streams for snow-making. The recent expansion of Vail Resort Ski Area into lynx habitat in the White River National Forest provides an example of how controversial ski area impacts can be (Thompson and Halfpenny 1991, Glick 2001). The White River National Forest currently has 18,212 ha under special use permits for skiing, and is contemplating plans for ski area expansion.

And the ski industry will continue to flourish in the Southern Rockies. Figure 3.4 predicts higher rates of growth in user days for activities like cross country skiing (242%), downhill skiing, and backpacking, and slower rates for hunting (22%), fishing (59%), snowmobiling, and off-road driving (54%)(Bowker et al. 1999), though presently off-road vehicle use is skyrocketing in popularity.

Summit County Colorado is growing the most rapidly of any county in the Southern Rockies and has the highest potential to provide additional capacity for skiing on National Forest lands. If growth rates stay the same, the combined daily capacity in 2010 will need to be 53,070 skiers at one time to meet the projected demand of 5,000,000 skiers per year. About 570 additional ha of National Forest lands are needed to meet this demand. By 2030, it is estimated an additional 3,424 ha would be needed to meet projected demand for skiing based on current growth rates. This would result in a total of 10,000 ha of National Forest land allocated to Summit County skiing (U.S. Forest Service 1999).

Though mechanized recreation is not projected to grow as fast as downhill and cross-country skiing over the next 50 years, it still has a significant and growing presence on the landscape. In Colorado, the number of registered all-terrain vehicles (ATVs) more than tripled during the 1990s, and snowmobile numbers increased by 64% (Finley 1999). Off-road vehicle use on fragile desert lands is of particular concern. Low elevation lands near population centers are under extreme pressure from ATVs and backcountry jeeping, while the higher elevations are increasingly subject to snowmobile activity. Mountain biking is a burgeoning sport with significant impacts to ecosystems on popular trails.

Even hiking and backpacking, seemingly low-impact activities, can have cumulative ecological impacts. Trails often traverse riparian areas and nesting areas and can harm native species and damage delicate natural communities. Heavy traffic in high elevation tundra can cause damage that takes years to repair. The Colorado Fourteeners Initiative is working throughout the state to improve trail systems and minimize human impact in fragile mountain ecosystems.

The main impacts associated with recreation on the public lands of the Southern Rockies are direct disturbance of wildlife, modification of habitat through vegetation damage, introduction of exotic species, erosion, and air and water pollution (Knight 1995).

## 4. The Cultural Landscape

Prior to the arrival of permanent Spanish and later Anglo settlements over 450 years ago, the Southern Rockies were home to native peoples who lived in both permanent habitations as well as on the land as hunters and gatherers. Beginning with the Paleo-Indians who inhabited the Southern Rockies around 11,000 years ago, the ecosystem gradually became home to the pueblo (village) peoples, who planted and harvested crops in valleys and on mesa tops, and later to hunting and gathering tribes, like the Apache, Ute, Comanche, Arapahoe and Cheyenne.

The most highly developed culture in prehistoric Colorado was the Anasazi culture, prevalent in the Four Corners region from the 11th through 13th centuries (Waldman 1999). The complexes they inhabited are popular National Parks and Monuments today: Mesa Verde, Chaco Canyon, Canyon de Chelly, and others. By 1300, the Anasazi had disappeared. The Pueblo Indians, descendants of the Anasazi, remain an important part of the region. In the 13th century, the Utes moved into the ecoregion from the Great Basin of Nevada, and ranged from northern New Mexico to southern Wyoming (Waldman 1999, Southern Ute website). Somewhere between the 9th and 14th Centuries, Athapascans arrived from the north to become the Apaches and Navajos, occupying northern New Mexico and southern Colorado (Waldman 1999).

The 17th century saw extensive colonization and missionary efforts by the Spanish. New Mexicans began to move into the Arkansas Valley in the early 1700s. This also was the beginning of Hispanic settlement of the San Luis Valley. French trappers and traders began moving into the region in the 1700s, as well, adding new dimensions to the ongoing conflicts between the Spanish and the Indians. Anglos began flooding the area with the Colorado Gold Rush of 1859. Trinidad became an important town, a point of contact between eastern Hispanic settlements and Anglo Coloradoans. Abbott et al. (1982:49) characterized the early history of this region as follows:

> *"Since the early 1700s, the southern Rockies, the San Luis basin, and the Arkansas Valley had been zones of contact among dissimilar peoples – Utes and Apaches, Comanches and Spaniards, Frenchmen and Spaniards, and above all, New Mexicans and Americans – competing for control of the same territory... [T]he lands of Colorado were one of the major frontiers of world history, a zone of interpenetration between the expansive societies of Hispanic and Anglo America."*

The Comanche began moving through southeastern Colorado in the 1700s (Waldman 1999). The Arapaho and Cheyenne arrived at the edges of the present Southern Rockies Ecosystem in the early 1800s (Waldman 1999). As these groups moved through the area they occasionally fought each other. Those plains tribes also kept the Ute tribes largely confined to the mountainous areas. In 1840, the Southern Cheyenne, Kiowa, and Comanche allied against the Crow, Shoshone, Pawnee, Ute, and Apache (Waldman 1999). Later, several of these tribes would become important fighters against white expansion into the Great Plains (Waldman 1999).

While the Southern Rockies share many physical characteristics that make it an ecoregion, socially and culturally, the landscape is quite diverse. Although county boundaries obviously do not conform to ecoregion boundaries, they offer the best approximation for the purposes of socioeconomic analysis. There are 64 counties in or near the Southern Rockies with significant socioeconomic ties to the region – 6 in south-central Wyoming, 48 in Colorado, and 10 in northwestern New Mexico. Using these county boundaries, the region encompasses 349,450 km$^2$. In terms of race and ethnicity, the Southern Rockies are predominantly white, but with more Hispanic and more American Indians than the nation as a whole. Table 3.2 compares the region's racial and ethnic composition with that of the United States.

Counties in New Mexico and southern Colorado had the highest percentages of Hispanic people. Six counties were more than half Hispanic — Mora, San Miguel, Rio Arriba, Costilla, Conejos, and Taos — and 20 were more than a quarter Hispanic. Thirty-two of the region's 64 counties – exactly half – were more than 90% white in 2000. San Juan County in New Mexico had the highest proportion of American Indians (36.88%), due to its overlap with the

**Table 3.2 Ethnic background of the Southern Rockies compared to the United States, 2002.**

| Ethnic Background | Southern Rockies | U.S. |
|---|---|---|
| White | 80.22% | 75.10% |
| Hispanic | 21.21% | 12.50% |
| Native Hawaiian or other Pacific Islander | 10.00% | 0.10% |
| Some other race alone | 8.64% | 5.50% |
| Black or African American | 3.43% | 12.30% |
| Two or more races | 3.03% | 2.40% |
| American Indian and Alaska Native | 2.54% | 0.90% |
| Asian | 2.05% | 3.60% |

Source: U.S. Department of Commerce, Bureau of the Census, 2000 Census of Population and Housing.

Navajo Reservation and the many Pueblos located there. Sandoval (16.28%), Rio Arriba (13.88%), and Montezuma (11.23%) counties, where the Colorado Ute Mountain Ute reservation is located, all had significant Indian populations, as well. The highest percentages of people with ancestry from Africa or Asia occurred in the urban counties along Colorado's Front Range and around Albuquerque.

In terms of educational attainment, Colorado, New Mexico, and Wyoming were all close to national averages (county data were not available). Only 11.48% of Americans over the age of 25 had a college degree in 2000. Wyoming and New Mexico were slightly below that average, at 10.57% and 10.35%, respectively, while Colorado was slightly higher, at 13.21%.

As a nation, the percentage of people below the poverty level was 13.3% in 2000. Colorado (10.2%) and Wyoming (12%) were both below the national average, while New Mexico, at 19.3%, was significantly higher.

Politically, the region was evenly split in 2001: 33.76% of registered voters were Democrats in 2001, while 35.75% were Republicans, and 30.49% were registered as Other. New Mexico was more Democratic (52.28%), Wyoming was more Republican (60.06%), and Colorado was more evenly split, but slightly more Republican (36.11% vs. 30.07% Democrat).

The most Democratic counties in the Southern Rockies coincide with concentrations of Latinos in New Mexico and southern Colorado, and with the Front Range urban areas. However, Republicans are dominant in El Paso County, Colorado (Colorado Springs), while the most highly concentrated Republican counties are located in southern Wyoming and northern Colorado.

## 6. Current Population Trends

The Southern Rockies are an attractive destination for migrants, as demonstrated by significant rates of population growth over the last century. In this section we look at population trends gleaned from the 2000 U.S. Census and examine the varied reasons for the region's longstanding popularity (Table 3.3).

**Table 3.3 Population growth from 1910 - 2050.**

| Area | 1910 | 1930 | 1950 | 1970 | 1990 | 2010 | 2050 |
|---|---|---|---|---|---|---|---|
| United States | 92,228,496 | 123,202,624 | 151,325,798 | 203,211,926 | 248,709,873 | 298,056,500 | 395,461,000 |
| Western states | 6,825,821 | 11,896,222 | 19,561,525 | 33,735,250 | 51,127,810 | 68,553,000 | 109,304,000 |
| Colorado | 799,024 | 1,035,791 | 1,325,089 | 2,207,259 | 3,294,394 | 4,650,500 | 6,208,000 |
| New Mexico | 327,301 | 423,317 | 681,187 | 1,016,000 | 1,515,069 | 2,158,000 | 3,364,000 |
| Wyoming | 145,965 | 225,565 | 290,529 | 332,416 | 453,588 | 604,500 | 863,000 |

1900 – 2000 Data Source: U.S. Census Bureau, Population Division, 2002.
2000 – 2050 Data Source: U.S. Department of Commerce, Bureau of the Census, 2000 Census of Population and Housing; Center of the American West, Western Futures Project, www.centerwest.org.

On April 1, 2000, the Southern Rockies county population was 5,408,152. Metropolitan Denver, the largest urban area, had roughly 2,000,000 people. Not surprisingly, the counties with the greatest populations are concentrated along the eastern slope of the mountains, where large cities like Fort Collins, Denver, Colorado Springs, Santa Fe and Albuquerque are located. The 12 counties in these areas account for 77% of the entire Southern Rockies population. Because county boundaries extend beyond the ecoregion boundaries, a significant portion of this urban population falls outside of the ecoregion boundary. The least populous counties in the region were mainly in the southwestern corner of Colorado.

The urban Front Range counties mentioned above ranked highest in population density, with Denver County leading the way at 2,090 people per km$^2$. The least densely populated counties were, again, in southwestern Colorado, with average densities of 0.4 people per km$^2$. Colorado's average population density was 16.4 per km$^2$, New Mexico's was 5.9 per km$^2$, and Wyoming's was 2 per km$^2$. Fifty-four of the 64 counties in the region had human densities lower than the national average of 31 per km$^2$.

It is important to note that population distribution statistics at the county level do not accurately demonstrate precise patterns of human population. In the Southern Rockies, population centers are typically concentrated in lower elevations and mountain valleys, so large portions of heavily populated counties may be relatively devoid of development.

Over the course of the 20th century, people came to the Southern Rockies in ever increasing numbers. Between 1900 and 2000, the U.S. population increased by 274%, the Southern Rockies by 710%, Colorado by 697%, New Mexico by 831%, and Wyoming by 434%. They came for many reasons. The gold rush, beginning in 1859, was typical of the hope for riches and a better life that attracted people to the region. The late 1800s and early 1900s saw significant population growth correlated with railroads, silver booms, "cattlemania," coal bonanzas, and town building efforts. Between 1900 and 1920, Colorado's population increased by 74%, but then dropped 20% in the next two decades because of economic down-turn (Abbot et al. 1982).

The population growth that occurred during and after World War II marks the beginning of the "New West" (Riebsame 1997). Commenting on the West as a whole, White (1991) attributes this westward movement to the federal bureaucracies, which devoted disproportionate shares of their resources to western development. The economy of the Southern Rockies remained strong after the war, due in large part to military expenditures on the Western Slope and in the Four Corners region (Wilkinson 1999). But people came to the region as much for quality of life as for jobs and the opportunity to make money.

The 1970s saw another population increase related to the energy boom when the Arab oil embargo resulted a focus on domestic sources: coal throughout the Intermountain region between New Mexico and Montana; oil and gas in the Overthrust Belt near the Wyoming-Utah border; and uranium in Colorado, New Mexico, and Wyoming. Later, oil shale and synfuels in northwest Colorado and southern Wyoming would contribute to the boom, which helped make Denver second only to Houston among energy capitals of the country (Riebsame 1997, Wiley and Gottlieb 1982). During the 1970s the human population of the Southern Rockies region as a whole increased by a third, nearly three times the national growth rate.

In the past decade, the Southern Rockies population has increased 28%. Colorado ranked third in the nation for growth over the past decade, while New Mexico and Wyoming increased their populations by 20.1% and 8.9%, respectively. This latest population boom was related to national prosperity fueled by a bullish market, the growing popularity of the region, and the continued growth of "footloose" industry – mostly in the high tech sector – that had begun in the late 1960s. This boom was characterized by wealth and spending on luxuries such as second homes in the mountains. Population projections for the region suggest that high growth rates and the aforementioned patterns of development will continue (Table 3.3).

## 7. Current Economic Trends

The Southern Rockies economy has evolved over the past 100 years from extraction of natural resources like gold, silver, oil, gas, uranium, coal, timber, and forage to service industries, retail trade, finance, insurance, and real estate. Today's economy is more diverse and complex, with significant portions of the region's income coming from small "footloose" businesses and non-labor sources.

Commonly used economic growth indicators are total personal income and employment (number of new jobs) as they relate to population growth over a period of time. In the Southern Rockies, total personal income has increased at rates significantly higher than in the U.S. overall, and job creation has stayed ahead of population growth (Table 3.4). While population nearly doubled in the Southern Rockies between 1969 and 1999, jobs in the region increased 187%, with 2,300,000 new jobs created during that period. In comparison, jobs in the United States as a whole increased by 80%.

In addition to number of jobs, an important economic indicator related to employment is the type of job being created, i.e., what percentage of the workforce is made up of wage and salary workers (those who work for someone else) versus proprietors (self-employed business owners). In the Southern Rockies, wage and salary earners dropped from 85% to 81% of the workforce, while proprietors increased from 15% to 19%. These changes closely mirror changes in the national workforce.

**Table 3.4 Percent increase in income, employment, and population across the U.S. and for the Southern Rockies from 1969-1999.**

| | United States | Southern Rockies | Colorado | New Mexico | Wyoming |
|---|---|---|---|---|---|
| Total Personal Income | 121% | 249% | 253% | 183% | 136% |
| Employment | 80% | 187% | 184% | 143% | 104% |
| Population | 35% | 87% | 87% | 72% | 46% |

Source: U.S. Department of Commerce, Bureau of Economic Analysis, Regional Economic Information System.

Twenty-two percent of new jobs created between 1969 and 1999 in the Southern Rockies were proprietors running their own businesses (compared to 20% in the U.S. as a whole). Stated another way, self-employment currently accounts for approximately one in five jobs in the Southern Rockies. Data from the Census Bureau's County Business Patterns does not include farm employment, self-employed people, railroad employees, or government employees.

Over half of the 164,280 Southern Rockies business establishments in 1999 employed less than five people, and

87% employed less than 20 people. This is indicative of a stable, diverse economy. When a community is dominated by one or two large employers (like in a mining town or a timber community), much of the economic risk is in the hands of one employer. With several smaller companies in a community, that risk is more dispersed (Rasker 1994).

A closer look at the type of proprietorship reveals that almost all new business owners are "nonfarm"-related, and many represent growing "footloose" industries. People who can take their businesses anywhere are increasingly choosing to relocate in the Southern Rockies for quality of life reasons (Power 1996). Many of these new, small businesses are in the service sector.

By such economic indicators, the Southern Rockies appear to be booming, both in terms of population and economics over the past 30 years. But, because the large majority of counties measured are in Colorado, the picture masks the 16 counties of Wyoming and New Mexico, many with less vibrant local economies. It is important to recognize the varying social and economic conditions in the different subregions of the Southern Rockies.

For example, while the counties in the Colorado Front Range are far removed from an extractive economy (defined here as farm, mining, agricultural services, forestry, fishing, "and other"), counties in southern Wyoming, southwestern Colorado, and northern New Mexico retain somewhat significant remnants of "Old West" economies. The four counties that had more than a quarter of their jobs in extractive industries in 1999 are Saguache (33%), Dolores (31%), and Conejos (27%) in southwestern Colorado and Mora (35%) in northern New Mexico.

## 8. Current Land Development Trends

Population and economic growth are inevitably accompanied by land development. Disturbingly, however, the physical expansion of residential housing in the Southern Rockies has occurred at a rate even faster than population growth, for three reasons: an increase in lower-density suburban development; the boom in exurban and "ranchette" rural development; and the growth in second homeownership in the Southern Rockies (twice the national average), which is not reflected in population statistics (Theobald 2000). Thus, the impact of urban sprawl and expansion of low-density housing developments into natural landscapes in the Southern Rockies and surrounding areas are even more significant than the high population growth rates suggest, and housing development is among the most significant agents of landscape change.

Moreover, the negative impact of housing expansion on ecosystems and species is actually much greater than the total area developed. Scattered, low-density development results in fragmented habitat. In many mountain valleys and foothill forests, low-density exurban developments often occur along public-private ownership boundaries, and may block wildlife movement. This insularizes wildlife habitat on surrounding public lands (Theobald 2000).

Developed areas also create a "disturbance zone" that extends beyond the actual development and into adjacent natural habitat. The ecology of this zone is affected by the spread of noxious weeds, predation by household pets (cats are particularly destructive), increases in human-adapted species (e.g., raccoons, skunks, or starlings), introduction of detrimental wildlife attractions (e.g., trash cans), and increases in recreational activity (Knight 1995). The extended zone of negative effect for songbirds and medium-sized mammals is similar around low-density housing development and high-density development; indeed, low-density housing may have greater overall impact due to the larger landscape area (Odell and Knight 2001). Moreover, important natural processes, such as fires and floods, are often suppressed around developed areas to protect houses and businesses. The generally close proximity of much of the region's forest land to private, developable land will restrict options for natural disturbance management on public lands, in particular the ability to allow natural and ecologically beneficial forest fires to burn (Shinneman et al. 2000, Theobald 2000).

Using housing-unit data from U.S. Census Block Groups, Theobald (2000, 2001) calculated historical and future spatial trends in development patterns for the region (Figure 3.6). Looking specifically at the Southern Rockies ecoregion (and not the county-defined region), land within urban (>1 housing unit per ha) and suburban (1 unit per 1-4 ha) development grew from roughly 42,000 ha in 1960 to 175,000 ha by 1990. This area is projected to grow to roughly 390,000 ha in 2020 and to 550,000 ha in 2050. Exurban development (1 unit per 4 - 16 ha), grew from roughly 190,000 ha to 600,000 ha between 1960 and 1990, and it is projected at roughly 850,000 ha by 2020 and 1,120,000 ha by 2050. Exurban, suburban, and urban developments collectively covered about 775,000 ha (4.6% of the ecoregion) in 1990 and are projected to grow to 1,670,000 ha (10% of the ecoregion) by 2050 (Table 3.5).

This development is mainly concentrated in mountain valleys, foothills, and lower elevation valleys (Shinneman et al. 2000, Theobald 2000). These areas often include valuable agricultural lands and species-rich wildlife habitat such as ponderosa pine forests, oak shrublands, montane grasslands, riparian, and wetland habitat (Shinneman et al. 2000).

Table 3.5 **Housing Density in the Southern Rockies.**

| Housing density | 1960 | 1990 | 2020 | 2050 |
|---|---|---|---|---|
| Urban/Suburban | 42,000 | 175,000 | 390,000 | 550,000 |
| Exurban | 190,000 | 600,000 | 850,000 | 1,120,000 |
| **Total** | **232,000** | **775,000** | **1,240,000** | **1,670,000** |

data source: Theobald (2000, 2001).

## 9. Conclusion

In sum, the Southern Rockies are booming, both in terms of population and economic growth; but the region retains significant wildlands. Indeed, it is the relatively pristine environment that attracts so many people and businesses to the region. As communities continue the shift from extractive economies to service economies and marketing natural amenities, they will be more likely to see the economic benefits of restored and intact ecosystems.

One of the biggest concerns is the astronomical population growth rate and attendant residential development. Because of proximity to the popular and highly traveled I-70 corridor, counties in west-central Colorado had high growth rates: Eagle County has grown by 90% and is ranked 9th in the country for the 1990s; Garfield County has grown by 46.1%. The tendency for this corridor to attract the second homes of wealthy people means that the effect on Nature is worse than the population numbers indicate.

Hinsdale and Mineral Counties (Southwest Colorado) had growth rates of 69% and 49% respectively, and both ranked in the top 26 fastest growing counties in the nation. Archuleta County ranked 14th in the nation, with an 85.2% growth rate. Granted, Hinsdale, Mineral, and Archuleta Counties were some of the most sparsely populated counties in the region to begin with, so even with high growth rates, their population density remains relatively low; but present trends are surely cause for concern.

Many people come to the Southern Rockies for their natural beauty and recreation opportunities. Eventually, growth will erode those amenities, particularly if scenic views from the porch are a more deeply held core value than functioning ecological processes. In addition, these changing demographics do not automatically translate to a change in power from the Old West to the New West. That may be true over the long-term, but in the short-term there may be heightened conflict between competing values (Glick 2001).

# SECTION II:

# THE APPROACH TO THE SOUTHERN ROCKIES WILDLANDS NETWORK VISION

*The thought of what was here once and is gone forever will not leave me as long as I live.*
*It is as though I walk knee-deep in its absence.*

-Wendell Berry

# 4 INTRODUCTION TO OUR APPROACH

Brian Miller, Dave Foreman

## 1. Introduction

The Southern Rockies ecoregion has enormous potential for conservation, yet it also faces many threats. We have touched briefly on the risks to biological integrity in Chapters 2 and 3, and we will supply more detail in Chapter 5 (Wounds). Chapter 6 will discuss our goals for healing those wounds. While the Southern Rockies Wildlands Network Vision proposes a long-term alternative to present patterns of exploitation, we also know that there is little time to waste. The momentum of current development plans results in the loss of natural areas, biodiversity, and native species each time we delay.

Wildlands network designs are based on site-specific proposals for cores, linkages and compatible-use areas that stretch across a landscape (Noss and Cooperrider 1994, Soulé and Noss 1998, Soulé and Terborgh 1999). The boundaries of the landscape are defined for each plan, and the network of protection is mapped in detail within those boundaries. The specific boundaries of this Wildlands Network Design encompass the Southern Rockies ecoregion. We chose the ecoregion as a base for the plan because an ecoregion is characterized by relatively homogeneous ecological patterns throughout its connected ecosystems (Shinneman et al. 2000). The Southern Rockies Wildlands Network Vision overlaps the New Mexico Highlands Vision to the south and the Heart of the West Vision to the northwest. This overlap ensures a relatively seamless transition between regions.

## 2. Principles of Reserve Design

In this chapter we outline our basic approach toward forming a wildlands network design. We follow the general pattern for reserve network design that is used in the Sky Islands Wildlands Network Conservation Plan (Foreman et al. 2000), the New Mexico Highlands Wildlands Network Vision, and the Maine Wildlands Network Vision (all three completed by the Wildlands Project and their cooperators), and we follow the reserve design principles of Noss and Cooperrider (1994), Meffe and Carroll (1997), Primack (1998), Soulé and Noss (1998), and Soulé and Terborgh (1999). "Rewilding the Rockies," a phrase proposed by Michael Soulé and Reed Noss, is a real possibility for this part of North America.

The weight that the Southern Rockies Ecosystem Project gives to the various strategies and principles of reserve design depends to some extent on our goals for this ecoregion. We all share the broader goals of conservation, but no one group can do everything. So, the primary problem that we will address in this plan is restoring the ecological role of large predators. Evidence is strong that without large predators, ecosystems are incomplete in both form and function (Soulé and Noss 1998, Terborgh et al. 1999, Oksanen and Oksanen 2000, Estes et al. 2001, Miller et al. 2001). Yet, while many groups emphasize biological diversity and ecosystem representation, large carnivores are often missing from the picture. Rewilding is, therefore, a complementary strategy to goals of biological diversity and ecosystem representation.

Restoring large carnivores is an essential step toward "rewilding" the landscape (Soulé and Noss 1998). The Southern Rockies Wildlands Network Vision embraces "rewilding," or protecting and restoring native ecosystems and large wilderness areas (Soulé and Noss 1998). Rewilding requires functioning keystone species and processes. Thus, we seek to protect native species, reintroduce extirpated species, and restore ecological processes. This includes wolves (*Canis lupus*) and grizzly bears (*Ursus arctos*).

Because of the large area needed to restore carnivores at numerical levels where they can fulfill ecological and evolutionary functions, we must plan at large temporal and geographic scales. The theory of island biogeography (MacArthur and Wilson 1967) was a major breakthrough for principles of reserve design. The concept of species-area relationships allowed people to see that no single park or reserve would be large enough to maintain viable populations of large carnivores (see Frankel and Soulé 1981, Newmark 1987 and Newmark1995, Noss and Cooperrider 1994, Woodroffe and Ginsberg 1998). And, even if a population could persist in a given park, albeit through heavy management and restocking, it would be little more than a taxonomic museum piece if it were unable to contribute to ecological and evolutionary function across its range (Leopold 1966). The restoration of wolves in Yellowstone National Park does not affect ecological processes in the Southern Rockies.

Thus we propose a wildlands network design that extends across the landscape in a system of core areas, connections, and compatible-use lands that allow movement of large carnivores throughout the ecoregion and beyond. In the following section, we touch briefly on this continental scale, but for more detail refer to the Sky Islands Wildlands Network Conservation Plan (Foreman et al. 2000), the New Mexico Highlands Wildlands Network Vision, the Maine Wildlands Network Vision, Noss and Cooperrider (1994), Meffe and Carroll (1997), Primack (1998), Soulé and Noss (1998), and Soulé and Terborgh (1999).

### Cores

Core reserves are essential to the reserve network (Noss and Cooperrider 1994). Human disturbance in core areas should be minimal (Noss 1992). Cores that represent wilderness areas are essential for the survival of many species, principally because of human behavior outside of those areas (Mattson 1997). In other words, cores must be allowed to function in their natural state, and they must be large to be effective (Soulé and Simberloff 1986, Wilcove and Murphy 1985, Noss 1992).

Larger cores will likely contain more habitats and more species. Deterministic or stochastic factors threaten small populations more than they do large populations (Frankel and Soulé 1981), and edge effects are more severe on smaller reserves because there is a larger perimeter to area ratio (Noss 1983, Harris 1984, Franklin and Forman 1987, Wilcove et al. 1986, Janzen 1986, Noss and Cooperrider 1994). Edge effects extend various distances into the heart of the reserve, depending on the type of effect studied (humidity changes, tree blow-down, nest predation, human poaching, etc.). All these effects can reduce viability of a species in a small reserve.

Newmark (1987, 1995) showed that more mammal species have disappeared from small national parks than have been lost from large parks. It is important, however, to not extrapolate size requirements across highly different ecoregions. For example, Gurd et al. (2001) estimated that a reserve of 5,000 $km^2$ would suffice to maintain any mammal species in eastern North America. Yet, an area of 5,000 $km^2$ in the Southern Rockies will not be enough space for an ecologically effective density and distribution of wolves.

We base the Southern Rockies Wildlands Network Design on a core system of federally protected Wilderness Areas ("wilderness" literally means *self-willed land*). Despite weaknesses and inconsistencies in the 1964 Wilderness Act, it has proven to be the most effective means of protecting large areas in the United States (Foreman 1995). In using National Wilderness Area designation as the cornerstone for a wildlands network, some basics about the Wilderness Act need to be understood.

Wilderness Areas are not human exclusion zones. A wide variety of non-motorized recreational activities is permitted, ranging from backpacking to hunting and fishing. However, Wilderness Areas are not solely recreational areas. In the various definitions of Wilderness Areas, both experiential and ecological values are prominent and considered compatible (Foreman 1998). The Wilderness Act has different criteria for candidate Wilderness Areas than for management of designated Wilderness Areas. For example, there is no requirement that a candidate area must be pristine or even roadless to be designated as Wilderness. "Pristine," which is an ultimate word like "unique," does not appear in the Wilderness Act. However, after designation, no permanent roads are allowed nor is use of mechanized equipment (except for certain administrative needs, usually of the emergency kind, Foreman 1998).

Section 2(c) of the 1964 Wilderness Act[3] sets the parameters:

> *A wilderness, in contrast with those areas where man and his works dominate the landscape, is hereby recognized as an area where the earth and its community of life are untrammeled by man,*

[3]The Wilderness Act. 1964. Public Law 88-577 (16 U.S. C. 1131-1136) 88th Congress, Second Session, September 3, 1964 in Watson, Jay, ed. 1998. The Wilderness Act Handbook third edition (revised). The Wilderness Society, Washington, D.C.

> *where man himself is a visitor who does not remain. An area of wilderness is further defined to mean in this Act an area of undeveloped Federal land retaining its primeval character and influence, without permanent improvements or human habitation, which is protected and managed so as to preserve its natural conditions and which (1) generally appears to have been affected primarily by the forces of nature, with the imprint of man's work substantially unnoticeable; (2) has outstanding opportunities for solitude or a primitive and unconfined type of recreation; (3) has at least five thousand acres or is of sufficient size as to make practicable its preservation and use in an unimpaired condition; and (4) may also contain ecological, geological, or other features of scientific, educational, scenic, or historical value.*

Note that this definition uses the phrases "earth and its community of life" and "protected and managed to preserve its natural condition" before the phrase "has outstanding opportunities for solitude or a primitive and unconfined type of recreation." Ecological concerns were clearly on the minds of the drafters of the Wilderness Act. Furthermore, the wording "which *generally* appears to have been affected *primarily* by the forces of nature, with the imprint of man's work *substantially* unnoticeable" (emphasis added) clearly shows that Congress did not believe candidate areas had to be pristine (Foreman 1998).

Designation of an area as Wilderness does not prevent management, such as reintroduction of wolves or beavers, restoration of natural fire, control of exotic species, or ecological restoration such as planting willow and cottonwood wands along degraded streams. Some Wilderness designation legislation has specifically called for restoration measures. In the 1999 Dugger Mountain (Alabama) Wilderness Act, for example, the Forest Service was directed to use equipment and an existing road to remove a fire tower. After removal, the road was to be closed. In other cases, areas have been designated as Potential Wilderness Additions to allow ecological restoration and removal of nonconforming structures or uses. After restoration, the area automatically becomes Wilderness, with roads closed and mechanized equipment banned.

Thus, conservationists should propose less-than-pristine areas for Wilderness designation as long as they acknowledge the intrusions (Soulé 1991). These include areas with roads, past logging, and other unnatural disturbances. Indeed, one of the most important criteria for large carnivores in the U.S. is not necessarily pristine habitat, but large areas of protected space. Ecological and experiential (recreational and aesthetic) justifications need to be made for proposing such areas, however. Resolution of conflicts over management in wilderness remains a lively topic (Soulé 2001).

In the past, both conservation groups and land management agencies have used standards of quality and purity to select candidate areas for protection and for drawing boundaries around such areas. Unfortunately, federal agencies sometimes have used the purity standard to ostensibly protect only those areas that appear to be without human impact. Boundary selection has often excluded portions of wild areas that are scenically of "lower quality". Often these "lower quality" areas were of greater ecological importance than the areas protected; commonly, they were at lower elevations with higher productivity. Purity has also been used as a subterfuge by the agencies to exclude areas with timber, minerals, or other exploitable resources.

Both the U.S. Forest Service and U.S. Bureau of Land Management have set standards of wilderness purity not required by the Wilderness Act (Cutler 1977). For example, in the Forest Service's roadless area review and evaluation (RARE), the Southwest Regional Forester decreed that areas to be evaluated for possible Wilderness designation had to be truly roadless. Consequently, tire tracks that remained visible into the next season excluded thousands of hectares from being identified as roadless. In 1972, the U.S. Forest Service proposed to remove several thousand hectares of the Gila Primitive Area from protection because of the faint sign of a long-abandoned airstrip.

In the early 1970s, the Forest Service stridently opposed designating Wilderness Areas in the eastern United States because of the agency's purity dogma. Members of Congress, including the champions of the 1964 Wilderness Act, made it clear that purity had not been their intent. Senator Frank Church (1973), the floor manager of the Wilderness Act, said that the Forest Service:

> *...would have us believe that no lands ever subject to past human impact can qualify as wilderness, now or ever. Nothing could be more contrary to the meaning and intent of the Wilderness Act. The effect of such an interpretation would be to automatically disqualify almost everything, for few if any lands on this continent-or any other-have escaped man's imprint to some degree.*

This is one of the great promises of the Wilderness Act. We can dedicate formerly abused areas where the primeval scene can be restored by natural forces.

## Connectivity

The degree of isolation affects species diversity. Isolated patches have less chance of genetic exchange with populations in other patches and less chance of being recolonized if a population is lost (MacArthur and Wilson 1967, Frankel and Soulé 1981). In addition, no single core area would be large enough to support large carnivores in the Southern Rockies, so carnivores must be able to move throughout the region (not just inside a given Wilderness Area) if they are to re-establish missing ecological and evolutionary functions. Therefore, connectivity is important for planning Wildlands Network Designs.

The antithesis of connectivity is fragmentation. Fragmentation isolates formerly continuous habitats and therefore changes the scale at which natural events operate. Connectivity is necessary to reconnect those fragments and restore the role of those natural events. Despite many complicating factors, connectivity in some form is essential for most species, especially large animals, which cannot maintain viable populations in small, isolated areas (Frankel and Soulé 1981, Noss and Harris 1986, Soulé 1991, Noss and Cooperrider 1994). Nevertheless, we should remember that while large animals may be excellent for estimating needed reserve size, they should not be the only choice for planning connections; this is because they can move across gaps in habitat that are inhospitable to smaller species. As an example, American martens (*Martes americana*) do not cross open expanses in winter much wider than 100 m (Koehler and Hornocker 1977), thus a very high degree of connectivity is necessary to provide viable habitat for these small to mid-size carnivores.

Such biological connections should be based on locally relevant processes, interactions, and the needs of particular species. Biological connections provide for natural dispersal of individuals within an area, seasonal migration of groups, genetic exchange between populations, and ability to shift natural ranges in response to climate change (Noss and Cooperrider 1994, Miller et al. 1998). Issues of scale come into play in planning connections, and issues of scale can be among the most difficult to understand.

If connections are designed for long distance dispersal, consideration should be given to connections wide enough to house residents of the focal species (Noss and Cooperrider 1994, Miller et al. 1998). Such connections more closely resemble historical conditions of connectivity. Many vertebrate species allow dispersing juveniles to pass through their territory. In addition, the typical dispersal pattern for many mammals is for females to remain fairly close to the area where they were raised, whereas males make long distance movements (Greenwood 1980, Dobson 1982). Areas wide enough to house residents would consequently allow females to disperse, and that could be important for natural restocking of extirpated colonies in a metapopulation (Miller et al. 1998). In addition, wide connections diminish the ratio of edge to interior.

The management of connectivity becomes progressively more complex as scale increases (Miller et al. 1998). Whereas connections within a single protected area may be relatively simple, movement that crosses agency, state, and international boundaries increases the number of managing partners. Connecting two protected areas that are already separated by roads and human settlements increases the number of social, economic, and enforcement dimensions. These considerations should not be taken lightly (Miller et al. 1998).

Interregional connectivity between wildlands network designs is important for many wide-ranging species. Connectivity between the Southern Rockies Wildlands Network and the New Mexico Highlands Network will be crucial for lynx (*Lynx canadensis*, a focal species for both designs) to successfully reestablish territories in northern New Mexico. Likewise, if the recently reintroduced Mexican wolf (*Canis lupus baileyi*) is to expand its currently limited range into suitable habitat further north, and reestablish a genetic cline with the northern gray wolf, connectivity between the Sky Islands, the New Mexico Highlands, the Southern Rockies, and the Heart of the West Wildlands Networks will be essential.

Connectivity for some species can be provided by a system of "stepping stone" habitats rather than by a discrete, linear connection. As a caveat, what passes as a system of stepping-stones for a generalist or large species can become an ecological sink (where mortality rates exceed birth rates) for a more specialized or smaller species (Simberloff and Cox 1987, Miller et al. 1998). Depending on how a species responds, various types of connections can be a travel conduit, a permanent home, a sink, an agent in disease transmission, a vehicle that promotes contact with an exotic competitor, or an avenue that provides increased contact with a predator (Miller et al. 1998).

Translocation of animals between isolated populations has been proposed as an alternative to connections. While it may be physically possible to move animals between sites, there may not be any functional benefit. Homing behavior, excessive movement from the release site, and conflict with residents has been a major problem in carnivore translocations, resulting in drastically reduced survival (see review by Linnel et al. 1997). Several mountain lions (*Puma concolor*) translocated over 400 km returned to their original territories (Logan et al. 1996, Logan and Sweanor 2001).

Most important, such an alternative can perpetuate existing patterns of habitat fragmentation. Thus, large animals may persist in patches (at least over the short term), but their numbers may remain too small for natural selection to act, and they would have little impact on ecosystem processes. Additionally, processes such as fire, nutrient cycling, grazing, and flooding would remain altered by isolation and reduced scale. At our present level of knowledge, protecting and restoring connections is a superior way to restore ecological integrity.

### Compatible-use Lands

Compatible use lands are areas managed for more human use than are Wilderness Areas, but still managed with strong conservation goals (Noss and Cooperrider 1994). In this plan, we recognize low, medium, and high compatible use levels (see Chapter 9). For example, low compatible use areas could include federal land bordering a designated Wilderness Area, thus extending the protection for the inhabitants of that Wilderness Area and greatly reducing edge effects. Areas managed for medium compatible use could form important connections allowing animal movement between areas. Even if areas of high compatible use are suboptimal for reproduction of a focal species, they can still provide temporary habitat for a dispersing individual waiting for a breeding area to open.

In this plan we propose compatible use designations only for federal and state lands. We recognize biologically important private land, but only in the context of providing information. Compatible use on private lands must be a cooperative measure, and often involves conservation easements, tax incentives and other measures (see Chapter 10).

## 3. The Three-Track Framework

To determine the location and size of cores, connections, and compatible use lands, we used the three-track approach pioneered by Reed Noss (Noss and Cooperrider 1994). One track embraces *representation* of habitats and vegetation types, within a network of core areas. A second track identifies and protects *special elements*, which are locations of threatened species, biodiversity "hotspots", and important places such as roadless areas. A third track identifies and protects key habitat for *focal species*. Focal species serve an important ecological function or indicate healthy, functioning systems.

Individually, each track may not provide complete protection. For example, representation analysis usually sets different protection goals for different types of vegetation; but representation alone does not deal with actual presence or absence of given species. Special element mapping does not account for spatial needs of many species. And focal species analysis usually assumes that the species being modeled can act as an adequate surrogate for many other, smaller-bodied species that use similar habitat. The optimal solution is to combine all three tracks into a comprehensive approach toward conservation planning. As we stated above, time and money often mean groups have to specialize instead of "trying to do it all." We employ all three tracks in the Southern Rockies Wildlands Network Design, but because The Nature Conservancy has drafted a plan for the region based on special elements and representation, and because of our emphasis on large carnivores, the focal species track receives more weight. In addition to the discussion below, see Chapter 7 for a more detailed explanation of our methods.

### Representation

The Nature Conservancy has called representation of habitats at a community level a "coarse filter" approach to biological conservation (Noss and Cooperrider 1994, Neely et al. 2001). A coarse filter approach is economically efficient. Broader vegetation schemes serve as a surrogate for data on each individual species within a given scheme, and these vegetation patterns are easier to map. In many cases such data already exist (though, they may be formatted differently by different agencies). The coarse filter approach assumes that species living in these vegetation classes can be saved by protecting the habitats (Noss and Cooperrider 1994). To that end, The Nature Conservancy estimates that 85% to 90% of all species in a region can be protected through a coarse filter approach (Noss and Cooperrider 1994). Over the long term, however, vegetation patterns change, and the associated species may or may not change in proportion (Noss and Cooperrider 1994).

### Special elements

While representation of broad vegetation types entails a coarse filter approach, *special elements* present a "fine filter" approach to conservation planning (Noss and Cooperrider 1994; Neely et al. 2001). It is an extension of representation, in that it highlights narrowly fitted endemics, rare species, unique ecosystems or other factors not well covered by representation of vegetative communities (Noss and Cooperrider 1994). The Nature Conservancy's Southern Rocky Mountains ecoregional plan used the coarse filter of habitat representation and the fine filter of rare species and rare ecosystem locations (Neely et al. 2001). The Southern Rockies Ecosystem Project has not duplicated The Nature Conservancy's excellent work; we have used roadless areas (both protected and unprotected), designated Wilderness, and Park Service lands for our special elements components. We are discussing cooperative options with The Nature Conservancy that would examine the complementary conservation plans and more fully blend all three tracks for the next round of planning.

### Focal species

Representations of vegetation types and special elements are excellent tools for biodiversity conservation, and they point to areas that should be considered in a reserve design. Those two tracks, however, may not provide enough spatial area for large carnivores and ecological processes to function (or even persist). So, the third track for wildlands network designs uses *focal species*. Focal species analysis identifies additional high-value habitats and addresses the questions, "How much area is needed?", "What is the quality of habitat?", and "In what configuration should we design habitat areas?" (Miller et al. 1998).

We use focal species in planning and protecting reserves because their requirements represent factors important to maintaining the natural state of the entire region (Lambeck 1997, Miller et al. 1998, Watson et al. 2001). Watson et al. (2001) demonstrated that a focal species approach was effective for planning conservation of woodland birds, and they speculated that it should be useful for guiding plans in other environments.

*Ecologically interactive* species, which include *keystone* species and *foundation* species, are better grounded in ecology than *flagship* species and *umbrella* species. However, both umbrella species and flagship species can be useful in implementation and outreach campaigns, where they serve educational and political functions. As an example, data are not yet available to use the reintroduced lynx as an ecologically interactive species, but we do propose lynx for a flagship species. Since lynx data are limited, bobcat (*Lynx rufus*) can be used as a surrogate species and may serve a useful role in monitoring and implementation. Studies of bobcat could be used to evaluate predicted impacts of roadways and roadway underpasses in areas where lynx are not yet present or exist in low densities (Crooks pers. comm.).

Ecologically interactive species (keystone and foundation species) are important focal species for building a wildlands network. Both have crucial interactions in an ecosystem that benefit other species associated with, or dependent upon, that system. The removal of an ecologically interactive species initiates changes in ecosystem structure by changing various interactions and processes. The result can trigger cascades of direct and indirect changes, including losses of diversity on more than one trophic level (Soulé and Noss 1998, Terborgh et al. 1999).

Keystone species enrich ecosystem function in a unique and significant manner through their activities, and the effect is disproportionate to their numerical abundance (Paine 1980, Terborgh 1988, Mills et al. 1993, Soulé et al. in press). Foundation species, however, enrich an ecosystem more in proportion to their numerical abundance (or perhaps by amount of biomass, Soulé et al. in press). Whether keystone or foundation, the interaction of a species with the system is the important point. Examples of these interactions include predation, competition, mutualism, and modifying habitats (Menge et al. 1994, Power et al. 1996, Jeo et al. 2000, Soulé et al. in press).

Important to the restoration of large carnivores, ecologically interactive species can affect the surrounding landscape through predation. Apex predators play a key role in ecosystems by maintaining ecosystem structure, diversity, and resilience via "top-down" effects that trickle through various trophic levels (Terborgh 1988, Terborgh et al. 1999, Estes et al. 2001, Miller et al. 2001, Soulé et al. in press). The disappearance of apex predators in a region can cause acute changes in that system, many of which can lead to loss of other species in the area. Most frequently this involves release of herbivores from predation pressure, which in turn exerts unnatural pressure on plant communities, often resulting in biotic simplification (Terborgh et al. 1999). Reduced diversity can also result from the "Paine effect" (Terborgh et al. 1999). Eliminating an apex predator dissolves the ecological boundaries that formerly held a dominant herbivore in check and allows that species to displace other herbivores (Paine 1966). Finally, the loss of an apex predator can cause "meso-predator release", which can result in noticeable declines of smaller prey species (Soulé et al. 1988, Palomares et al. 1995, Crooks and Soulé 1999).

Another ecological interaction is via "ecological shapers" or species that significantly modify their habitat in a way that benefits associated species. Examples of species in this category include beavers (*Castor canadensis*), prairie dogs (*Cynomys* spp.), and bison (*Bison bison*, Naiman et al. 1988, Detling 1998). The habitat changes of ecosystem shapers create a mosaic of habitat types, allowing for greater diversity in an area.

In addition, species can be interactive through mutualism. An example is the relationship between the five-needled pines, and some members of the corvid family—particularly Clark's nutcracker (*Nucifraga columbiana*). The five-needled pines feed the Clark's nutcracker, which cracks the cones and deposits the seeds in caches. In return, the five-needled pines depend on the corvid to disperse their seeds. If this relationship were broken, many other species would suffer, as many mammals also feed on the pine nuts.

Another type of focal species that is grounded in ecology is the *indicator species*. Indicator species are tightly linked to specific biological elements, processes, or qualities, and are sensitive to ecological changes. They are essentially surrogates for the system they use, and they must function as an early warning system to be effective at that task (Noss and Cooperrider 1994). Some indicator species are defined quite narrowly and may reflect a habitat quality like stream temperature or contaminant level. Other indicators can represent broad spatial or temporal scales, although broader goals can be more difficult to ground in the choice of indicator.

A particularly broad type of indicator species is a *wilderness-quality indicator*, implying that the species chosen is wilderness-dependent. While many species use and benefit from wilderness, they may not be dependent upon it (Hendee and Mattson 2002). Typically, species that use wilderness primarily or exclusively are species that are directly persecuted or are associated with habitats that humans convert to other uses (Mattson 1997). In other words, such species are vulnerable to human behavior. Many species may be able to adjust to humans at some level if human behavior does not affect them adversely. But without significant changes in human values and culture, these species need habitat that is remote from negative human behavior if they are to survive in viable numbers (Mattson 1997). To quote Hendee and Mattson (2002: 324):

> *The distribution, numbers, diversity, and behavior of wildlife species can be a measure of that naturalness and solitude of a wilderness. Wildlife reflects ecological conditions and their changes over time, so wildlife can serve as indicators of wilderness character and quality—in fact, as well as in human perception.*

Nine focal species were used to plan for cores and connections in the Southern Rockies Wildlands Network Design: American marten, beaver, bighorn sheep (*Ovis canadensis*), black bear (*Ursus americanus*), grizzly bear, lynx, pronghorn (*Antilocapra americana*), cutthroat trout (*Oncorhynchus clarki*), and gray wolf. Justifications for inclusion of each species and full species accounts are listed in the Appendix. We also make management recommendations for the selected focal species.

Through dynamic modeling, Carroll et al. (2003) evaluated whether the Southern Rockies could support metapopulations of wolves over time. We also used models to propose linkages between cores for wolves and black bears, and assess habitat suitability models for pronghorn and cutthroat trout. We chose focal species for models that would be complementary for habitat needs. Choices during the first round were limited by time and money, but we plan to expand the list of focal species on the second iteration of the plan. During implementation, indicator species and flagship species (*e.g.* the lynx) will be important for monitoring success and for outreach efforts.

## 4. Precautionary Principle

We advocate the *Precautionary Principle*, which proposes that it is better to err on the side of protecting "too much" habitat than too little. Nature is not predictable over time and space. To assume that we can accurately predict a minimum population size for a species is arrogant and dangerous. Our perspectives of time are too short and our knowledge of ecology too minuscule. The stochastic nature of the communities and systems dictates that we err on the side of caution, and that we use wide margins of safety. If we do not, sooner or later our plan will fail to protect biodiversity or ecological processes.

In addition, this report is not the end of the process. We advocate adaptive management (Hollings 1978) and intend future iterations of this Wildlands Network Design. This draft is a series of hypotheses to test. As they are tested, the plan will be refined accordingly.

## 5. Conclusion

The principles of reserve design provide a foundation for a wildlands network design. We offer only a brief description of those principles here, and specifically the principles that directly affect our goals. For more detail on principles of reserve design and conservation biology, refer to the literature cited at the beginning of this chapter. In general, we seek broad and well-accepted goals outlined by Noss and Cooperrider (1994). These goals include representing all types of functioning ecosystems in protected areas, maintaining viable populations of native species in natural patterns of abundance and distribution, and supporting natural ecological and evolutionary processes. A specific goal of this design is to support the return of large carnivores, particularly wolves, to the Southern Rockies by identifying and protecting the optimal network of roadless core areas and linkages.

Fremont Cottonwood @ Horseshoe Canyon Evan Cantor

*Now it seems as though our mother planet is telling us, "My children, my dear children, behave in a more harmonious way. My children, please take care of me."*

-Dalai Lama

*One of the penalties of an ecological education is that one lives alone in a world of wounds.... An ecologist must either harden his shell and make believe that the consequences of science are none of his business, or he must be the doctor who sees the marks of death in a community that believes itself well and does not want to be told otherwise.*

-Aldo Leopold

# 5 ECOLOGICAL WOUNDS

Dave Foreman, Doug Shinneman, Brian Miller, Robert Howard, Matt Clark

## 1. Introduction

Aldo Leopold (1972) wrote,

> *The last word in ignorance is the man who says of an animal or plant: "What good is it?" To keep every cog and wheel is the first precaution of intelligent tinkering.*

Despite Leopold's (1966) half-century old advice on "intelligent tinkering" we have not kept "every cog and wheel." Finely tuned interactions among species, physical environments, and ecological processes form the webs of life on our planet. Each web is not static, but varies continuously within certain bounds, and the species and systems have adapted over time to the range of variability in their particular region (Noss 1999). When "cogs or wheels" are lost, a system can fluctuate outside of the bounds to which it has adapted. Depending on which parts are lost, and the rate of loss, the pressure on a given system can exceed its ability to respond. Once such a vortex is entered, runaway positive feedback can make escape difficult, as altered structure and function can cause secondary waves of extinction that further heighten the instability.

The human-caused wounds to the land are tied directly to the current extinction crisis - the most serious since the end of the dinosaurs 65,000,000 years ago (Wilson 1992, Soulé 1996). Wilcove et al. (1998) list the current causes of extinction in the United States in order of importance: (1) habitat destruction; (2) non-native (alien) species; (3) pollution; (4) overexploitation; (5) disease. Reading and Miller (2000) show that overexploitation was a more important cause of extinction historically in North America and remains a highly important cause in much of the world today. Reading and Miller include disease with exotic species, since most diseases that have catastrophic population effects are exotic.

There are several ways to categorize ecological wounds. We follow a medical diagnosis approach of differentiating between wounds or illnesses (pathology) and the causes (etiological agents that perturb natural systems). As an analogy, cigarette smoking is not a human illness, but it can be a cause of several illnesses, including emphysema, lung cancer, mouth and throat cancer, and heart disease. Likewise, grazing by domestic livestock is not a wound to the land in itself, but it can contribute to several landscape wounds. The Southern Rockies Wildlands Network Vision identifies actual wounds to the land and then considers the anthropogenic causes for each. Just as a medical doctor seeks not only to treat the symptoms and the disease, but also tries to understand the root cause(s) of the illness, so do ecological "doctors" look to healing the wounds of the land through a therapeutic approach with specific therapies to return to a healthy functioning system.

In recent years, ecological and historical researchers have greatly improved our understanding of the ecological wounds in the Southern Rockies. Unfortunately, they are finding, even in the best-protected areas, pre-existing wounds may continue to suppurate. Efforts to protect the land and create a sustainable human society in the Southern Rockies ecoregion will come to naught without understanding these wounds and their underlying causes, and then attempting to heal them.

## 2. Wounds to the Land

From the beginning of human habitation of the Southern Rockies, our species has inflicted wounds of varying severity on terrestrial and aquatic communities. Many natural communities are greatly reduced or disrupted in extent, composition, and function due to activities such as logging, fire suppression, overgrazing, housing development, agricultural conversion, water use/dams, and pollution (Figure 5.1). Habitats particularly at risk include old-growth ponderosa pine forest, old aspen forest, low-elevation riparian communities, sagebrush shrublands, montane grasslands, and most wetlands and aquatic systems (Noss et al. 1995, Shinneman et al. 2000, Neely et al. 2001).

This chapter contains a discussion of the major wounds in the Southern Rockies ecoregion and is a summary or review of more detailed discussions in Chapters 2 and 3. Chapter 6 will present the goals and objectives of the Southern Rockies Wildlands Network Vision which are designed to heal the wounds. These wounds, although listed separately, are interconnected and overlap each other.

The major wounds in the Southern Rockies are:

- Loss and decline of native species
- Loss and degradation of terrestrial and aquatic ecosystems
- Loss and alteration of natural processes
- Fragmentation of wildlife habitat
- Invasion by exotic plant and animal species
- Pollution and climate change

The Southern Rockies Ecosystem Project identified this configuration of wounds over several years beginning with a workshop held in May 2000. Additional research and publication of *The State of the Southern Rockies Ecoregion* (Shinneman et al. 2000) led to further refinement. The list presented here is drawn from these prior sources, and, in addition, has incorporated appropriate materials from the New Mexico Highlands Wildlands Network Design for the overlapping portions of the two study areas.

### *Wound 1: Loss and Decline of Native Species*

During the past 200 years, native carnivores such as wolves (*Canis lupus*), grizzlies (*Ursus arctos*), lynx (*Lynx canadensis*), wolverines (*Gulo gulo*), river otters (*Lontra canadensis*), ungulates including bison (*Bison bison*) and pronghorn (*Antilocapra americana*), and important rodents like beavers (*Castor canadensis*) and prairie dogs (*Cynomys* spp.), as well as numerous birds, amphibians and reptiles have been eliminated entirely or greatly reduced in numbers. Some of the causes are commercial and recreational trapping, hunting and collecting; competition from domestic livestock; diseases introduced by settlers and domestic livestock; livestock fencing; predator and rodent control programs; loss and transformation of natural habitats for human uses; recreation; contamination of riparian areas; and ecological imbalances caused by the loss of ecologically important species and natural processes (See Chapter 2 and Chapter 3).

Trappers entered the Southern Rockies in the 1820s, and by the 1840s, beavers were functionally extinct (Pollock and Suckling 1998). Market and hide hunters killed off the southern herd of bison in the 1870s (Matthiessen 1987, Cook 1989). With the loss of bison and the defeat of the resident tribes, the land was overstocked with livestock, precipitating the first of several grazing collapses in the early 1880s (Flores 1996). Mining camps sprang up in the late 1800s, drawing market hunters who slaughtered elk (*Cervus elaphus*), pronghorn, deer (*Odocoileus* spp.), bighorn sheep (*Ovis canadensis*), and turkey (*Meleagris gallopavo*) to feed the miners (Matthiessen 1987).

With their natural prey gone, wolves, grizzlies, and mountain lions (*Puma concolor*) turned to cattle and sheep for food (Mackie et al. 1982, Bogan et al. 1998). The U.S. Department of Biological Services' (forerunner of the U.S. Fish and Wildlife Service) Division of Predatory Animal and Rodent Control (PARC) used traps, guns, and poison to try to completely exterminate predators, including bobcats (*Lynx rufus*), lynx, mountain lions, wolverines, and coyotes (*Canis latrans*, Dunlap 1988). Prairie dogs were functionally exterminated by a taxpayer-sponsored poisoning program that continues today (Miller et al. 1996).

As a result of direct killing and loss of habitat, today four species native to the Southern Rockies are globally extinct and at least 14 are considered extirpated (see sidebar, Shinneman et al. 2000, Neely at al. 2001). At least four species of birds that historically bred in the region no longer do so, including the marbled godwit (*Limosa fedoa*), harlequin duck (*Histrionicus histrionicus*), merlin (*Falco columbarius*), and ring-billed gull (*Larus delawarensis*, Andrews and Righter 1992, Neely et al. 2001).

Twenty species of plants, invertebrates, and vertebrates from the region are currently listed as Threatened or Endangered under the U.S. Endangered Species Act (U.S. Fish and Wildlife Service 2001a). At least 101 species native to the Southern Rockies are globally imperiled (ranked G1-G2), and nearly 300 other species are considered of special concern due to restricted ranges, population declines, and other vulnerability factors (Neely et al. 2001).

This level of loss from the Southern Rockies is shocking, but it is deeper than just missing parts. For example, the

## LOSS OF SPECIES IN THE SOUTHERN ROCKIES

### EXTINCT

Yellowfin cutthroat trout *(Salmo clarki macdonaldi)*

Eskimo curlew *(Numenius borealis)*

Carolina Parakeet *(Conuropsis carolinensis)*

New Mexico sharp-tailed grouse *(Tympanuchus phasianellus hueyi)*

### EXTIRPATED

Grizzly bear *(Ursus arctos)*

Gray wolf *(Canis lupus)*

Bison *(Bison bison)*

Black-footed ferret *(Mustela nigripes)*

Canada lynx *(Lynx canadensis )*

Wolverine *(Gulo gulo)*

Rio Grande bluntnose shiner *(Notropis simus simus)*

Rio Grande silvery minnow *(Hybognathus amarus)*

American eel *(Anguilla rostrata)*

Freshwater drum *(Aplodinotus grunniens)*

Sturgeon *(Scaphirhynchus platorynchus)*

Blue sucker *(Cycleptus elongatus)*

Rio Grande shiner *(Notropis jemezanus)*

Speckled chub *(Macrhybopsis aestivalis aestivalis)*

(Shinneman et al. 2000, Neely at al. 2001)

loss of large carnivores is more complex than the simple absence of a species. It creates an ecological imbalance, which is discussed in more detail under Wound 3.

Black-footed ferret *(Mustela nigripes)*

## *Wound 2: Loss and Degradation of Terrestrial and Aquatic Ecosystems*

Streams, wetlands, and riparian forests have been severely damaged by loss of beavers and the introduction of livestock grazing; dams (Figure 5.2) and water diversions, groundwater pumping, draining of wetlands and agricultural clearing, and general watershed damage from a variety of human activities. Forest systems are degraded because of historical cutting for mine timbers, railroad ties, and firewood; industrial silviculture operations; exotic tree pathogens; fire management (largely suppression); and grazing impacts on seedling establishment. Grazing by domestic livestock and fire suppression have disrupted and degraded high mountain meadows, intermountain grasslands, and woodlands.

### *Aquatic and riparian ecosystems*

A detailed discussion of human-caused changes to aquatic and riparian ecosystems is found in Chapter 2.6. The near-extermination of beavers by 1840 began the degradation of watersheds and riparian areas. Beaver dams had created extensive wetlands, controlled floods, and stored water for slow release throughout the year. With the loss of beaver dams, wetlands shrunk and seasonal floods went unchecked; fish, invertebrates, and other forms of wildlife that depended on beaver ponds were affected. Although beavers have recovered somewhat, they are still absent in many areas (Knight 1994, Bogan et al. 1998, Pollock and Suckling 1998).

Domestic cattle and sheep grazing also has been a major cause of watershed and stream destruction, as well as degrading grassland ecosystems (U.S. General Accounting Office 1988, Belsky et al. 1999). Human development and water use have destroyed or dramatically altered most species-rich aquatic and riparian ecosystems in the Southern Rockies. Wetlands drained for agricultural use or human habitation, lowered groundwater tables resulting from irrigation and domestic water pumping, and altered hydrological cycles from dams and water diversions all affect the integrity and naturalness of aquatic and riparian ecosystems. (See Chapter 3.3 Water Use for more details).

### *Forests*

Historically, many forests were cut to supply railroad ties, mine timbers, housing materials and firewood. Today, commercial logging, a continuing cause of loss and degradation of forest ecosystems, varies from forest to forest and from year to year in the Southern Rockies. Timber harvests declined in all forests between 1987 and 1997, but there is great concern that the trend is toward more commercial logging in the early 21st century (Figure 5.3).

Bogan et al. (1998: 547) summarize forest health in New Mexico as follows:

> *Fire suppression, commercial forestry practices, and overgrazing have pervasively altered the structure and species composition of most southwestern forests (U.S. Forest Service 1993, Covington and Moore 1994). Old-growth forests have been greatly reduced by high-grading and even-aged management practices that targeted the most valuable old trees, especially ponderosa pine. In addition, until 20 years ago, snags (dead trees) were systematically removed as fire and forest health hazards, while extensive road networks aided those who poached snags for fuel wood. Hence, most managed forests now lack desired numbers of large-diameter snags, which serve important ecological roles such as cavity-nesting sites for many breeding birds (Thomas et al. 1979, Hejl 1994) and probably for many bats as well (Green, pers. comm., Bogan, unpublished data).*
>
> *After 100 and more years of intensive grazing, fire suppression and commercial logging, today's forests are characterized by unnaturally dense stands of young trees, a variety of forest health concerns, increasing potential for widespread insect population outbreaks (Swetnam and Lynch 1993), and unnatural crown fires (Covington and Moore 1994, Sackett et al. 1994, Samson et al. 1994).*

See Chapter 2 for details of concerns about each forest ecosystem type.

#### *Grasslands and shrublands*

The grassland and shrubland ecosystems in and adjacent to the Southern Rockies have largely been converted to farmland and urban landscapes, or subjected to heavy livestock grazing and fire suppression. In addition to direct loss of these habitats, invasion by exotic plants, loss and decline of native species such as bison, prairie dogs and grassland birds, and replacement of palatable forbs favored by wildlife by unpalatable woody plants have all taken their toll (see Chapter 2.6 for more details).

### *Wound 3: Loss and Alteration of Natural Processes*

The removal or decline of large carnivores has lessened or eliminated top-down regulation of ecosystems. Natural fire, which is vital to the health of forest, woodland, and grassland ecosystems in the Southern Rockies, has been largely eliminated by more than 100 years of livestock grazing and fire suppression. Natural over-the-bank stream flooding, which is key for the reproduction, maintenance, and recovery of riparian communities, has been lost because of dams, stream diversions, and flood-control structures. Simberloff and his co-authors (1999) identify predation, fire, and hydrology as the ecologically most essential natural processes to restore. All have been severely disrupted in the Southern Rockies.

#### *Predation and top-down regulation*

Campaigns to eliminate carnivores have caused severe ecosystem dysfunction. Carnivores play an important role in regulating ecosystems, and predation can affect flora and fauna that seem ecologically distant from the carnivore (Terborgh 1988). Through predation, carnivores directly reduce numbers of prey (Terborgh 1988, Terborgh et al. 1997, 2001, Estes et al. 1998, Schoener and Spiller 1999). Carnivores also cause prey to alter their behavior by choosing different habitats, different food sources, different group sizes, different time of activity, or reducing the amount of time spent feeding, so they are less vulnerable (Kotler et al. 1993, Brown et al. 1994, FitzGibbon and Lazarus 1995, Palomares and Delibes 1997, Schmitz 1998, Berger et al. 2001a).

By reducing the numerical abundance of a competitively dominant prey species (or by changing its behavior), carnivores erect and enforce ecological boundaries that allow weaker competitors to persist (Estes et al. 2001). If a predator selects from a wide-range of prey species, the presence of the predator may cause all prey species to reduce their respective niches and thus reduce competition among those species. Removing the predator will dissolve the ecological boundaries that check competition. As a result, prey species may compete for limited resources, and superior competitors may displace weaker competitors, leading to less diversity through competitive exclusion (see Paine 1966, Terborgh et al. 1997, Henke and Bryant 1999).

The impact of carnivores thus extends past the objects of their predation. Because herbivores eat plants and their seeds, predation of herbivores influences the structure of the plant community (Terborgh 1988, Terborgh et al. 1997, 2001, Estes et al. 1998). The plant community, in turn, influences distribution, abundance, and competitive interaction within groups of birds, mammals, and insects. There is ample evidence to support these ideas (see Estes et al. 1978, 1989, 1998, Pastor et al. 1988, McLaren and Peterson 1994, Estes and Duggins 1995, Krebs et al. 1995, 2001, Estes 1996, Terborgh et al. 1997, 2001).

Large carnivores also directly and indirectly affect smaller predators, and therefore the community structure of small prey (Terborgh and Winter 1980, Soulé et al. 1988, Bolger et al. 1991, Vickery et al. 1994, Palomares et al. 1995, Sovada et al. 1995, Crooks and Soulé 1999, Henke and Bryant 1999, Schoener and Spiller 1999). Small prey distribution and abundance affects ecological factors like seed dispersal, disease epizootics, soil porosity, soil chemistry, plant biomass, and plant nutrient content (Whicker and Detling 1988, Hoogland 1995, Detling 1998, Keesing 2000).

When large carnivores are extirpated, ungulate numbers increase, sometimes by a factor of five to seven times (Crête and Manseau 1996, Crête 1999, Berger et al. 2001b). The loss of some species thus creates an overabundance of other species, to the detriment of the ecosystem. High elk numbers negatively affect the growth of aspen (*Populus tremuloides*, Kay 1990, Kay and Wagner 1994, White et al. 1998, Ripple and Larson 2000). Similarly, Berger et al. (2001b) showed that high numbers of moose (*Alces alces*) decreased the quality and quantity of willow (*Salix* spp.); thus, neotropical migrant birds fared better in areas with wolves and bears than in areas where moose were released from predation. Overabundance of white-tailed deer (*Odocoileus virginianus*) has been shown to reduce numbers of native rodent species, cause declines in understory nesting birds, obliterate understory vegetation in some forests, and even eliminate regeneration of the oak (*Quercus* spp.) canopy (Alverson et al. 1988, 1994, McShea and Rappole 1992, McShea et al. 1997).

Managing carnivores without considering the indirect effects that will cascade through a system when we change their numbers will undoubtedly continue to alter the structure and function of an area (see Terborgh et al. 1999). Management policies based on reducing carnivore numbers have caused, and will continue to cause, severe harm to many other organisms that seem distantly removed from the apex trophic layer (see Terborgh 1988, Terborgh et al. 1999). While short-term control and hunting restrictions may be necessary when a system is highly perturbed or fluctuating outside its normal bounds of variability, such tactics only address a symptom.

In the Southern Rockies, extirpation of the wolf and grizzly bear, and the decline of mountain lions have disrupted ecological integrity through the behavioral and numerical release of prey animals, as exemplified by Colorado elk herds that exceed the carrying capacity of the land. Carnivore eradication and reduction has simplified systems and reduced biodiversity, largely by eliminating their keystone role of ungulate predation.

### *Fire*

While Native Americans may have set fires, fire frequencies were probably controlled by climate and fuel dynamics, rather than by source of ignition (Allen et al. 1998). Although ecosystems such as spruce-fir likely burned in stand replacing fires every few hundred years, many ecosystems in the Southern Rockies coevolved with frequent

fire. Before about 1900, most montane forests burned in accordance with the two-to-seven-year wet-dry cycles associated with the El Niño-Southern Oscillation (Swetnam and Betancourt 1998). Incomplete understanding of the ecological role of natural fire in these ecosystems led the Forest Service and other land managers to aggressively try to put out fires from about 1906 on. In addition to fighting fires, the Forest Service deliberately encouraged overgrazing by cattle and sheep to eliminate grass that carried the natural, cool, ground fires (Savage and Swetnam 1990, Swetnam 1990, Bogan et al. 1998). Bogan et al. (1998: 547) conclude, "Fire suppression over the past pervasively affected many southwestern ecosystems (Covington and Moore 1994)." Increasing numbers of scientists recognized fire's important role by the 1960s, but many foresters and ranchers were notably unreceptive to these findings. See Chapter 2 for a more detailed discussion of natural fire regimes for various forest types.

Thinning techniques (logging) and prescribed fire are methods used to return these altered forests to their pre-settlement structures and composition, especially around residential development (US Forest Service 2000). However, these approaches may be misguided in areas where the forests were naturally dense and experienced stand-replacing fires, especially in mesic sites and upper elevations (Shinneman and Baker 1997, Veblen et al. 2000). Furthermore, any restoration effort must actually retain remaining old trees and not increase road densities or edge habitat, destroy interior habitat and roadless areas, alter landscape structure, or aid the spread of weedy species (Shinneman and Baker 2000). In addition, the ability of land managers to allow fires to burn unimpeded is increasingly restricted as people continue to build mountain homes, especially in lower elevation lands, such as ponderosa pine/Douglas-fir forests, and in the wild-urban interface.

### *Flooding*

Riparian woodlands were naturally maintained by

flooding from snowmelt in the spring and thunderstorms in the summer. Degradation of watersheds, loss of the functional role of beavers, and increase of dams, water diversions, groundwater pumping, channelization, and flood control structures have largely eliminated this natural disturbance regime. Simberloff et al. (1999) discussed the many benefits of naturally fluctuating water levels in streams. Elimination of natural flooding has allowed exotic tamarisk (*Tamarix* spp.) to out compete native cottonwood (*Populus* spp.) and willows in parts of the region. Dams and diversions also alter sediment loads, oxygen levels, and water temperatures. In some cases in the Southern Rockies, streams have been channeled, resulting in drastically altered stream ecosystems, destroyed riparian habitat, and diminished populations of native species (Osmundson et al. 1995).

## Wound 4: Fragmentation of Wildlife Habitat

Wildlife habitat in the region has been fragmented by highways, roads, and vehicle ways; dams, irrigation diversions, and dewatering of streams; destruction and conversion of natural habitat, and other works of civilization, such as urban, suburban, and exurban (ranchette) development. Fragmentation has severed historic wildlife migration routes and has potentially isolated wide-ranging species such as wolf, mountain lion, lynx, pronghorn, and bighorn sheep, and even more localized species, in nonviable habitat islands.

### *Roads*

Roads and travelways have cut into river valleys and wild areas alike, connecting human settlements and providing high-speed transportation corridors. They cut the land into smaller and smaller pieces, increasing edge habitat and presenting barriers to wildlife movements. Interstate highways 25, 70 and 80 are particularly formidable for many kinds of wildlife. The Colorado Division of Wildlife characterized I-70 in central Colorado as the Berlin Wall for wildlife. See Chapter 3.2, Roads and Infrastructure, for a detailed description of the effects of the more than 122,840 km of primary and secondary roads and uninventoried routes in the Southern Rockies (Shinneman et al. 2000).

The Jemez Mountains provide an example of how habitat fragmentation from roads has increased in New Mexico. There, road length increased from 719 km in 1935 to 8,433 km in 1981, and the estimated total area of road surfaces grew from 0.13% of the map area in 1935 (247 ha) to 1.67% in 1981 (3,132 ha, Allen et al. 1998). In Wyoming, a study on a 29,600 ha parcel of the Medicine Bow National Forest reported that roads and the associated edge effects may cover as much as 20% of the land area (Reed et al. 1996, Figure 5.4).

### *Recreation*

In many parts of the Southern Rockies, the effects of recreation and tourism on wildlife habitat and connections between core areas are at least as serious as extractive uses of the land. Backcountry travel, summer and winter off-road vehicle use, and ski-area development are of great concern (Knight 2000).

The burgeoning use of mechanized transportation, including dirt bikes, ATVs, extreme jeeps, snowmobiles, jet skis, and "personal watercraft" exert pressures on habitat previously unknown and often unanticipated. Between 1990 and 2000, ATV and dirt bike registrations in Colorado increased six-fold, to a total of 62,000. Although owners represent less than 2% of Colorado's population, their mobility spreads the impact faster and farther than many other types of recreation. Off road vehicles (ORVs) have tremendous ecological effects, including soil compaction and erosion, reduced water infiltration rates, spread of invasive weeds, and various negative effects on birds and vegetation (Meffe and Carroll 1997, Primack 1998). The Southern Rockies Conservation Alliance made suggestions to encourage responsibility where ORVs are used. They are available at http://www.southernrockies.org/motorized_recreation_contacts.htm. Major population centers adjacent to the eastern edge of the Southern Rockies, as well as hundreds of smaller cities and communities within the ecoregion, bring residents and visitors alike in unprecedented numbers to the forests and mountain valleys. Hiking, mountain biking, motorcycling, skiing, boating, fishing, photography, climbing, hunting, wildlife viewing, horseback riding, scenic driving, and almost every other imaginable recreational pursuit are increasing, and contributing to further direct loss and fragmentation of wildlife habitat and natural areas. See Chapter 3.3, Recreation, for a more detailed discussion.

### *Dams and diversions*

Dams on rivers and headwater streams, irrigation diversions, and dewatered and degraded stretches of once-perennial streams have fragmented the habitat for native fish, amphibians, and aquatic invertebrates (Bogan et al. 1998, Crawford et al. 1993). In the arid West, year-round access to water for agriculture and domestic use comes from storage reservoirs, as well as ground water. Over 5 million acre-feet of water in the Southern Rockies are stored in more than 1,000 dams, with a network of diversions, trans-basin pipelines, canals and ditches to deliver water to its human-designed destination (Shinneman et al. 2001). Aquatic species are locked into limited and degraded stretches of rivers and streams, and as the current drought cycle continues, alteration of aquatic and riparian habitats is likely to increase.

### *Destruction and conversion of habitat*

Destruction or conversion of natural habitats to other uses, in addition to direct destruction, also results in fragmenting the larger landscape. Remaining natural habitats are separated from adjacent patches, and are interrupted by lands devoted to livestock, clear-cuts, human settlements, and recreation complexes (such as large ski resorts). Much of the Medicine Bow National Forest in Wyoming and portions of the Rio Grande National Forest in southern Colorado offer good examples of logging levels that led to extensive habitat fragmentation (Reed et al. 1996, U.S. Forest Service 1998, Shinneman et al. 2000). The Medicine Bow is now the most heavily clearcut and roaded forest in the Rocky Mountains and is criss-crossed by more than 3,500 miles of roads (Biodiversity Conservation Alliance 2003). In addition, inappropriate logging practices have inhibited stand regeneration (Reed et al. 1996, U.S. Forest Service 1998, Shinneman et al. 2000).

Because of prolonged drought, "logging for water" has been proposed in Colorado in 2003. In order to increase water supply, as much as 25 to 40 % of the forest must be clear-cut, resulting in a highly fragmented landscape like this one in the Medicine Bow (Figure 5.4). Colorado Trout Unlimited notes that a recent literature survey produced no direct evidence that thinning trees would increase baseflow (Rhodes and Purser, 1998). One of the articles they cited (Marvin 1996), a mid-1990's review of 30 studies, included data from a Colorado watershed where 40% of the trees were selectively logged, yet there was no increase in annual water yield (Colorado Trout Unlimited 2002).

### *Development*

As population continues to grow throughout the Southern Rockies, primary and second homes, ski areas, golf courses, and transportation infrastructure expand to service human wants and needs. Population in the Southern Rockies grew six-fold between 1900 and 1995, from .5 million to 3 million, and the population projection for 2020 is nearly 4 million (Shinneman et al. 2001). In some areas, especially near major urban centers, wildlife habitat is an island in a sea of humanity. This extreme fragmentation and loss of habitat drives species to other areas and often to their demise.

## *Wound 5: Invasion of Exotic Species*

Native species, once held in a delicate balance by natural interactions, are displaced by aggressive and disruptive exotic plant and animal species. Some of these destructive invaders were deliberate introductions; some escaped from cultivation; others hitchhiked in on shoes, tires, livestock, and the wind. Most do well in disturbed habitats. Once established, exotic species can displace natural species and disrupt ecological processes (Wilcove et al. 1998, 2000).

Colorado has about 500 known exotic plants, and at least 78 exotic fish species (Shinneman et al. 2000). Based on a rate of spread of 14% per year, the BLM estimates that 1,840 ha per day are infected by exotics in the western U.S. (Shinneman et al. 2000). New Mexico now has about 390 established alien plant species or 11 % of the state's flora (Cox 2001). Even Rocky Mountain National Park has 21 exotic plant species deemed to be "species of special concern" for their ability to cause ecological damage (Shinneman et al. 2000, Table 5.1).

Tamarisk (salt cedar), a native of the Middle East, provides a graphic illustration of the problem. Planted as an ornamental, it spread through cattle-damaged riparian areas. It benefits from dams and flood-control levees, which prevent natural cycles of drying and flooding with which native cottonwoods and willows evolved. Tamarisk is not a major habitat or food source for native species, although it does provide interim nesting habitat for the southwestern willow flycatcher (*Empidonax traillii extimus*), an Endangered species, in a few areas where native vegetation has been lost. Tamarisk is a phreatophyte, sucking up large amounts of water through its roots and transpiring this moisture into the air, thereby drying up springs and streams upon which native species depend.

Various factors and conditions in the Southern Rockies have contributed to the invasion of exotic species: The elimination of seasonal flooding by upstream dam construction has allowed non-native salt cedar and Russian olive (*Elaeagnus angustifolia*) to dominate large areas (Dick-Peddie 1993, Farley et al. 1994). Heavy livestock grazing has helped to spread non-native weeds, such as Kentucky bluegrass, Russian thistle (*Salsola tragus*), and cheatgrass, among the many weeds of serious concern (Fleischner 1994, Weddel 1996). The eastern bullfrog (*Rana catesbeiana*) threatens native frogs (Rosen and Schwalbe 1995). Colorado has intentionally introduced the moose and mountain goat (*Oreamnos americanus*) for hunting opportunities. Accidentally introduced mammals in Colorado include the house mouse (*Mus musculus*) and Norway rat (*Rattus norvegicus*).

Additionally, throughout the Southern Rockies, many streams and lakes have been stocked with non-native fish species, which have altered the food chain, and prey upon, out-compete, or hybridize with native fishes. These exotic fish species threaten native Rio Grande (*Oncorhynchus clarki virginalis*), greenback (*O. c. stomias*), and Colorado River cutthroat trout (*O. c. pleuriticus*, Young 1995) in their respective historical ranges in New Mexico, Colorado, and Wyoming. At least 75 exotic species of fish are present in New Mexico (Bogan et al. 1998; Boydstun et al. 1995). Murray (1996) discusses the ecological impacts of stocking high mountain lakes, which typically did not have natural fish populations, with exotic trout. Aquatic invertebrates extirpated from such lakes after stocking of fish include beetles, midges, dragonflies, water striders, mayflies, water fleas, copepods, and sideswimmers. Stocking non-native trout, a practice that continues today, has compromised native strains of cutthroat trout (see focal species account). Particularly egregious examples of management for exotic species are the dams built in the Pecos Wilderness to raise the water level of natural lakes for the sole reason of supporting exotic trout (Murray 1996).

#### *Pathogens*

Many 5-needled pine forests face the threat of blister rust (*Cronartium* spp.), which has arrived from Eurasia. Blister rust is anticipated to eliminate 98% of the 5-needled pines before it runs its course (Mitten pers. comm. 2001). Clark's nutcrackers (*Nucifraga columbiana*) are the prime seed dispersal agents for the 5-needled pines. With the arrival of West Nile virus (*Flavivirus* spp.) in the Southern Rockies, a virus particularly virulent for corvids, the mutualistic relationship between jays and 5-needled pines may be impaired, costing the pines their method of dispersal and the jays and nutcrackers a critical food source.

The prairie dog ecosystem faces a severe threat from sylvatic plague (*Yersinia pestis*). Approximately two-thirds of the range of the black-tailed prairie dog (*Cynomys ludovicianus*) has been affected by plague, and the disease is spreading from southwest to northeast (U.S. Fish and Wildlife Service 2000). Considering that much of the eastern part of the prairie dog range has already been converted to cropland, the picture is dire. The only states where prairie dogs are largely free from plague are North and South Dakota, but the northeastward movement of the disease means that outbreaks of plague may occur there in the future (U.S. Fish and Wildlife Service 2000). At present, a third of the remaining black-tailed prairie dogs exist in seven complexes that are larger than 4,000 ha. Because plague has the potential to eliminate entire complexes (as does poisoning), seven independent events of disease or poisoning could eliminate 36%

**Table 5.1. Exotic plants of concern in Rocky Mountain National Park.**

| Species | Common Name |
|---|---|
| *Elymus repens* | Quackgrass, couchgrass |
| *Bromus inermis* | Smooth brome |
| *Bromus tectorum* | Cheatgrass, downy brome, downy chess |
| *Carduus nutans* | Musk thistle, nodding plumeless thistle |
| *Centaurea diffusa* | Tumble knapweed, diffuse knapweed |
| *Centaurea biebersteinii* | Spotted knapweed |
| *Leucanthemum vulgare* | Oxeye daisy, white daisy |
| *Cirsium arvense* | Canada thistle, creeping thistle |
| *Cirsium vulgare* | Bull thistle, spear thistle |
| *Convolvulus arvensis* | Field bindweed, morning glory |
| *Dactylis glomerata* | Orchard grass, cocksfoot |
| *Euphorbia esula* | Leafy spurge |
| *Gypsophila paniculata* | Babysbreath |
| *Hypericum perforatum* | St. John's wort, goatweed, Klamath weed |
| *Linaria dalmatica* | Dalmation toadflax |
| *Linaria vulgaris* | Butter and eggs, yellow toadflax, white snapdragon |
| *Lythrum salicaria* | Purple loosestrife, purple lythrum |
| *Melilotus alba* | White sweetclover, honey clover |
| *Melilotus officinalis* | Yellow sweetclover |
| *Phalaris arundinacea* | Reed canarygrass |
| *Poa pratensis* | Kentucky bluegrass |
| *Sonchus arvensis* | Meadow, perennial, or creeping sowthistle |

of the black-tailed prairie dogs.

Brucellosis (*Brucella abortus*), which causes fetus abortion in cattle, was transmitted to bison and elk in the past. Now, Wyoming ranchers and state officials are concerned that bison and elk may pass the disease back to cattle. For example, Wyoming has a vigorous bison control program for bison when they leave Yellowstone National Park. The inoculation for brucellosis in cattle is effective, but it has not yet proved effective in bison.

Whirling disease (*Myxobolus cerebralis*) is caused by a protozoan parasite introduced from Europe, and it affects newly hatched trout. At first, the disease was thought to only affect hatchery trout, but hatchery trout have now passed the parasite to wild trout populations (Stohlgren 1998). Yet, hatchery releases continue in Colorado. Whirling disease may affect the native greenback cutthroat trout (Stohlgren 1998).

Chronic Wasting Disease (CWD) is related to scrapie in sheep, bovine spongiform encephalopathy in cattle (Mad Cow Disease) and Creutzfeldt-Jakob disease of humans. The causative agent is believed to be a modified protein (prion). Chronic Wasting Disease infects elk, white-tailed deer, and mule deer (*Odocoileus hemionus*), but is not known to infect livestock or humans at present. No treatment is known and the disease is typically fatal. It is not known how the disease is transmitted between animals, although direct contact between infected and non-infected animals and contamination of soil by excreta may be routes. The disease was originally described in captive animals 35 years ago in Colorado, and over the last five years, it has been found in wild herds of southeastern Wyoming, northeastern Colorado, and southwestern Nebraska. In 2002, it was also found in New Mexico. It is of increasing concern for wildlife managers across North America for its effect both on the herds and on a major recreational hunting industry (Animal and Plant Health Inspection Services, www.aphis.usda.gov/vs/nahps/cwd/).

### *Wound 6: Pollution and Climate Change*

Pollution of the air, waterways, and land, does not respect political or ecosystem boundaries. Many sources are distant from the Southern Rockies, and likewise Rocky Mountain sources may spread afar. Mines, feedlots, factories, smelters, power plants, agricultural and forestry biocides, automobiles, and urban areas have spread heavy metals, toxic wastes, and chemicals in the air, land, and water - all affecting species, ecosystems, and climate.

Many streams and mountain lakes are polluted by pesticides, herbicides, and excess nutrients from agricultural sources; acid and heavy metal mine drainage; increased sedimentation from land uses such as logging, road-building, and recreation; and acid deposition from power plants and urban areas. For example, concentrations of DDT (dichlorodiphenyltrichloroethane) and other biocides remain from their heavy application for forestry and agricultural purposes in the 1950s and 1960s (Bogan et al. 1998), as illustrated by the 516,664 kilograms of DDT that was sprayed on the 478,055 ha of the Santa Fe and Carson National Forests between 1955 and 1963 (Bogan et al. 1998).

Colorado has nearly 9,000 abandoned mines that still release toxic material at some level (Shinneman et al. 2000). Water quality in the Arkansas River headwaters is compromised by heavy metal depositions, and the Alamosa River is barely recovering from the effects of the Summitville mine.

Air pollution from the Front Range and coal-burning power plants harms mountain systems. Colorado snow packs hold high concentrations of sulfate and nitrate in the mountains near Denver (Turk et al. 1992 in Stohlgren 1998). High elevation sites in the Front Range receive 9 times more nitrogen per hectare (4.7 kg per ha.) than remote areas of the world (Stohlgren 1998).

Various pollutants lead to ozone depletion and contribute to global climate change. Increases in atmospheric ultraviolet transmission may result in direct physiological damage, damage to DNA, and increased susceptibility to pathogens, predators, and competition (Caldwell, et al. 1998, Licht and Grant 1977). In a recent study, global warming is suggested as a possible cause of the apparent extirpation of 7 of 25 American pika (*Ochotona princeps*) populations in the Great Basin (Beever et al. 2003).

## 3. Conclusion

The wounds to the Southern Rockies are pervasive, complex and interrelated, and they must be considered as a whole system. To deal with one in isolation may only exacerbate some other wound – using herbicides to control exotic species risks serious damage from toxic chemicals, for example. Good science and careful planning are needed to inform both the Wildlands Network Vision and public policy.

*Whatever you can do, or dream you can, begin it. Boldness has genius, power, and magic in it.*

-Johann von Goethe

*I hold the most archaic values on earth ... the fertility of the soul, the magic of the animals, the power-vision in solitude, ...the love and ecstasy of the dance, the common work of the tribe.*

- Gary Snyder

# 6 MISSION AND GOALS

Dave Foreman

*(This chapter, originally written for the New Mexico Highlands Wildlands Network Vision, was modified with the author's permission to reflect differences in ecoregion characteristics and Southern Rockies Wildlands Network Vision conservation goals.)*

## 1. Mission

The mission of the Southern Rockies Wildlands Network Vision is to protect and rewild the regional landscape. The vision justifies and provides references for design decisions for a wildlands network made up of core protected areas, wildlife movement and riparian linkages, and compatible-use areas. The vision proposes (1) to protect all remaining natural habitats, native species, and natural processes; and (2) to heal the region's ecological wounds by developing and implementing a conservation vision for the region.

## 2. Healing the Wounds Approach to Ecological Restoration

Until recently, conservation has focused primarily on protecting wildlands and wildlife from development and further wounding. A hallmark of recent conservation, however, is ecological restoration. Unfortunately, much of what is called ecological restoration today falls far short of the mark. Soulé (1996) warned against "restoration" that seeks only to put back the process, but not the community. Soulé wrote that "it is technically possible to maintain ecological processes, including a high level of economically beneficial productivity, by replacing the hundreds of native plants, invertebrates and vertebrates with about 15 or 20 introduced, weedy species." The contributors to *Continental Conservation* cautioned that *"process and function are no substitute for species"* (Simberloff et al. 1999: 67).

"Regional and Continental Restoration" (Simberloff et al. 1999), Chapter 4 in *Continental Conservation* (Soulé and Terborgh 1999), provides state-of-the-art guidance for wildlands restoration. In setting goals, the question of which point in time is referenced for "the full range of native species and ecosystems" must be answered. Simberloff and his co-authors explain that restoration can never achieve an exact reproduction of a system that existed at some previous time. Instead, they recommend:

> *Thus, restoration should be aimed at returning to the point on this trajectory of change that would have obtained in the absence of human disturbance, rather than trying to replicate the precise system that once was present (Simberloff et al. 1999:66).*

Therefore, we work to put all the pieces back into an ecosystem instead of trying to recreate a poorly understood ecosystem at some arbitrary point before significant human disruption.

In addition, restoration needs to be done on a landscape level because wide-ranging species require large areas; and ecological disturbance (such as fire) can only be restored in a large area. The "dynamic, nondeterministic character of natural communities requires restoration of large areas in order to promote the long-term viability and adaptability of populations and communities" (Simberloff et al. 1999: 69). Less-than landscape-scale restoration produces "ecological museum pieces — single representatives of communities that, although present because of unusually large restoration and maintenance investments, do not exist in any ecologically meaningful way" (Simberloff et al. 1999: 71). A medical analogy would be keeping an otherwise terminally ill patient permanently on life support at high cost.

The Southern Rockies Wildlands Network Vision calls for ecological restoration, and the specific restoration approach adopted by our vision is based on healing the ecological wounds discussed in the previous chapter. In general, the Southern Rockies Wildlands Network Vision follows the direction from *Continental Conservation*: "Restoration methods for wildlands can be divided into three categories: control of invasive nonindigenous species; reestablishment of natural abiotic forces; and reintroduction or augmentation of native species" (Simberloff et al. 1999:72).

The Southern Rockies Network Vision is also based on *rewilding*, which calls for the recovery of large carnivores and protection of their core habitats and the connectivity between cores. Ecological restoration and rewilding are closely related. Loss of large carnivores is a common result of human disruption. Not only are species lost, but also the important ecological process of top-down regulation through predation is lost. Rewilding, then, is a way of healing the wounds caused by loss of large carnivores.

Ecological restoration is a growing science that will play a crucial role in healing many of the ecological wounds identified in the previous section. The Society for Ecological Restoration (SER) has developed project policies and guidelines that provide the methodologies for implementing a successful restoration project (http://www.ser.org/reading.php?pg=primer2). Their Project Policies include information on project planning, contending with exotic species at project sites, integration of a project into a larger landscape, planting of regional ecotypes, local stewardship, and project evaluation. The Project Guidelines give detailed information on the process of an ecological restoration project from start to finish, including: conceptual planning, preliminary tasks, installation planning, installation planning tasks, post installation planning tasks, and evaluation.

Ecological restoration within designated Wilderness Areas may be necessary in special circumstances to restore ecological integrity. Mechanized equipment may be necessary in certain cases (Sydoriak et al. 2000). However, Crumbo (unpublished) cautions:

> *All management decisions affecting wilderness, including restoration or visitor use, should conform to the minimum requirement concept derived from the Wilderness Act (Section 4(c)). In wilderness, any management action must be based on the minimum intervention necessary to achieve wilderness conservation goals. Wilderness's minimum requirement concept simply comprises the most rigorous interpretation of a general precautionary approach applicable to all public lands.*

Unfortunately, even the largest protected Wilderness Areas are spatially inadequate to maintain native biodiversity (Newmark 1995, Noss and Cooperrider 1994). Many reserves, including Wilderness, may require some degree of active management since they are simply too small and isolated for essential ecological processes to operate. But management actions in wilderness, if deemed necessary at all, must be viewed as interim measures and the minimum required to achieve the long-term goal of a wild, self-sustaining wilderness (Noss et al. 1999).

## 3. Healing the Wounds Goal-Setting

A conservation strategy is more likely to succeed if it has clearly defined and scientifically justifiable goals and objectives. Goal setting must be the first step in the conservation process, preceding biological, technical, and political questions of how best to design and manage such systems. Primary goals for ecosystem management should be comprehensive and idealistic so that conservation programs have a vision toward which to strive over the decades. They should address both preventing additional wounding and healing existing wounds. A series of increasingly specific objectives and action plans should follow these goals and be reviewed regularly to assure consistency with primary goals and objectives (Noss 1992).

The order of implementation steps to achieve goals for the Southern Rockies ecoregion may be complex and interdependent. For instance, the reintroduction of predators into certain regions may first require the restoration of vegetation and prey before the ecosystem is able to support them. Strategies to increase the biomass food base may include revitalizing the vegetation through the reintroduction of fire, reversing the encroachment of woody vegetation into former grasslands/savannas, and replacing livestock with grazers that are available to predators. However, reintroduction of formerly extirpated predators like the wolf (*Canis lupus)* may ultimately help restore native vegetation through top-down regulation of grazers and browsers. Thus, goals and the objectives for achieving them are integrally connected, although it is not always clear what action should be the prerequisite for another because of the complexity of ecosystems.

## 4. Goals and Objectives

### *Goal 1: Protect and Recover Native Species*

Protect extant native species from extinction or endangerment and recover all native species to the region.

Some of these species are listed in our focal species group, and others are not. That is only because we tried to select focal species based on complementary habitat needs, and we did not list species that duplicated needs.

*Objectives*

1. Maintain viable populations of focal and other key species through protected large core areas and functional landscape connectivity, allowing for redundancy in the system in anticipation of future natural and anthropogenic changes.

2. Protect, recover, or reintroduce declining or extirpated native carnivores, including but not limited to, wolf, grizzly bear (*Ursus arctos*), river otter (*Lontra canadensis*), lynx (*Lynx canadensis*), wolverine (*Gulo gulo*), American marten (*Martes americana*), and black-footed ferret (*Mustela nigripes*).

3. Protect, recover, or reintroduce other declining or extirpated native species, including, but not limited to, bison (*Bison bison*), bighorn sheep (*Ovis canadensis*), pronghorn (*Antilocapra americana*), beaver (*Castor canadensis*), prairie dog (*Cynomys* ssp.), and cutthroat trout (*Oncorhynchus clarki*).

4. Protect federally listed Threatened and Endangered species and other species throughout the region, including but not limited to northern goshawk (*Accipiter gentilis*), Mexican spotted owl (*Strix occidentalis*), burrowing owl (*Athene cunicularia*), sage grouse (*Centrocercus urophasianus*), mountain plover (*Charadrius montanus*), and willow flycatcher (*Empidonax traillii*).

### *Goal 2: Protect and Restore Native Habitats*

Protect and restore all habitat types from further degradation and loss.

*Objectives*

1. Protect remaining roadless areas as National Wilderness Areas, National Parks/Monuments, or special management units with high protection.

2. Identify, protect, and restore riparian forests, wetlands, watersheds, and watercourses to maintain habitat integrity and connectivity.

3. Reduce erosion and restore eroded areas.

4. Expand key private lands under protective management through purchase, conservation easements, and other mechanisms.

5. Protect native forests (old-growth and other generally intact forests) and restore large areas of previously logged or degraded forests to recover old-growth characteristics.

6. Protect native grasslands and restore areas previously overgrazed and degraded.

7. Encourage ecological grazing management that allows for restoration of natural forest and grassland conditions and processes on private ranches.

8. Protect or restore native species that have important roles in maintaining native habitats, such as large ungulates and keystone species.

### *Goal 3: Protect, Restore and Maintain Ecological and Evolutionary Processes*

Protect functioning ecological and evolutionary processes, and restore and maintain disrupted ecological and evolutionary processes.

*Objectives*

1. Restore native predators to their historical range, when and where appropriate, to maintain predation and its top-down ecological functions.

2. Restore natural fire within the special restrictions of Wilderness Areas management and constraints of the wild-urban interface; and reduce or eliminate the disruptive role of livestock grazing on natural fire cycles.

3. Implement management policies that allow natural insect and disease outbreaks to take their course.

4. Restore flooding and hydrologic cycles by allowing natural flooding to occur where feasible, mim-

icking seasonal flooding cycles below reservoirs, and removing unnecessary dams and rip-rap from rivers.

### *Goal 4: Protect and Restore Landscape Connectivity*

Protect the land from further fragmentation, and restore functional connectivity for all species native to the region.

*Objectives*

1. Identify and protect terrestrial, riparian, and aquatic linkages and areas important for wildlife movement.

2. Develop management standards and legal protection for such linkage areas.

3. Prevent road construction, logging, off-road vehicle use, mining, and other disruptive activities in Forest Service and BLM roadless areas.

4. Promote the closure, removal and complete rehabilitation of old logging roads and other roads and ORV routes that no longer serve a legitimate purpose.

5. Promote modification of barriers (highways, *etc.*) to allow the safe movement of wildlife.

### *Goal 5. Control and Remove Exotic Species*

Prevent further spread of exotic species, and eliminate or control present exotic species.

*Objectives*

1. Implement a comprehensive program to remove, control, and mitigate exotic species, including non-native pests and disease organisms.

2. Prevent or minimize new introductions of other exotic plants, animals, and disease organisms.

### *Goal 6: Reduce Pollution and Restore Areas Degraded by Pollution*

Prevent or reduce further introduction of ecologically harmful pollutants into the region, and remove existing pollutants.

*Objectives*

1. Close and/or remediate polluting mines and restore river ecosystems affected by mining activities.

2. Promote clean-up of polluted sites (including nuclear waste), especially those that affect surface and ground water.

3. Discourage additional mining and oil and gas development in ecologically sensitive areas.

4. Reduce sedimentation loads in streams and rivers to natural levels.

## 5. A Prescription for Healing the Wounds

The Southern Rockies planning team believes that a healing-the-wounds approach is an excellent way to analyze conservation problems and to accomplish visionary but achievable goals across a landscape. Healing the wounds is also a powerful metaphor that can move conservationists to action and inspire the public. Healing ecological wounds can change people from conquerors to plain citizens of the land community (Leopold 1966). Unless we heal the wounds, we will have a continent "wiped clean of old-growth forests and large carnivores"; we will "live in a continent of weeds" (Terborgh and Soulé 1999). However, if we can achieve our goals to heal the wounds, we can restore integrity to ecological systems and safeguard the rich biodiversity of the Southern Rockies. To succeed at this will lead to a healthy and more sustainable relationship between natural and human communities.

While the goals and objectives of a conservation vision should be bold, even audacious, they should also be achievable. The Southern Rockies Wildlands Network Vision specifies realistic implementation tactics to address the wounds of the Southern Rockies. Action plans will be developed for each implementation step (see Chapter 10).

EC
99

*Glance at the sun.*
*See the moon and the stars.*
*Gaze at the beauty of earth's greenings.*
*Now, think.*

- Hildegard Von Bingen

*WITH PHILOSOPHY*
*HE CONTEMPLATES*
*THE MOUNTAIN…*
*OLD PROFESSOR FROG*

-Issa

# 7 METHODS FOR CREATING THE WILDLANDS NETWORK

Michelle Fink, Kurt Menke, Doug Shinneman

The Southern Rockies Wildlands Network Design used computer modeling and expert opinion to craft a design that includes the best available data and best meets the overarching goals of rewilding and ensuring persistence of native biodiversity in the ecoregion. Computer models that were used include SITES and least cost path analysis, while expert opinion was incorporated from literature, several expert workshops, individual review by SREP's science team, and regional and national experts.

## 1. The SITES Model

A wildlands network design asks two basic questions: "where should the network components (especially cores and linkages) be located?" and, "how large should the network be?" One tool to find the answers is the site selection optimization program SITES v1.0 (Andelman et al. 1999). Goals for each target component (special elements, representation, and focal species) are all stated quantitatively as inputs into SITES, and the model highlights the most important areas for a wildlands network (Noss and Harris 1986).

The SITES model was developed for The Nature Conservancy (TNC) by land-use planning experts, and has been used by TNC to develop ecoregional plans for nine different ecoregions. The SITES model attempts to minimize the "cost" of a conservation network while maximizing attainment of conservation goals, usually in a compact set of core areas. This set of objectives constitutes the "objective cost function," in which:

*Total Cost = (Cost of selected planning units) + (Penalty cost) + (Weighted boundary length)*

The user of the SITES model determines what the costs are, which may include literal interpretations such as actual cost of acquiring and protecting lands for conservation. In our case the *cost of selected planning units* was determined by an index that represents the relative cost of restoring these areas back to wild land. Another way to look at this index is that it represents the relative level of human-caused ecological degradation of an area. The *penalty cost* is an additional cost for failing to meet stated target goals. SITES automatically calculates a penalty cost for each target that estimates how much more it would cost, in terms of planning unit cost plus weighted boundary length cost, to meet the stated goal. If desired, each target goal can be assigned a unique weight to apply to the penalty cost that reflects its greater or lower intrinsic value to the overall network design. However, we weighted all target elements equally. The *weighted boundary length* is the cost of the spatial dispersion of the selected sites as measured by the total boundary length of the network design multiplied by an arbitrary modifier. Increasing this boundary length modifier has the effect of clumping areas chosen in order to minimize the perimeter to area ratio, and thereby reduce fragmentation of the network.

The entire study area was divided into 1,000-hectare hexagonal areas that were used as the individual planning units so that SITES could create an initial optimized design irrespective of jurisdictional boundaries. A hexagonal shape was selected over other shapes or entities (*e.g.* square cells, watersheds) because the unit size remains constant (planning units that vary widely in size can present problems for the SITES algorithm), it approximates a circle (which has a low edge to area ratio), and provides a relatively smooth output (as compared with similar sized square cells). We chose a planning unit size of 1,000 ha because this was larger than the resolution of the coarsest input data set and provided output that had sufficient resolution for the purposes of this study. Because other Wildlands Network Designs in the Spine of the Continent MegaLinkage are also using the unit size of 1,000 ha, it will facilitate collation of the regional plans into a continental plan.

The SITES algorithm uses a process termed "simulated annealing" (Andelman et al. 1999). Through iterations, this technique gradually "hones in" on a set of planning units that best meet the target conservation goals, while minimizing cost. The annealing process starts with a random set of planning units and then adds and discards planning units via an iterative process in an attempt to maximize conservation target goals while minimizing costs. This method does not necessarily find the "perfect" solution each time, but instead returns a near optimal solution. Thus several runs of the model are recommended in order to determine the best selection output. Each time we ran the model, we did it in sets of ten runs, each with one million iterations. SITES then selected the run that best met the target goals with the least cost as the "best" run for that set of ten. SITES also creates a summary file, which indicates the number of times (out of the ten) that a particular planning unit was included in the final set of planning units. Planning units that are selected in multiple runs of the set, even though they do not necessarily appear in the "best" (lowest cost) solution, may highlight potential linkages between core wild areas and/or areas with more intrinsic value to the overall design (Noss et al. 2002). Therefore, planning units that were chosen at least five out of the ten times and were in groups of at least four adjacent units were used to delineate the final SITES output.

Details of the methods used are below. Preliminary drafts of the focal species inputs and SITES analysis were reviewed during three expert workshops, with participation from the scientific, academic and conservation communities (see list of participants in Acknowledgements). Contributions were evaluated and incorporated in a second run of the SITES model. These data were then made available again for further review. We then used these results, along with the expert opinion of local conservation groups, as the basis of the Network Design.

### *Planning Unit Cost*

To determine the cost of selecting each planning unit, each 1,000 hectare hexagon was assigned a 'cost score' which was compiled from three data inputs: land cover (level of human disturbance), housing density, and road density/edge effect (Figure 7.1). The three inputs were given compatible relative scores and then combined for an overall score that represents the relative degree to which these areas have been removed from their natural state. These scores are relative and subjective and imply natural values rather than dollar amounts. General assumptions of this method for assessing planning unit cost include:

- That a measure of natural value is more relevant to a Wildlands Network Vision than are monetary or political costs.
- That the input data used are sufficient to reflect the relative region-wide pattern of natural values.
- That preserving natural areas before they are degraded is generally more desirable than restoring already degraded areas.

The land cover component of the planning unit cost represents the degree to which each hexagonal unit has been developed or otherwise modified by man. Land cover data is based on The Nature Conservancy's *Southern Rocky Mountain Terrestrial Ecological Systems* map, which was in turn derived from U.S. GAP Analysis land cover information (Neely et al. 2001). We modified the map to account for differences between The Nature Conservancy's Southern Rocky Mountains ecoregion boundary and our ecoregion boundary, which is approximately 500,000 ha larger. Additional data from the Gap Analysis Projects of Wyoming, Colorado, and New Mexico were added to fill in the missing areas. Each category within this layer was then re-classified to a general land cover category (Table 7.1). A cross tabulation was performed to derive a sum of the hectares of each land cover category within each hexagonal planning unit. Each hexagon was then assigned a corresponding cost score according to its perceived level of human disturbance (Table 7.2).

**Table 7.1 Land cover reclassification.**

| GAP Layer Classification | Land cover category |
|---|---|
| Agriculture categories | Agriculture |
| Mining and recent clearcuts | Disturbed |
| All native vegetation community categories | Natural Cover |
| Water (including reservoirs) | Open Water |
| Urban | Urban |

**Table 7.2 Land cover cost.**

| Land cover category | Criteria | Cost Score |
|---|---|---|
| Natural | Hexagon area>= 80% Natural Cover+ Open Water | 0 |
| Semi-natural | all other combinations | 10 |
| Agriculture | Hexagon area >= 50% Agriculture | 15 |
| Disturbed | Hexagon area >= 33% Disturbed | 15 |
| Developed | Hexagon area >= 50% Agriculture + Urban | 25 |
| Urban | Hexagon area >= 50% Urban | 100 |

Housing density was based on 1990 census data compiled and analyzed by Theobald (2001). Future revisions will use 2000 census data instead. The number of housing units was originally recorded at the block group level, and then converted to units per hectare. Housing density data were intersected with the hexagonal planning units to derive a composite estimate of housing density per hexagon. Levels of housing density were then separated into density classes and given a cost score (Table 7.3).

**Table 7.3 Housing density cost.**

| Category | Units/ha | Cost Score |
|---|---|---|
| Undeveloped | 0 | 0 |
| Rural | 0 – 0.062 | 5 |
| Exurban | 0.062 – 0.25 | 25 |
| Suburban | 0.25 – 1.25 | 50 |
| Urban | > 1.25 | 100 |

Roads were based on the 2000 census TIGER/Line data. Some obvious coding errors in the data were corrected prior to analysis. This dataset clearly under-represents unpaved roads on federal lands. However, it is more complete and up to date than USGS Digital Line Graph (DLG) road data, so we decided to use it. Roads were classified as Primary (interstates and major highways), Secondary (other paved roads), or Primitive (unpaved). Each classification was then assigned a weight and a level of road/edge influence. These parameters were then used in a kernel line density function (ESRI 2001) to determine the relative road/edge effect contained within each hexagon. This function treats the weights given to each road type as a smooth contour of diminishing value with distance from the road. The full value of the weighting factor is used at the center of the road itself, with the value decreasing to zero at the edge of the stated distance of road edge influence (Table 7.4).

**Table 7.4 Road density weights and edge influence.**

| Classification | Weight | Road edge influence (meters from road) |
|---|---|---|
| Primary | 100 | 2000 |
| Secondary | 33 | 800 |
| Primitive | 11 | 300 |

Each hexagon was then given a score that reflected its relative level of road density plus edge effect (Table 7.5).

The three scores for land cover, housing density, and road density/edge effect were then added within each hexagon, to give each planning unit a relative cost score ranging from 0 to 300.

**Table 7.5 Road density and edge effect cost.**

| Relative Road Density plus Edge Effect (weighted km/sq.km) | Cost Score |
|---|---|
| 0 | 0 |
| 0 – 12.2 | 5 |
| 12.2 – 24.4 | 10 |
| 24.4 – 48.8 | 20 |
| 48.8 – 97.6 | 35 |
| 97.6 – 195.2 | 50 |
| 195.2 – 390.4 | 65 |
| 390.4 – 780.8 | 80 |
| 780.8 – 1561.6 | 100 |

### *Weighted Boundary Length*

Unlike many previous algorithms, which often neglected the configuration of sites and resulted in fragmented conservation designs that are difficult to manage, the simulated annealing algorithm employed by SITES includes a parameter, the boundary length modifier, which allows planners to achieve a compact design by forcing the clustering of selected sites through weighting of the total boundary length (Andelman et al. 1999, Possingham et al. 2000). Total boundary length is defined as the sum of the perimeters of all planning unit clusters that were selected. A boundary length modifier of 0 results in no influence over clumping, whereas increasing the modifier value gives a relatively greater importance to boundary costs and results in greater clumping. A very high boundary length modifier value would create the extreme of a single clump of planning units in the shape of a circle. Minimizing the perimeter to area ratio helps retain the ecological integrity of protected areas by decreasing the amount of edge effect and decreasing fragmentation. However, an extreme design of a single large circular protected area would not adequately achieve most conservation goals, such as representation of all natural community types. Thus, the appropriate boundary length modifier must compromise between meeting most conservation target goals and minimizing the spatial scattering of selected planning units. We tested several boundary length modifiers and evaluated their effects, as discussed in the Results chapter (see Chapter 8).

### *Special Elements*

We used roadless areas, National Wilderness Areas, and Park Service lands as special elements (Figure 7.2). Roadless areas cover 3,840,000 ha, but only about 40% of these roadless lands are designated Wilderness Areas. Congressionally designated Wilderness Areas comprise 1,556,931 ha or 9.3% of the land area of the Southern Rockies ecoregion. The unprotected 60% of roadless areas (2,301,368 ha) represents 13.8% of all the land in the Southern Rockies ecoregion(Shinneman et al. 2000).

National Wilderness Areas and National Park Service lands were given a target goal of inclusion of 100%, while roadless areas had a target of 75%. We would have preferred a higher target goal for unprotected roadless areas. However, data currently available for roadless area boundaries are incomplete and currently undergoing revision by citizen groups for most of the National Forest lands within the region. We expect some of these changes to be substantive, and so chose not to place too much emphasis within the model on the aerial extent of the currently available data. The final Wildlands Network Design does not rely solely on the output of SITES, but also incorporates information from citizen proposals and local expert opinion, so that we are confident the design includes those areas most valuable and in need of conservation. Future iterations of the Design will make use of the latest inventory results.

Based on the known ecological values of roadless areas (Hitt and Frissell 1999, Wilcove et al. 2000, DeVelice and Martin 2001, Strittholt and DellaSala 2001), we used this category of lands as the focus for special elements. There are other options for special elements, such as old growth or locations of rare and imperiled species. The Nature Conservancy recently used best available data from state Natural Heritage Programs, regional experts, and other sources to include over 600 terrestrial and aquatic ecosystems, plant communities, and individual species, with an emphasis on rare and imperiled species and communities as target elements of their conservation vision for the Southern Rockies (Neely et al. 2001). We chose not to duplicate this tremendous effort, but rather to emphasize the large roadless wild areas that are needed for rewilding and for focal species habitat needs. However, we recognize that future iterations of the Southern Rockies Wildlands Network Design should include a comprehensive old-growth component, among other special elements. This is a difficult assignment because only a few of the National Forests in the region have even attempted an old-growth inventory.

### *Representation*

Special elements and focal species are specific areas of emphasis in our conservation planning. However, the main goal is to preserve the integrity of the Southern Rockies ecoregion as a functioning whole. Representation of all distinct natural communities within conservation landscapes and protected area networks is a long-standing goal of biodiversity conservation (Noss 1987). To that end, all reasonable effort must be made to retain all unique components of the ecoregion in sufficient amounts to promote their persistence over time. It is impossible to account for every species, assemblage, and community with SITES or any other reserve design algorithm. Instead, broad categories of community and ecosystem types are chosen with the assumption that these broad classifications will include most of the biodiversity within the ecoregion.

The Nature Conservancy spent close to two years working with other organizations, agencies, and area experts to derive their target goals for the Southern Rockies region, including representation goals for both terrestrial and aquatic ecological community types (Neely et al. 2001). In this initial version, we have included general target goals for terrestrial communities only. Future revisions will incorporate more specific terrestrial and aquatic community goals, possibly through a cooperative effort with TNC.

We based most of the vegetation community representation target goals for our design on TNC's terrestrial ecological systems representation goals as reported in Appendix 14 in Neely et al. (2001). The goals reported in that appendix represent simple aerial extent within the ecoregion and do not take into account TNC's efforts to disperse target elements evenly through subregions of the total ecoregion. Our vegetation community GIS coverage is based on TNC's *Southern Rocky Mountain Terrestrial Ecological Systems* map, which in turn is based on GAP land cover information (Neely et al. 2001). We modified the map to account for differences between TNC's Southern Rocky Mountains ecoregion boundary and our Southern Rockies ecoregion boundary, which is approximately 500,000 ha larger. Our design includes 30 terrestrial vegetation communities (Table 7.6).

The representation goals for 23 of these communities were based directly on the aerial extent goals used by TNC. Refer to the "Conservation Goals" section of Neely et al. (2002) for an explanation of how these goals were derived. The remaining seven were the three riparian communities

**Table 7.6 Vegetation community representation goals.**

| Community Type | Available Area (Ha) | Representation Goal (%) | Goal (Ha) |
|---|---|---|---|
| Active sand dune & swale complex | 10,494.9 | 38% | 4,000 |
| Alpine dry tundra & moist meadow | 680,381.3 | 28% | 191,103 |
| Alpine substrate - ice field | 206,565.0 | 30% | 61,969 |
| Alpine tundra - dwarf shrub & fell field | 125,341.6 | 38% | 47,556 |
| Aspen forest | 1,336,482.8 | 30% | 399,827 |
| Bristlecone - limber pine forest & woodland | 77,709.7 | 30% | 23,312 |
| Douglas fir - ponderosa pine forest | 383,707.9 | 17% | 66,585 |
| Foothills riparian woodland & shrubland | 5,283.0 | 66% | 3,487 |
| Gambel's oak shrubland | 641,881.9 | 33% | 210,190 |
| Greasewood flat & ephemeral meadow complex | 180,650.0 | 32% | 58,457 |
| Intermontane - foothill grassland | 837,424.4 | 35% | 290,272 |
| Juniper savanna | 312,702.2 | 30% | 93,811 |
| Lodgepole pine forest | 1,108,411.7 | 30% | 332,450 |
| Lower montane - foothills shrubland | 759,921.5 | 32% | 246,402 |
| Marsh & wet meadow | 19,000.6 | 66% | 12,500 |
| Montane - foothill cliff & canyon | 21,055.1 | 29% | 6,142 |
| Montane grassland | 293,271.6 | 33% | 96,775 |
| Montane mixed conifer forest | 616,665.3 | 28% | 172,856 |
| Montane riparian shrubland | 13,253.7 | 66% | 8,747 |
| Mountain sagebrush shrubland | 1,339,986.6 | 30% | 398,787 |
| North Park sand dunes | 342.4 | 30% | 103 |
| Piñon - juniper woodland | 1,726,695.7 | 30% | 518,009 |
| Ponderosa pine woodland | 1,985,826.5 | 33% | 663,227 |
| Sagebrush steppe | 272,676.8 | 30% | 81,803 |
| San Luis valley winterfat shrub steppe | 141,259.4 | 35% | 50,047 |
| South Park montane grasslands | 221,107.2 | 33% | 72,317 |
| Spruce-fir forest | 2,251,858.6 | 30% | 674,148 |
| Stabilized sand dune | 38,335.6 | 29% | 11,162 |
| Upper montane riparian forest & woodland | 19,380.9 | 66% | 12,791 |
| Winterfat shrub steppe | 131,048.1 | 30% | 39,314 |

(foothills riparian woodland & shrubland, montane riparian shrubland, and upper montane riparian forest & woodland), winterfat shrub steppe, sagebrush steppe, piñon-juniper, and juniper savannah.

Riparian communities were under represented in TNC's terrestrial ecological community goals, probably because they treat riparian areas as special elements that they specifically mapped as element occurrences. The GAP-based mapping efforts are at too coarse a scale to include most riparian communities in the ecoregion. Finer scale mapping of riparian areas is currently underway in Colorado by the Colorado Division of Wildlife, and we hope to be able to use this and similar data in future revisions of the Network Design. However, in the interest of time, this initial version of the design includes target goals of 66% (to match the percentage used for the marsh and wet meadow community type) of the aerial extent of each of the riparian vegetation communities as currently mapped in the ecoregion.

Our ecoregional boundary includes the Gunnison River valley, as well as larger portions of the upper Canadian and upper Pecos basins. This results in our boundary containing substantially more piñon-juniper woodland, ponderosa pine woodland, winterfat shrub steppe, mountain sagebrush shrubland, sagebrush steppe, intermontane-foothill grassland, and juniper savannah than TNC's boundary (see Table 7.6). Our vegetation community goals were therefore adjusted to include at least 30% of the available aerial extent of these particular communities.

One of the primary goals for the next iteration of this document will be to more thoroughly study representation needs and to revise the target goals to better reflect viability of the native diversity within the ecoregion over time.

### *Focal Species*

Our Wildlands Network Design concentrates more heavily on focal species, which provides an interesting comparison and important complement to the map produced by The Nature Conservancy, which did minimal analysis of focal species and did not emphasize connectivity. The suite of focal species was selected so as to achieve a balance of both habitat quality indicators and keystone species representing all the principal community types within the Southern Rockies. This suite was selected by the science team, with justification for each described in Appendix 1. However, not all of these species are appropriate for inclusion as inputs in the SITES selection model, for various reasons explained in Appendix 1. Therefore, the final list of focal species chosen for analysis was reduced to the following species:

We compiled known location and suitable habitat data for gray wolf (*Canis lupus*), black bear (*Ursus americanus*), pronghorn (*Antilocapra americana*), and cutthroat trout (*Oncorhynchus clarki*). For the cutthroat trout we included the greenback (*O. c. stomias*), Rio Grande (*O. c. virginalis*), and Colorado River (*O. c. pleuriticus*) subspecies. Black bear was used as a surrogate for grizzly bear (*Ursus arctos*). The needs of the two bear species, while substantially different in the Northern Rockies, are considered to be essentially the same within the Southern Rockies (T. Beck pers. comm., L. Craighead pers. comm., and S. Cain pers. comm.). Spatial data representing this information were primarily derived from existing data sources, with the exception of black bear, which had to be created (see below). Data gathered and created were then modified based on expert opinion from several workshops and meetings.

GIS data for the cutthroat trout subspecies came primarily from the Biodiversity Conservation Alliance, which had collected locations of genetically pure populations of each subspecies. Added to these were areas identified by the Colorado Division of Wildlife for the overall known range of the Rio Grande and Colorado River subspecies. Equivalent data were not available for the greenback, and so this subspecies is probably under represented in the data. We took the original stream segments that represented individual trout populations and created subwatersheds for each using 82 meter resolution digital elevation data to derive slope and flow accumulation. The subwatershed polygons, not the linear stream segments, were used as input for SITES (Figure 7.3).

GIS data for gray wolf were provided by Carroll et al. (2003) and used with permission by Carlos Carroll. Wolf data received represented probability of occurrence at a resolution of 500 $km^2$ over the extended Southern Rockies ecoregion and supporting areas as modeled by PATCH, a dynamic wildlife population modeling software. Wolf core areas are areas suitable for potential reintroduction of wolf populations in the ecoregion and are those areas that Carroll et al. (2003) had identified as likely reintroduction areas, based on a combination of habitat suitability and current land management. Suitable wolf habitat outside of cores were those areas with a 60% or greater probability of occurrence as assessed by Carroll et al. (2003), plus other areas from a static habitat model based on prey densities by Martin et al. (1999, Figure 7.4 [a] and [b]).

Pronghorn data came from the state wildlife/Gap program data from Wyoming, Colorado, and New Mexico. This data was then slightly modified based on expert opinion (Figure 7.5).

We created GIS data representing core areas for bears because of a lack of suitable existing data. The data were created in consultation with Tom Beck, wildlife biologist and regional bear expert. Bear habitat suitability was modeled using the vegetation community layer and unweighted road density. The addition of prey data, such as ungulate concentration areas, was considered but not included because currently existing ungulate GIS data are not considered to be accurate enough (T. Beck, pers. comm.). The following vegetation communities were considered to be primary habitat for bears, based on expert input and literature review (T. Beck, pers. comm., Carroll et al. 1998):

- Gambel's oak shrubland
- Aspen forest
- Riparian communities
- Piñon-juniper woodland
- Ponderosa pine woodland

In addition, all other forest types and all grassland types that were adjacent to each other were considered to be secondary habitat. Primary habitat patches were given a score of 5 and secondary habitat a score of 3. Road density was based on 2000 census TIGER/Line roads. Simple road density was calculated per hectare as averaged over a 3 $km^2$ area. The 3 $km^2$ average was used to represent the mean daily movement of adult female black bears (Carroll et al. 1998), which is assumed to be the biologically relevant scale for this analysis. Road density was then classified and scored (Table 7.7).

Road density scores were then subtracted from the vegetation scores to derive a habitat suitability layer, with suitability scores ranging from –5 to +5. These data were smoothed using 10 iterations of a majority filter function (ESRI 2001), to help smooth edges and remove isolated patches. Those patches with a suitability score greater than 1 were then selected out as suitable black bear habitat. T. Beck (pers. comm.) considers black bear harvest statistics from the Colorado Division of Wildlife (CDOW) to be strongly correlated with bear population densities. Therefore, a ten-year average of these harvest statistics was compared against the modeled suitable habitat layer. Selecting out suitable habitat patches that are 600,000 ha or larger gave the best match to the CDOW harvest data. These large patches then became the core areas for black bear. These core areas were later only slightly modified by expert opinion at workshops and meetings (Figure 7.6).

**Table 7.7 Road density scores for bear habitat modeling.**

| Road density (m/ha) | Score |
|---|---|
| 0 | 0 |
| >0 – 5 | 1 |
| >5 – 10 | 2 |
| >10 – 25 | 3 |
| >25 – 50 | 4 |
| >50 | 5 |

Target goals for focal species are meant to represent minimum areas necessary for viable populations of each species to persist over time. Ideally one would use data that identify the size and location of interconnected subpopulations necessary to maintain a metapopulation within the ecoregion and surrounding lands indefinitely. However, at this time the data needed to represent each focal species to this level of detail are not available. Again, this vision is a work in progress and should be considered a hypothesis to test. Future iterations will include more detailed data and research into viability over time. For this initial study, however, the target goals chosen were at least 50% of the available suitable habitat, as represented by the spatial data gathered for each species (Table 7.8).

**Table 7.8 Focal species target goals (% available suitable habitat).**

| Focal Species | Goal (%) |
|---|---|
| Black/grizzly bear core areas | 75 |
| Wolf core areas | 100 |
| Other suitable wolf habitat | 50 |
| Pronghorn suitable habitat | 50 |
| Colorado River cutthroat subwatersheds | 100 |
| Greenback cutthroat subwatersheds | 100 |
| Rio Grande cutthroat subwatersheds | 50 |

## 2. Least Cost Path Models for Wolf and Bear Dispersal

Least cost path models were constructed for the wolf and grizzly bear, in part to help identify best linkages between predicted core areas. Methods used were adapted from Singleton et al. (2001). First, we created a 'permeability' layer for each species, representing the relative ease with which an individual can travel through the landscape. Then

**Table 7.9 Land cover permeability scores for wolves and bears.**

| Land cover | Wolf | Bear |
|---|---|---|
| Active sand dune & swale complex | 8 | 3 |
| Agriculture - dry | 8 | 3 |
| Agriculture - irrigated | 5 | 3 |
| Alpine dry tundra & moist meadow | 10 | 10 |
| Alpine substrate - ice field | 7 | 1 |
| Alpine tundra - dwarf shrub & fell field | 7 | 1 |
| Aspen forest | 10 | 10 |
| Bristlecone - limber pine forest & woodland | 10 | 10 |
| Douglas fir - ponderosa pine forest | 10 | 10 |
| Foothills riparian woodland & shrubland | 10 | 10 |
| Gambel's oak shrubland | 10 | 8 |
| Greasewood flat & ephemeral meadow complex | 10 | 5 |
| Intermontane - foothill grassland | 10 | 5 |
| Juniper savanna | 10 | 5 |
| Lodgepole pine forest | 10 | 10 |
| Lower montane - foothills shrubland | 10 | 8 |
| Marsh & wet meadow | 8 | 10 |
| Mining operation | 3 | 1 |
| Montane - foothill cliff & canyon | 5 | 2 |
| Montane grassland | 8 | 5 |
| Montane mixed conifer forest | 8 | 10 |
| Montane riparian shrubland | 10 | 10 |
| Mountain sagebrush shrubland | 10 | 8 |
| North Park sand dunes | 8 | 3 |
| Piñon - juniper woodland | 8 | 10 |
| Ponderosa pine woodland | 10 | 10 |
| Recent clearcut conifer forest | 7 | 3 |
| Sagebrush steppe | 10 | 8 |
| San Luis Valley winterfat shrub steppe | 10 | 8 |
| South Park montane grasslands | 8 | 5 |
| Spruce-fir forest | 10 | 10 |
| Stabilized sand dune | 8 | 3 |
| Upper montane riparian forest & woodland | 8 | 10 |
| Urban | 1 | 1 |
| Water | 1 | 1 |
| Winterfat shrub steppe | 10 | 8 |

a weighted cost distance layer was calculated from the permeability layer to represent the cost of traveling through an area such that, the poorer the dispersal habitat, the greater the effective distance an individual would have to travel to get through that area. Finally, we executed a least cost path analysis for both wolves and bears using the weighted cost distance layer to determine the most probable dispersal linkages each species would take to travel between core population areas. It is important to recognize that the results of this analysis represent the probability of successful dispersal between two areas and not necessarily what an individual animal will choose to do. For example, a wolf may choose to disperse through agricultural areas, but its probability for success is low.

The permeability layers were derived from four separate GIS data layers: land cover (categorical), population density (people/km$^2$), weighted road density (km/km$^2$), and slope (%). Scores were assigned to each component layer and then combined into the overall permeability layers for each species. Permeability values were adapted from values given for gray wolves and grizzly bears in Singleton et al. (2001), with input from the science team. We used ArcGIS 8.2 for all spatial computations. The data resolution was 100 meters.

Land cover was based on the vegetation community data. We gave each land cover type a relative score for wolves and bears that represents the ease with which an individual can move through an area. Scores ranged from 1-10, with 1 = *extremely difficult* and 10 = *no difficulty*. Land cover categories were not identical with those used by Singleton et al. (2001), therefore some interpretation was required (Table 7.9).

Human population density was based on 1990 census data compiled and analyzed by Theobald (2001). Future revisions will use 2000 census data instead. Population data were recorded at the block group level; this was then converted to population density per square kilometer.

**Table 7.10 Human population density permeability scores for wolves and bears.**

| Population density (people/km$^2$) | Wolf | Bear |
|---|---|---|
| 0– 26 | 10 | 10 |
| 26 – 65 | 5 | 5 |
| 65 – 130 | 3 | 3 |
| 130 – 260 | 2 | 2 |
| > 260 | 1 | 1 |

Permeability parameters used were adapted from Singleton et al. (2001), who determined from their research that wolves and grizzly bears responded in the same way to human population densities (Table 7.10).

Road density was based on 2000 census TIGER/Line data and was calculated using a kernel line density function in ArcGrid (ESRI 2001) to derive weighted kilometers of road per square kilometer. Roads were weighted based on the classifications of *primary* (Interstates and major highways), *secondary* (other paved roads), and *primitive* (unpaved), with the assumption that the higher the road classification, the greater and faster the traffic, and therefore the more difficult it is for an individual animal to successfully cross. Primary roads were given a weight of 5 and secondary roads were weighted 2. This has the practical effect of increasing the road density by a factor of 5 and 2, respectively. Primitive roads were not weighted. Permeability parameters used were adapted from Singleton et al. (2001), although it should be noted that the authors of this paper did not weight roads in their study. Our approach represents a higher relative level of sensitivity to the presence of roads (Table 7.11).

**Table 7.11 Road density permeability scores for wolves and bears.**

| Road density ($km/km^2$) | Wolf | Bear |
|---|---|---|
| 0 – 0.6 | 10 | 10 |
| 0.6 – 1.2 | 8 | 8 |
| 1.2 – 2.5 | 5 | 5 |
| 2.5 – 3.7 | 5 | 3 |
| 3.7 – 6.2 | 2 | 2 |
| 6.2 – 30 | 1 | 1 |
| > 30 | impassible | impassible |

We calculated percent slope from a composite Digital Elevation Model (DEM) for the ecoregion with a cell resolution of 82 meters. The DEM data originated from the U.S. Geologic Survey. Cell size of the resulting slope grid was recalculated to 100 meters to match all other permeability input grids. According to Singleton et al. (2001), slope does not play a factor in bear dispersal. However, there are certain canyons within the ecoregion that do pose a dispersal barrier because of their sheer drop. Some of these are listed under the Gap layer category of "Cliff and Canyon" but others, most notably the Black Canyon of the Gunnison, are not. Therefore, we decided to incorporate slope for bears, but coded it so that only the sheerest drops posed a barrier to movement. The slope permeability scores for wolves were taken directly from Singleton et al. (2001, Table 7.12).

The permeability layers for each species were then created by multiplying the individual component layers and then dividing by 10,000 in order to scale permeability between 0 – 1.0. Permeability as a probability function was then transformed into a weighted cost distance function that represented distance traveled in terms of weighted meters. Singleton et al. (2001) used a linearly weighted cost distance, such that, when combined with the cell width in a cost distance function, values ranged from actual cell width to 100x cell width. The equation is:

**Table 7.12 Slope permeability scores for wolves and bears.**

| Slope (%) | Wolf | Slope (%) | Bear |
|---|---|---|---|
| 0 – 20 | 10 | 0 – 75 | 10 |
| 20 – 40 | 8 | 75 – 100 | 6 |
| > 40 | 6 | > 100 | 1 |

$$WCD = C * (100 - (100 * V)$$

where WCD = the weighted cost distance, C = the cell width (100 m in this case) and V = the permeability value (range 0-1.0). Permeability values of 1.0 would actually result in a cost distance of 0, but are manually adjusted to 100 to reflect actual cell width. This linear function was tried, but the results appeared unnecessarily restrictive and not reflective of a realistic behavioral response. For example, using the linear model, the hypothetical dispersing individual would travel miles out of its way to avoid crossing even isolated dirt roads. This happens because the linear model allows for no tolerance of minor disturbances or barriers. In the absence of concrete behavioral response data, we decided to use a non-linear cost distance function instead, on the assumption that animal behavior in general follows a response curve that allows minor disturbances to be buffered by adaptive responses. The function used is a power function:

$$WCD = C * (V)^{-2}$$

The resulting weighted cost distance retains the approximate range of the linear function (Figure 7.7).

## 3. Limitations

### *Limitations of SITES and Least Cost Path Analysis*

Regardless of the model used, output is only as good as the input data. Because nature is complex and available resources are scarce, data can never be wholly complete or without error. Ideally, input data should first be validated with field research, but this is rarely the case due to limitations of time and money. Static models such as SITES, particularly when based on static input data, do not take into account local population and meta-population dynamics and changing environmental conditions over time. The initial design solutions provided by SITES must therefore be reviewed and modified using local knowledge and professional judgment in order for the results to be relevant. The simulated annealing algorithm used by SITES trades a guaranteed optimal solution for computational speed and software accessibility. There are other algorithms that can guarantee optimal solutions, but they require much more computing power, time and cost, particularly with large study areas (Andelman et al. 1999, Pressey et al. 1996). Another limitation to simulated annealing algorithms is a sensitivity to input parameters. Simpler algorithms are more robust, but do not always provide as optimal a solution (Possingham et al. 2000).

There are several limitations to the base cost layer used in our SITES analysis. Not all relevant factors, such as grazing pressure, were incorporated. The three inputs used are not exclusive of each other and are in fact strongly correlated. However, because each input layer is coarse-scale and comes from a different data source, none are wholly complete and accurate. By using all three inputs in a relative scoring scheme, we hope to reach a more representative picture of the costs at the scale of the entire ecoregion. The cost layer as a whole is correlated to some degree to the focal species suitable habitat inputs, *i.e.*, suitable wildlife habitat usually occurs in the more natural and undeveloped lands. We did a simple correlation test of the cost layer with both wolf and black bear suitable habitat inputs. The correlation coefficient was $r = -0.248$ for the wolf suitable habitat layer, and $r = -0.372$ for the bear suitable habitat layer. Therefore, these suitable habitat inputs are only partially redundant to the cost layer. Finally, SITES is sensitive to initial input values (Possingham et al. 2000), so that modifying the planning unit cost in the future will produce different results, even if the relative degree of cost is maintained. Further research on model sensitivity, method assumptions, and confounding effects should be undertaken in future versions of this design.

Least cost path analysis has no direct bearing on what an individual animal (of any species) is actually going to do when trying to get from point A to point B. The analysis makes many untested assumptions about animal behavior and appropriate scale, such as using the non-linear versus linear cost distance functions. In addition, dispersing individuals frequently do not know ahead of time where they are heading or what obstacles lie in their path, whereas the computer model has the advantage of viewing the entire area at once. Results of such an analysis should not be used to make important management decisions, but should rather be used as a way to identify and prioritize further research needs. For a good discussion of least cost path limitations and assumptions as pertains to large mammal dispersal, see Walker and Craighead (1997).

### *Limitations of expert opinion*

This Network Design relies heavily on the expert opinion of area biologists, ecologists, and local activists who are well acquainted with their area of interest. However, as the term implies, expert opinion is based on opinion, and may reflect a personal bias as to what is important. The tendency of most people, including established scientists, is to draw conclusions based on what they have personally observed, rather than what can be concluded through substantiated, empirical studies. In the absence of such studies, however, expert opinion can be a valuable source of information.

## 4. Synthesis of the Wildlands Network Design

The Wildlands Network Design was created from a combination of the SITES output, least cost path analysis for wolves and bears, expert opinion about additional high value areas from local scientists and activists, and citizen proposed forest management plans from local conservation groups. We also referred to the network design used by the New Mexico Highlands project and the draft SITES results for the Heart of the West project for those areas that overlapped with the Southern Rockies ecoregion. The end result is a collection of core wild areas connected by other areas of various levels of compatible management. Each area was assigned a 'network unit classification' based on its level of recommended protection and management. The network unit classifications are defined in Chapter 9.

Three workshops were held within the region to gather local expert opinion: December 16, 2002 in Denver with 25 participants; January 14, 2003 in Carbondale; and

January 16, 2003 again in Denver. The January meetings were identical and had 31 participants total. There were 44 different people in the three meetings, representing scientists and conservation advocates, with a few agency participants. These workshops were used to gather information about areas of importance not included in the models, and the type of threats present in each area, as well as to gather initial feedback about the methods employed.

Citizen proposed management plans, which use the U.S. Forest Service management prescription codes, were translated into the various unit classifications based on the perceived level of protection and type of use (Appendix 2). Areas considered to be high use areas were not included in the network design. In areas where citizen plans were not available, unit classifications were decided initially by using available road and land use data, and then local conservation groups and regional experts reviewed them.

## 5. Conclusion

The results of these methods are discussed in Chapter 8. They become the basis for the Southern Rockies Wildlands Network Design discussed in Chapter 9.

*This is the most beautiful place on earth. There are many such places.*

-Edward Abbey

# 8 RESULTS

Michelle Fink, Kurt Menke, Doug Shinneman

## 1. Results of SITES Analysis

### *Multiple scenario testing*

Various iterations of the SITES model were run in order to test the effects of various parameter settings. We tested the outcome of adjusting planning unit boundaries to take into account currently protected area boundaries, adjusting the boundary length modifier, and three different target goals for bear suitable habitat.

SITES can be programmed to force inclusion or exclusion of certain planning units, regardless of how this affects the objective cost function. Strictly protected lands are the foundation of a Wildlands Network Design, so we tested the model outcome using both the base 1,000 ha hexagon planning units and a target goal of 100% inclusion of all National Wilderness Areas and National Park Service lands, and by locking in these areas to their exact boundaries. The science team decided it was preferable to ensure inclusion of these protected areas to their actual boundaries, so the final SITES run used modified planning units and forced the inclusion of these areas into the design. This changed the number of planning units from 17,361 whole hexagons to 19,192 units, 3,660 (19%) of which were less than 1,000 ha in size. The protected areas that were locked into the design consisted of 2,576 planning units, totaling 1,645,000 ha, or approximately 10% of the ecoregion.

Another variation tested was the effect of various boundary length modifiers. After evaluating different boundary length modifiers, it quickly became apparent that a large modifier value was unnecessary, because the base cost layer created much the same clustering effect. The science team decided upon a fairly small boundary modifier of 0.005, which resulted in smoother clusters and fewer single, outlying planning units, while still retaining a reasonable amount of flexibility in meeting target goals.

The final parameter tested was the target goal for suitable bear (*Ursus* spp.) habitat. As the top predator being proposed for eventual reintroduction, the grizzly bear (*Ursus arctos*) is of enormous influence to the final configuration of the network design. Therefore, the science team wanted to see how various goals for this species (using black bear [*Ursus americanus*] data as a surrogate) affected the outcome of the model, and so the model was run using 50%, 75%, and 100% of the aerial extent of suitable bear habitat. After reviewing the results, the science team decided to use the 75% target goal as the best balance with the other model inputs while still emphasizing the importance of this focal species.

### *Target goals*

Our final SITES output configuration included 9,929 planning units that cover 8,244,100 ha (49%) of the ecoregion (Figure 8.1). The model included six focal species, two special elements, and thirty vegetation communities. Of these input elements, the majority (87%) had their target goals of inclusion met or exceeded (Table 8.1). Two of the five elements whose goals were not met – core habitat for wolf (*Canis lupus*) and montane mixed conifer forest – were short of the stated goal by no more than two tenths of a percent, and can for all practical purposes be seen as meeting the goal. The remaining three elements that are not adequately represented in the SITES output were the Colorado River

Table 8.1 Target goals met for each element input with the final SITES analysis.

| Conservation Element<br>Focal Species | Available (ha) | Target (ha) | Target met | Proportion of target met | Value (ha) | Proportion of available |
|---|---|---|---|---|---|---|
| Colorado River Cutthroat Trout | 56,190 | 56,190 | no | 95% | 53,183 | 95% |
| Greenback Cutthroat Trout | 55,723 | 55,723 | no | 96% | 53,438 | 96% |
| Rio Grande Cutthroat Trout | 456,896 | 228,493 | yes | 102% | 233,547 | 51% |
| Wolf Core Habitat | 1,007,314 | 1,007,314 | no | 99.8% | 1,004,956 | 99.8% |
| Wolf Secondary Habitat | 5,651,101 | 2,825,550 | yes | 151% | 4,260,411 | 75% |
| Bear Core Habitat | 3,519,784 | 2,639,840 | yes | 103% | 2,714,021 | 77% |
| Pronghorn Habitat | 3,844,834 | 1,922,420 | yes | 104% | 1,994,390 | 52% |
| **Special Elements** | | | | | | |
| Roadless Areas | 2,978,613 | 2,233,960 | yes | 102% | 2,272,915 | 76% |
| Wilderness & Park lands* | 1,645,040 | 1,645,040 | yes | 100% | 1,645,040 | 100% |
| **Vegetation Communities** | | | | | | |
| Active sand dune & swale complex | 10,495 | 4,000 | yes | 262% | 10,495 | 100% |
| Alpine dry tundra & moist meadow | 680,381 | 191,103 | yes | 235% | 448,769 | 66% |
| Alpine substrate - ice field | 206,565 | 61,969 | yes | 304% | 188,569 | 91% |
| Alpine tundra - dwarf shrub & fell field | 125,342 | 47,556 | yes | 234% | 111,328 | 89% |
| Aspen forest | 1,336,483 | 399,827 | yes | 235% | 941,306 | 70% |
| Bristlecone - limber pine forest & woodland | 77,710 | 23,312 | yes | 168% | 39,171 | 50% |
| Douglas fir - ponderosa pine forest | 383,708 | 66,585 | yes | 200% | 133,095 | 35% |
| Foothills riparian woodland & shrubland | 5,283 | 3,487 | no | 96% | 3,355 | 64% |
| Gambel's oak shrubland | 641,882 | 210,190 | yes | 213% | 447,030 | 70% |
| Greasewood flat & ephemeral meadow complex | 180,650 | 58,457 | yes | 106% | 61,894 | 34% |
| Intermontane - foothill grassland | 837,424 | 290,272 | yes | 115% | 335,165 | 40% |
| Juniper savanna | 312,702 | 93,811 | yes | 115% | 108,107 | 35% |
| Lodgepole pine forest | 1,108,412 | 332,450 | yes | 128% | 426,547 | 38% |
| Lower montane - foothills shrubland | 759,922 | 246,402 | yes | 132% | 326,309 | 43% |
| Marsh & wet meadow | 19,001 | 12,500 | yes | 105% | 13,092 | 69% |
| Montane - foothill cliff & canyon | 21,055 | 6,142 | yes | 271% | 16,652 | 79% |
| Montane grassland | 293,272 | 96,775 | yes | 110% | 106,213 | 36% |
| Montane mixed conifer forest | 616,665 | 172,856 | no | 99.9% | 172,745 | 28% |
| Montane riparian shrubland | 13,254 | 8,747 | yes | 106% | 9,305 | 70% |
| Mountain sagebrush shrubland | 1,339,987 | 398,787 | yes | 106% | 422,287 | 32% |
| North Park sand dunes | 342 | 103 | yes | 191% | 196 | 57% |
| Piñon - juniper woodland | 1,726,696 | 518,009 | yes | 157% | 812,430 | 47% |
| Ponderosa pine woodland | 1,985,827 | 663,227 | yes | 128% | 848,289 | 43% |
| Sagebrush steppe | 272,677 | 81,803 | yes | 152% | 124,305 | 46% |
| San Luis Valley winterfat shrub steppe | 141,259 | 50,047 | yes | 120% | 60,088 | 43% |
| South Park montane grasslands | 221,107 | 72,317 | yes | 113% | 81,373 | 37% |
| Spruce-fir forest | 2,251,859 | 674,148 | yes | 237% | 1,595,035 | 71% |
| Stabilized sand dune | 38,336 | 11,162 | yes | 163% | 18,173 | 47% |
| Upper montane riparian forest & woodland | 19,381 | 12,791 | yes | 106% | 13,519 | 70% |
| Winterfat shrub steppe | 131,048 | 39,314 | yes | 194% | 76,351 | 58% |

* locked in to guarantee 100% representation

cutthroat trout (*Oncorhynchus clarki pleuriticus*), the greenback cutthroat trout (*O. c. stomias*), and the foothills riparian woodland & shrubland vegetation community. Even so, the SITES output included at least 95% of the target goal aerial extents of each of these elements. Subwatersheds for the Colorado River cutthroat trout are recognized as already being under represented in this model, and so this deficiency is the greatest concern. The riparian vegetation communities are also poorly represented in general because of the scale of the input data, and so it is difficult to know how much of these plant community types is actually included in the model output.

## 2. Results of Least Cost Path Analysis

The results of the least cost path analysis for dispersal linkages between core habitat areas for wolves and bears are shown in Figures 8.2 and 8.3. The contours represent decreasing probability of successful dispersal with increasing distance from the central least cost path, expressed in terms of increasing weighted cost distance. Constrictions of the dispersal linkages around major roads are evident for both species. That wolf dispersal behavior was coded in the model as being much less sensitive to agricultural land use than was bear dispersal behavior is evident in comparing the respective response of the model to the San Luis Valley. The San Luis Valley is an area heavily devoted to agriculture, mostly potatoes and grains such as barley and wheat. The least cost model shows that wolves going between the Vermejo Ranch area and the San Juan Mountains would be likely to cross directly through the valley, close to the New Mexico-Colorado state line. Bears, on the other hand, are more likely to avoid the valley entirely, traveling instead along the Sangre de Cristo Mountains and Cochetopa Hills, taking a much longer, but presumably safer, route.

These likely dispersal linkages were overlaid onto the SITES output to highlight areas of likely carnivore movement not otherwise covered by the initial network design. Together, these two components formed the basis of the network design.

## 3. The Wildlands Network Design

The next step was to refine this design based on current land management and citizen proposed management plans for federal lands. The results of this refinement are displayed on the Southern Rockies Wildlands Network Design map (Figure 9.1).

The Wildlands Network Design was created from a combination of the SITES output, least cost path analysis for wolves and bears, expert opinion about additional high value areas from local scientists and activists, and citizen proposed forest management plans from local conservation groups. We also referred to the network unit classifications used by the New Mexico Highlands project and the draft SITES results for the Heart of the West project for those areas that overlapped with the Southern Rockies ecoregion. Areas identified through the above means were assigned a network unit classification as described in Chapter 9, section 2. Citizen proposed management plans, which use the U.S. Forest Service management prescription codes, were translated into the various unit classifications based on the level of protection and type of use. Areas considered to be high use areas were not included in the Network Design. In areas where citizen plans were not available, unit classifications were decided initially by using available road and land use data and then reviewed by local conservation groups.

The completed design covers 10,429,615 ha (25,772,037 ac), or 62% of the ecoregion, plus a few areas that go slightly beyond the ecoregion boundary because of ownership boundaries. Core areas— Core Agency, Core Private, and Core Wilderness (both designated and proposed)— comprise 4,330,241 ha, or 42% of the design (26% of the ecoregion). Table 8.2 shows the total area of each network unit classification in the design. In comparison, currently protected areas— National Wilderness Areas, National Park Service lands, and other congressionally protected areas— only cover about 1,716,000 ha, or 10% of the ecoregion.

**Table 8.2 Summary of Network Design by unit classification.**

| Unit Classification | Hectares | Acres |
|---|---|---|
| Core Agency | 613,796 | 1,516,717 |
| Core Private | 372,410 | 920,242 |
| Core Wilderness | 3,344,035 | 8,263,257 |
| Low Use Compatible | 1,662,800 | 4,108,851 |
| Medium Use Compatible | 1,986,599 | 4,908,973 |
| Private/Tribal High Value | 559,228 | 1,381,876 |
| Study Area | 915,257 | 2,261,639 |
| Wildlife Linkage | 975,492 | 2,410,482 |
| **Total** | **10,429,615** | **25,772,037** |

For general comparison purposes, the Wildlands Network Design was compared against The Nature Conservancy's conservation portfolio for the Southern Rockies (Neely et al. 2002). The conservation portfolio covers 48% of the ecoregion (using SREP's ecoregion boundary) and overlaps with our Wildlands Network Design by 63%. Of the target elements used by The Nature Conservancy, the Network Design includes approximately 80% of these, compared to 97% included in the conservation portfolio (based on generalized counts of target elements only, regardless of whether these were presented as points, lines, or polygon features). A further breakdown of target elements covered by general taxonomic or ecological group is provided in Appendix 3. Further detail about specific rare species and endemic populations covered is difficult because of complications of data sensitivity and ownership. An analysis using Colorado Natural Heritage Program low resolution element occurrence data is also included in the Appendix. These data are for the Colorado portion of the ecoregion only and have deliberately obscured location data to protect the sensitive species mentioned. Therefore, this analysis of inclusion should only be regarded as a general indication of which sensitive species and assemblages are included in the network design. SREP is currently working in conjunction with The Nature Conservancy for further, higher accuracy, analysis and comparison.

Because the completed network design tends to follow lines of land ownership, unlike the results of the SITES analysis, the specific targets used in our SITES analysis are met to different degrees in the Network Design (Appendix 3). A few goals, most notably for the various riparian vegetation communities, are now far from being met, pointing to the need to work with private landowners to include more lowland and riparian corridors in future iterations and implementation of the design.

# SECTION IV:

# THE CONSERVATION VISION AND IMPLEMENTATION ACTION

*Climb the mountains and get their good tidings. Nature's peace will flow into you as sunshine flows into trees. The winds will blow their own freshness into you, and the storms their energy, while cares will drop off like autumn leaves.*

-John Muir

# 9. A CONSERVATION VISION FOR THE SOUTHERN ROCKIES

Mark Pearson, Dave Foreman, Brian Miller, Jean Smith, Tim Hogan, Michael Soulé

## 1. Southern Rockies Wildlands Network Design Overview

The Southern Rockies Wildlands Network Design emphasizes large wild core areas, landscape connectivity, and protecting habitat and linkages for large carnivores (Figure 9.1). It is strategically located in the central part of the Spine of the Continent MegaLinkage. Today, federal public lands include a variety of designated and proposed Wilderness Areas and other protected areas in the ecoregion. However, existing land ownership, development, and management patterns leave critical gaps in protected areas and bottlenecks that currently impede linkages for wildlife movement.

### *Regional Overview*

The Southern Rockies ecoregion includes parts (or all) of 10 major watersheds: the North Platte, Yampa-White, South Platte, Upper Colorado, Gunnison, Dolores, Upper Arkansas, San Juan, Canadian, and Rio Grande (Upper and Middle, Figure 9.2). The federal government owns 55% of the land, private ownership covers 37.8% of the land, the states own about 3.8%, and Tribal lands cover 3% of the region (Shinneman et al. 2000, Figure 9.3 and Table 3.1). The 41.4% of the land owned by the Forest Service is found on the Medicine Bow, Routt, Arapaho-Roosevelt, White River, Pike-San Isabel, Rio Grande, Grand Mesa/Gunnison/Uncompahgre, San Juan, Carson and Santa Fe National Forests.

There are six National Parks and National Monuments (Table 9.1) representing 194,950 ha, and 49 federally designated Wilderness Areas (Table 9.2, Figure 7.2) representing 1,556,931 ha or 9.3% of the land area. The largest is the Weminuche/Peidra Wilderness Area with 222,885 ha. Roadless areas cover 3,840,000 ha, but only about 40% of these roadless lands are represented in the federally designated Wilderness Areas that are listed in Table 9.2. The unprotected 60% of roadless areas (2,283,069 ha) represents 13.7% of all the land in the Southern Rockies ecoregion and is therefore very important for improving the conservation of nature. The 10 largest roadless areas are listed in Table 9.3, but there are only a few areas located more than 3.2 km from a road (Shinneman et al. 2000).

**Table 9.1 National Parks and Monuments.**

| Name | Size (ha) |
|---|---|
| Rocky Mountain NP | 107,000 |
| Great Sand Dunes NP | 61,000 |
| Bandelier NM | 13,000 |
| Black Canyon of the Gunnison NP | 11,400 |
| Pecos NHP | 150 |
| Florissant Fossil Beds NM | 2,400 |
| **Total** | **194,950** |

U.S. Forest Service and Bureau of Land Management lands are fairly continuous throughout the region and offer excellent opportunities for core wild areas, compatible use lands, and linkages (Figure 9.1). Gaining protection for the unprotected roadless areas on federal lands is a key objective to meet the Southern Rockies rewilding goals. For example, an unprotected roadless area in the Routt National Forest provides an important link between the Mount Zirkel Wilderness Area and the Sarvis Creek Wilderness Area. This critical area falls within lands identified by the combined SITES analysis and is also potential wolf habitat that supplements the identified wolf cores (Figure 9.4).

Overall, 62% of the roadless areas are below 3,077 meters (10,000 ft.) elevation compared to 31% of GAP Status 1 (Wilderness, National Parks) lands below 3,077 meters (Shinneman et al. 2000). Thus a fair proportion of unprotected roadless lands are in the lower elevation, richer habitat. If unprotected roadless areas were elevated to Wilderness Area status, then 10 of 13 major ecotypes in the Southern Rockies would be 10% or better protected. As an example, 12,000 ha of Douglas-fir forest are covered with GAP Status 1 level of protection, whereas another 80,000 ha (22% of all Douglas-fir) fall in unprotected roadless areas. In addition, there are 136,000 ha of ponderosa pine forest, 200,000 ha of piñon-juniper forest, and 372,000 ha of aspen forest in unprotected roadless areas.

Large protected wild areas are central to our bold vision for the Southern Rockies. Other public lands should be managed as wildlife movement linkages and low to medium compatible-use lands. Private ranches managed for conservation by their owners are also recognized as important wildlife habitat. By expanding and linking Wilderness Areas, and by improving stewardship of wounded areas, ample Southern Rockies habitat will be protected for focal species, including wide-ranging carnivores, sensitive native birds and fish, and other species.

## 2. Wildlands Network Unit Classification and Management Guidelines

While National Parks and National Wilderness Areas have clear guidelines for management, the wildlands network type of vision we discuss here is still being developed and refined. The land unit classification and guidelines below from the Sky Islands Conservation Vision and the New Mexico Highlands Wildlands Network Vision are based on Reed Noss' original classification system (Noss 1992). We use the same classification and guidelines to facilitate connection among the Southern Rockies and adjacent regional wildlands networks.

### ***Core Wild Areas*** *(Noss et al. 1999)*

#### *Study Areas (SA)*

Public land areas that need additional fieldwork to determine if an area has wilderness value, or the level of compatible use that should be allowed. An entire study area will not necessarily be proposed for Wilderness designation; it may be a combination of recommended Wilderness, linkage, and/or compatible-use lands.

#### *Federal Core Wilderness Areas (CW)*

Existing or proposed Wilderness Areas on National Forests, Bureau of Land Management lands, National Park units, and National Wildlife Refuges. Also included are Wilderness Recovery Areas, where significant restorative management is needed.

*Prohibited Uses:* These areas will be managed in accordance with the 1964 Wilderness Act, specifically with no permanent roads, no use of motorized/mechanized equipment (including bicycles, hang gliders), no commercial logging, and no new mining claims. Predator control and trapping should be prohibited, unless necessary for restorative management or recovery of extirpated native species.

*Permitted Uses:* Human use should be managed to protect the ecological integrity of the area. In general, permitted uses include traditional wilderness recreation (hiking, backpacking, horse packing, canoeing, river running), scientific study and research conducted under wilderness principles, and recreational hunting and fishing that does not degrade the ecological integrity of the area or jeopardize the reintroduction of carnivores.

*Restorative Management* (Simberloff et al. 1999): In addition, these areas should be managed to restore and protect natural ecological conditions. Allowed restorative management might include:

- Manual thinning of fire-suppressed (and artificially dense) stands of naturally open-structured forest types (e.g., ponderosa pine). This should be done to facilitate the reintroduction of fire (but only in areas not yet designated as Wilderness except in extreme circumstances). Only the minimum tools necessary should be employed. No commercial sales of timber should be allowed. Reintroduction of fire, either by allowing natural fires to burn or by prescribed fires that mimic natural fires in intensity, frequency (return interval), and seasonality. Fire management activities such as cutting down large snags and prophylactic clearing of fire lines should not be allowed.
- Road closures and, where necessary, revegetation and recontouring.
- Soil inoculation with mycorrhizal fungi, where necessary to reestablish native vegetation.
- Control or, where possible, removal of exotic species, including non-native game fish, game birds and game

**Table 9.2 Federally protected Wilderness Areas.**

| Wilderness Area | Size (ha) | Agency | Wilderness Area | Size (ha) | Agency |
|---|---|---|---|---|---|
| 1. Weminuche/Piedra* | 222,885 | USFS | 25. Lizard Head | 16,717 | USFS |
| 2. Sangre de Cristo/Great Sand Dunes | 105,863 | USFS/USNPS | 26. San Pedro Parks | 16,646 | USFS |
| 3. Flat Tops | 95,188 | USFS | 27. Fossil Ridge | 12,763 | USFS |
| 4. Pecos | 90,380 | USFS | 28. Huston Park | 12,379 | USFS |
| 5. Maroon Bells-Snowmass | 73,455 | USFS | 29. Bandelier/Dome | 11,520 | USNPS, USFS |
| 6. West Elk | 71,392 | USFS | 30. Platte River | 9,507 | USFS |
| 7. Collegiate Peaks | 67,750 | USFS | 31. Greenhorn Mountain | 9,250 | USFS |
| 8. Mount Zirkel | 64,723 | USFS | 32. Never Summer | 8,535 | USFS |
| 9. South San Juan | 64,260 | USFS | 33. Latir Peak | 8,094 | USFS |
| 10. Eagles Nest | 53,949 | USFS | 34. Wheeler Peak | 7,957 | USFS |
| 11. La Garita | 52,147 | USFS | 35. Roubideau Area* | 7,952 | USFS |
| 12. Holy Cross | 49,729 | USFS | 36. Cruces Basin | 7,284 | USFS |
| 13. Lost Creek | 48,477 | USFS | 37. Spanish Peaks | 7,226 | USFS |
| 14. Mt. Massive/Hunter-Fryingpan | 45,554 | USFS, USFWS | 38. Gunnison Gorge | 7,163 | USBLM |
| 15. Uncompahgre | 41,570 | USFS, USBLM | 39. Tabeguache Area* | 6,977 | USFS, USBLM |
| 16. Mt. Evans | 30,109 | USFS | 40. Mt. Sneffels | 6,704 | USFS |
| 17. Indian Peaks | 29,751 | USFS, USNPS | 41. Black Canyon of the Gunnison | 6,313 | USNPS |
| 18. Rawah | 29,570 | USFS | 42. Savage Run | 6,041 | USFS |
| 19. Commanche Peak | 27,029 | USFS | 43. James Peak | 5,666 | USFS |
| 20. Raggeds | 26,464 | USFS | 44. Vasquez Peak | 5,255 | USFS |
| 21. Powderhorn | 24,892 | USBLM, USFS | 45. Ptarmigan Peak | 5,097 | USFS |
| 22. Chama River Canyon | 20,356 | USFS | 46. Encampment River | 4,097 | USFS |
| 23. Sarvis Creek | 18,288 | USFS | 47. Neota | 4,016 | USFS |
| 24. Buffalo Peaks | 17,567 | USFS | 48. Cache La Poudre | 3,747 | USFS |
| | | | 49. Byers Peak | 3,607 | USFS |

Acreages are from the National Wilderness Preservation System website (http://nwps.wilderness.net/) 7/2003. Contiguous areas are lumped together.

* Piedra, Roubideau, and Tabeguache are not designated Wilderness Areas, but are Congressionally protected areas managed for wilderness values.

mammals. Where there is no other practical alternative for removal of exotic plants, judicious use of herbicides may be allowed.

- Phasing out of domestic livestock grazing, especially in riparian and other sensitive areas.
- Restoration of damaged watersheds and watercourses through willow and cottonwood wand planting, loose-rock gabions, and reintroduction of beaver.
- Reintroduction of extirpated native species, including large carnivores.

All restorative management should be conducted under "minimum tool" (The Wilderness Society 1998) and "precautionary principle" (see Chapter 1) standards and should be sensitive to maintaining a sense of wilderness.

*Core Agency Non-Wilderness Protected Public Areas (CA)*

National Parks, National Wildlife Refuges, and some State Parks, State Wildlife Areas, city and county open space, and other protected public lands that are not designated or proposed as Wilderness Areas.

*Prohibited Uses*: No commercial logging, livestock grazing, vehicle use off designated roads, trapping, or predator control, except when absolutely needed for endangered or threatened species recovery.

*Restorative Management*: The same principles as for Core Wilderness should apply.

*Permitted Uses*: Core Agency areas might have roads

**Table 9.3 Ten largest roadless areas.**

| Name | Size (ha) | % Protected |
|---|---|---|
| 1. Weminuche/Piedra Area | 315,900 | 75% |
| 2. Flat Tops Area | 150,400 | 62% |
| 3. Collegiate Peaks Area | 145,300 | 46% |
| 4. Pecos Area | 130,400 | 67% |
| 5. West Elk Area | 129,400 | 55% |
| 6. Northern Rocky Mountain NP/Commanche Peak | 113,700 | 68%* |
| 7. Mt. Zirkel Area | 106,900 | 61% |
| 8. Southern Rocky Mountain NP/Indian Peaks | 106,600 | 83%* |
| 9. La Garita Area | 103,600 | 49% |
| 10. South San Juan Area | 99,500 | 65% |

*Rocky Mountain National Park, although not designated as Wilderness, is managed with a high degree of protection.

(though the overall road density should be no more than 0.3 km/km$^2$), constructed campgrounds, visitor centers, *etc.*

*Private Reserves and Conservation Ranches (CP)*

Private Cores include private nature reserves such as those managed by The Nature Conservancy, National Audubon Society, and land trusts. These are areas not generally afforded levels of protection less than wilderness. Also included here are large private ranches with controlled road access that are managed for biodiversity conservation purposes.

*Core Private Wilderness (CPW)*

Other Private Cores include private land protected essentially as wilderness with conservation easements. Such areas include large private ranches where large carnivores, prairie dogs and other sensitive and imperiled species are accepted.

## Compatible Use Lands *(Groom et al. 1999)*

Compatible-use lands have important ecological functions: They ameliorate edge effects on core wild areas by insulating core wild areas from intensive land use; they provide a suitable habitat matrix for animals to move between core wild areas (*i.e.*, enhance connectivity); they provide supplemental habitat for populations of many native species inhabiting core wild areas, and stabilize population dynamics; they protect adjacent developed areas from any adverse impacts by large mammals that reach relatively high densities in core wild areas. In general there are three classes of compatible-use lands owned by the government (low, medium, and high). These include federal, state, county, and some city open-space lands. Private lands with conservation easements are listed under a separate classification.

*Low Use Compatible-Use Lands (UL)*

We suggest that such lands have a low road density (no more than 0.3 km/km$^2$) and low-intensity uses. Uses might include:

- Primitive recreation, including mountain bike and vehicle use on designated dirt roads only, with no vehicle use off-road. Mountain bikes may be allowed on designated trails.
- Low impact, small, developed campgrounds accessible by vehicle, and some dispersed camping areas.
- Hunting and fishing, in so far as these are compatible with the full range of biological diversity.
- Ecologically sensitive and predator-friendly livestock grazing, except in riparian areas or other highly sensitive areas.
- Limited low-intensity silviculture, such as light selective cutting of previously logged forest followed by road obliteration and closure, and restoration thinning. Cutting of large trees should be prohibited and the goal should be to restore old-growth conditions and natural fire regimes.
- Limited habitat manipulation for focal plant and animal species.
- Restorative management, including those measures listed for Wilderness Areas, but without wilderness restrictions.
- No road construction, vehicle use, or resource extraction in roadless areas of 400 ha or larger.

*Moderate Use Compatible-Use Lands (UM)*

Such areas have a higher road density than UL lands (but still no more than 0.66 km/km$^2$) and more intensive use. In addition to those listed for UL areas, uses in UM zones might include:

- Larger developed campgrounds and heavier recreational use, including dispersed camping and hunter camps, but with motorized vehicles and mountain bikes still restricted to designated routes.

- Habitat manipulation to favor focal wildlife species, but with the goal of returning areas to self-regulated functioning.
- No road construction, vehicle use, or resource extraction in roadless areas of 400 ha or larger.

*Transportation Compatible-Use Lands (UT)*

These lands are found along roads dividing adjacent core wild areas. Management should prevent and modify barriers to wildlife movement. In many cases data on wildlife crossings need to be gathered. Information on road kills, often kept by state game and fish departments, and analysis of snow-tracks can provide clues.

*Private Compatible-Use Lands (UP)*

Private lands voluntarily managed to protect wildlife and restore ecosystems. These include working ranches under ecologically oriented management. Biologically important areas in the Wildlands Network are mapped without regard to ownership for SITES and focal species. If important areas fall on private or tribal lands, then of course those owners make the decisions as to how the land is managed. However, we hope to alert them to the conservation value of their lands, and it may be worthwhile for all parties to explore options for cooperation.

### Landscape Linkages (Wildlife Connections)

*(Dobson et al. 1999)*

There are three classes of linkages in a Wildlands Network: Riparian Linkages, Wildlife Movement Linkages, and Dispersal Linkages. Linkages have several primary functions. One, they provide dwelling habitat, as extensions of core wild areas. Two, they provide for seasonal movement of wildlife. Three, they provide for dispersal and genetic interchange between core wild areas (tie metapopulations together). Four, they allow for latitudinal and elevational range shifts with climate change. Finally, they allow for uninterrupted flows of natural processes (*e.g.*, fire, flood, wind). Linkages are critically important, and we suggest the following management criteria:

- Road density no more than 0.15 $km/km^2$.
- Very few and strictly limited developed sites (campgrounds, *etc.*).
- When intersecting main-traveled roads, linkages should include wildlife land bridges or underpasses, tunnels, bridges, viaducts, speed bumps, and other structures that allow wildlife to cross roads safely.
- No trapping or predator control, except when necessary to protect sensitive species.
- No logging, except thinning to prepare for restoration of fire regime.
- Restorative management as appropriate.
- No motorized vehicles off designated roads.
- No mechanized vehicles, including bikes, off designated routes.
- Seasonal closures of activity where necessary to protect wildlife (birthing, breeding, nesting, *etc.*).
- No road construction, vehicle use, and resource extraction in roadless areas of 400 ha or larger.

*(Note: these guidelines do not apply to Dispersal Linkages.)*

*Riparian Linkage*

Riparian linkages are found along rivers, including National Wild, Scenic, and Recreational Rivers. The primary purpose is to protect continuous habitat for aquatic species, including native fish, beaver, river otter, and invertebrates, and for riparian woodland-dependent species such as birds.

*Wildlife Movement Linkage*

These provide terrestrial linkages for wildlife seasonal movement, dispersal, and movement between cores. Although wildlife movement linkages may hold habitat needed by a given species that is inferior to the habitat available in a core area, the wildlife movement linkages may provide areas that can support sub-adults until a territory opens in the core area. Areas should be managed primarily for movement by specific terrestrial species, with management guidelines based on the needs of those species.

*Dispersal Linkage*

Areas of federal, state, private, or mixed land that may not provide good habitat, but are generally safe for wildlife dispersal from one core habitat to another. In other words, a given species may not choose to live in a dispersal linkage for even a short period of time, but they would cross the area freely. Such dispersal linkages are thus important for genetic exchange. With road closures and restoration, a dispersal linkage may become a wildlife movement linkage.

## 3. Watersheds as Organizational Regions for Wildlands Network Units

The Wildlands Network Design and proposed unit classifications were derived from the computer modeling and expert opinion described in previous chapters. They are drafts in progress, and as time and fieldwork proceed, proposals for wildlife cores, linkages, and low and medium compatible-use lands will be refined. The completion of this work will clarify the value of the study areas shown on the map and listed below. For example, in the two years since releasing the Sky Islands Wildlands Network, the Sky Island Alliance, New Mexico Wilderness Alliance, and Arizona Wilderness Coalition have greatly refined the design through fieldwork. Iterations of the Southern Rockies map and unit descriptions will be published periodically as fieldwork and final proposals are completed.

For this iteration, existing citizens' conservation plans were taken into account. The San Juan Citizens Alliance, Biodiversity Conservation Alliance, White River Conservation Project, and the Upper Arkansas and South Platte Project conservation visions are based on extensive fieldwork and a solid understanding of the scientific data and principles. To translate the citizens' plans into the Wildlands Network Design classification schema, each Forest Service management prescription was assigned a unit code and subclass. These codes match the Wildlands Network Design unit description to the U.S. Forest Service Region 2 definition for management. As an additional note, in the Southern Rockies, we used the category of high compatible use infrequently. In general, we propose that all public land be managed at low or medium levels of compatible use. Unfortunately, some public lands are highly altered by ski resorts or off-road vehicle trails, and some private lands may be severely impacted by industrial or agricultural uses. Thus high compatible use is a category left more to the discretion of local groups.

Watersheds are listed below with a brief description of their ecological values, current protection status, some general recommendations, a justification for protection, and some of the threats to the watershed. The various recommendations for unit classifications are listed by name, and these are displayed on the Southern Rockies Wildlands Network Design (Figure 9.2).

### *3.1 Middle Rio Grande / Canadian Watershed*

#### *Description and Ecological Values*

The Rocky Mountains terminate in northern New Mexico in two high ranges on either side of the Rio Grande: the San Juan-Jemez and the Sangre de Cristos. Most of the wildlands network here is in the Santa Fe and Carson National Forests and two large private ranches, Philmont Boy Scout Ranch and Vermejo Ranch. Additional parts of the wildlands network are owned and managed by the U.S. BLM, several pueblos, the Jicarilla Apache tribe, the New Mexico Department of Game and Fish, and private landowners.

Existing Wilderness Areas are the Pecos, Wheeler Peak, Latir Peak, Bandelier, San Pedro Parks, Chama River Canyon, and Cruces Basin. Taos Pueblo manages Blue Lake as a wilderness. The South San Juan and Weminuche Wilderness Areas are just across the Colorado border in the Upper Rio Grand Watershed. The New Mexico Wilderness Alliance proposes several U.S. BLM Wilderness Areas and is developing Wilderness Area proposals for Forest Service roadless areas. Two complexes of compatible-use lands would link the Sangre de Cristos to the San Juans across the Rio Grande.

The Rockies bring the topography, forests, and wildlife of Canada and Alaska south into New Mexico with the southernmost habitat for wolverine (*Gulo gulo*), Canada lynx (*Lynx canadensis*), American marten (*Martes americana*), white-tailed ptarmigan (*Lagopus leucurus*), and boreal owl (*Aegolius funereus*). There are also areas of high sagebrush steppe with good pronghorn (*Antilocapra americana*) habitat. The area contains old-growth forests, recovering Great Plains grassland, sagebrush steppe, and many streams. American marten, black bear (*Ursus americanus*), mountain lion (*Puma concolor*), and major elk (*Cervus elaphus*) populations are present. Bighorn sheep (*Ovis canadensis*) are found in Columbine Hondo and were recently reintroduced into Latir Peaks. Owners of the 235,200 ha Vermejo Ranch are interested in restoring bison (*Bison bison*), black-tailed prairie dogs (*Cynomys ludovicianus*), Rio Grande cutthroat trout (*Oncorhynchus clarki virginalis*), and black-footed ferrets (*Mustela nigripes*).

Private ranches hostile to wildlife, proposed subdivisions, highways, other roads, and abused lands present existing and future challenges to wildlife movement and connectivity throughout the New Mexico Southern Rockies subregion. However, the expanse of National Forest and BLM lands, private lands, and state lands provide a wildlife movement linkage throughout the watershed.

A reintroduction area for gray wolves (*Canis lupus*) is largely defined by the Carson National Forest (6,000 km$^2$), Santa Fe National Forest (6,400 km$^2$), Vermejo Park Ranch, and several other large tracts of private land each of which encompasses over 100 km$^2$. These private lands include 268 km$^2$ protected under conservation easements. The Taos Pueblo lands encompass 391 km$^2$ of which 230 km$^2$ are managed as wilderness by the tribe (Shinneman et al. 2000). Most of these lands are undeveloped and contiguous, thus forming a large block of habitat that would serve as a secure core area for wolves. This region also contains the Bosque del Oro, Urraca, Elliot Baker and Colin Neblitt State Wildlife Areas.

*Status*

In addition to the existing Wilderness Areas and the Taos Pueblo Blue Lake reserve, the New Mexico Wilderness Alliance is developing Wilderness Area proposals for Wheeler and Latir Peaks additions, Columbine Hondo, and possibly for areas in Valle Vidal. Road closures and a good vehicle management plan for Valle Vidal have created several large roadless areas. Vermejo is well managed for conservation, wildlife, and ecological restoration. New Mexico Department of Game and Fish areas are generally protected.

Much of the federal lands in the area are unprotected. The Pecos Wilderness covers 90,380 ha but there could easily be another 60,000 ha added. The Pecos River, Chama River, and Rio Grande River are wild and scenic rivers, but the Chama River is fragmented by dams and diversions in spots.

*Recommendations*

Develop and campaign for new Wilderness Area and Wild & Scenic River designations on National Forests (e.g. additions to the Pecos Wilderness Area and Cruces Basin). Gain controlled release of gray wolves in Vermejo. Continue research for river otter (*Lontra canadensis*) reintroduction. Restrict vehicle use to designated roads and address issues of logging and restoration of natural fire regimes. This will probably require thinning near human settlements. Grazing issues are important, and some may be addressed with buy-out strategies, particularly in the high country. Restore low elevation habitat for cutthroat trout (*Oncorhynchus clarki*).

*Justification*

Contains vital large, relatively intact habitat area for focal species. In particular, the Carson National Forest/Vermejo complex is a core area for wolf reintroduction. The expanse of federal land offers excellent opportunities for wildlife linkages throughout the watershed and among neighboring watersheds.

*Further Study*

Develop Wilderness Area and Wild & Scenic River proposals. Study potential for grizzly bear (*Ursus arctos*) and river otter reintroduction. Prepare for wolf reintroductions. The Valle Vidal area is particularly important to study for value to a wolf reintroduction in the Vermejo and surrounding areas. It would be important to develop wilderness proposals for Canjilon Mountain, Bull Canyon, and Sierra Negra, and additions to the Pecos, Cruces Basin, and Chama Wilderness Areas.

*Vulnerability*

Ski area and resort development, ranchettes (particularly near Santa Fe), small paved roads, off-road vehicle use, logging threats, risk of catastrophic fire, mine pollution affecting streams (especially Red River), and oil and gas exploration and extraction are threats to the watershed. There is a potential for catastrophic crown fire because of altered fire regime. There is also a threat from poorly conceived (or politically motivated) logging schemes to reduce the risk of fire.

### Middle Rio Grande / Canadian Watershed Unit List

*Core Wilderness*

Bandelier National Monument Wilderness Area (BNM)
Chama National Wild and Scenic River
Chama River Canyon Wilderness Area /National Wild and Scenic River (Carson & Santa Fe NFs)
Cruces Basin Wilderness Area (Carson NF)
Dome Wilderness Area (Santa Fe NF)
Latir Peak Wilderness Area (Carson NF)
Pecos Wilderness Area (Carson & Santa Fe NF)
Rio Grande National Wild and Scenic River (BLM)
San Pedro Parks Wilderness Area (Santa Fe NF)
Wheeler Peak Wilderness Area (Carson NF)

*Proposed Core Wilderness*

Cerro de la Olla Proposed Wilderness
Chama River Canyon Proposed Additions (Carson & Santa Fe NFs & BLM)
Rincon del Cuervo Proposed Wilderness (BLM)
Rio Grande Gorge Proposed Wilderness
Rio San Antonio Proposed Wilderness (Carson NF & BLM)

*Core Wilderness Study*

Caballo Mt./Turkey Ridge Study Area (SFNF & Valle Grande National Preserve)
Camino Real Study Area (Carson NF)
Canada del Oso Study Area (Carson NF)
Canjilon Mt. Study Area (Carson NF)
Capulin Peak Study Area (Carson NF)
Columbine-Hondo WSA (Carson NF)
Corral Canyon Study Area (Santa Fe NF)
Cruces Basin Additions (Carson and Rio Grande NFs)
Dome Wilderness Area Addition (Santa Fe NF)
East Fork Jemez River Study Area (SFNF & VGNP)
Frijoles/Cerro Pelon Study Area (Santa Fe NF)
Addition to Pecos Wilderness Area (Santa Fe NF)
La Cueva Study Area (Carson NF)
Lagunitas/Jawbone Study Area (Carson NF)
Latir Peak Additions (Carson NF)
Little Costilla Peak Study Area (Carson NF)

Polvadera/Cañones/Cebolla Study Area (SFNF & VGNP)
Rio de la Oso Study Area (Santa Fe NF & BLM)
San Pedro Parks Additions (Santa Fe NF)
Santa Barbara Additions to Pecos Wilderness Area (Santa Fe NF)
Santa Fe/Glorieta Addition to Pecos Wilderness Area (Santa Fe NF)
Shuree Study Area (Carson NF)
Sierra de los Valles Study Area (Santa Fe NF & VGNP)
Sierra Negra/Bull Canyon Study Area (Carson NF)
Valle Vidal
Valle Vidal Study Area (Carson NF)
Wheeler Peak Additions (Carson NF)
White Rock Canyon (Santa Fe NF)

*Core Agency*

Bandelier National Monument (USNPS)
Colin Nesbit State Wildlife Area (NMGFD)
East Fork Jemez National Wild and Scenic River
Elliot Barker State Wildlife Area (NMGFD)
Pecos National Monument
Pecos National Wild and Scenic River
Urraca State Wildlife Management Area (NMGFD)

*Core Private*

Blue Lake Wilderness (Taos Pueblo)
Philmont Scout Ranch (Boy Scouts of America)
Vermejo Park Ranch (Turner)
Vermejo Ranch Cimarron Section (Turner)
Vermejo Ranch Greenwood Section (Turner)

*Core Wilderness Study/Compatible Use Low*

Caja del Rio Plateau (Santa Fe NF & BLM)
Cisneros Study Area (Carson NF & BLM)
Elephant Rock Study Area (Carson NF)
Elk Mt./Barillas Peak Study Area to Pecos Wilderness Area (Santa Fe NF)

*Core Wilderness Study/Compatible Use Low/Medium*

Naciemento/Jemez River Study Area (Santa Fe NF)
Peralta Canyon Study Area (Santa Fe NF)

*Compatible Use Lands (Low)*

Edward Sargent State Wildlife Management Area (NMGFD)
El Vado-Heron-Rio Chama State Recreation Areas
Humphries State Wildlife Management Area (NMGF)
Los Pinos State Recreation Area (NMGFD)
San Antonio Mountain (Carson NF & BLM)
Santa Clara Creek (Santa Clara Pueblo)
Valle Grande National Preserve (Santa Fe NF)
Wolf Draw/Pollywog (Santa Fe NF)

*Compatible Use Lands (Low/Medium)*

Glorietta Mesa (Santa Fe NF)
Jarosa (Santa Fe NF)
Las Tampas (Carson NF & BLM)

*Compatible Use Lands (Medium)*

Arroyo Punche (BLM)
Burned Mountain (Carson NF)
El Rito (Carson NF & BLM)
Gallina (Santa Fe NF)
Lama (Carson NF)
Las Viejas Mesa (Carson NF)
Los Pinos North (BLM & Carson NF)
Mogore Ridge (Carson NF)

*Transportation Compatible Use Area*

Arroyo Hondo (Carson NF, Taos Ski Valley)
Chama River (Santa Fe NF)
Cowles/Pecos (Santa Fe NF)
Tres Piedras Road Network (Carson NF)

*Riparian Linkage*

Cañon del Rio Grande (BLM & private)

*Dispersal Linkage*

El Vado-Humphries-Sargent Dispersal Linkage (Jicarilla Apache Game Ranches)

## 3.2 Upper Colorado-Dolores Watershed

*Description and Ecological Values*

The Dolores River and its primary subwatershed of the San Miguel drain the westernmost reaches of the San Juan Mountains. Both rivers arise at Lizard Head Pass. Three 4,300 m peaks dominate the river's headwaters, and numerous peaks over 4,000 meters occur at these higher reaches. The single most compelling feature of these watersheds may be the sweeping, continuous aspen forests that characterize the middle elevations. The largest contiguous tracts of aspen (*Populus tremuloides*) probably occur in the Dolores River watershed near Rico and Dunton. The forested tracts are managed primarily by the San Juan and Uncompahgre National Forests.

Only two relatively small Wilderness Areas have been designated in the watershed, Lizard Head and Mount

Sneffels, both alpine landscapes centered on craggy peaks. Lower elevations tend to be plateaus and mesas that are extensively roaded and logged, although several significant areas of roadless aspen forests still remain.

The Dolores watershed is home to large herds of mule deer (*Odocoileus hemionus*) and elk. Gunnison sage grouse (*Centrocercus minimus*) inhabit the far western edge of the watershed. Now extirpated Columbian sharp-tailed grouse (*Tympanuchus phasianellus columbianus*) were previously found in Glade Park. Reintroduced lynx make significant use of higher elevation spruce forests, and Lizard Head Pass functions as an important landscape corridor for lynx. River otters have been reintroduced successfully to the lower Dolores River.

### *Status*

The Lizard Head and Mount Sneffels Wilderness Areas are located in the basin headwaters. The BLM's Tabeguache special management area protects a major tributary canyon to the San Miguel and originates on the Uncompahgre Plateau. A new state park, Lone Mesa, has been acquired from about 4,800 ha of ranchland, ponderosa pine forest, and aspen groves near Glade Park. The lower Dolores River below McPhee Reservoir was studied and recommended for Wild and Scenic River designation in the 1970s, but has not yet been designated. The Nature Conservancy manages three major preserves along more than 16 km of the San Miguel River. Several large National Forest roadless areas remain unprotected at middle and upper elevations.

### *Recommendations*

Develop and campaign for new Wilderness Area designations on National Forests, including Stoner Mesa, Storm Peak, and San Miguel. Advocate for Wild and Scenic River studies of the upper Dolores and San Miguel Rivers, and major tributaries such as Bear Creek and Fish Creek. Utilize existing proposals of such state and local conservation organizations as Colorado Wilderness Network and San Juan Citizens Alliance. Support ongoing reintroduction of lynx. Advocate for similar reintroduction program for wolverine in the near future. Pursue recovery of Columbian sharp-tailed grouse. Longer-term goals include wolf reintroduction. Research need for supplemental river otter reintroduction. Restore natural fire regimes to low elevation ponderosa pine forests. Restore heavily clearcut spruce forests through road closures and replanting as necessary.

### *Justification*

Contains the largest stands of aspen forest in Southern Rockies. The region is important habitat for wolf and bear (Figures 7.4 [a] and 7.6). Dolores watershed has key landscape linkages to the lower elevation lands of the Colorado Plateau.

### *Further Study*

Identify locations for Columbian sharp-tailed grouse recovery. Prepare for wolverine and wolf reintroductions. Determine most appropriate actions to restoring natural fire regime in ponderosa pine ecosystems.

### *Vulnerability*

The most serious threat is sprawling residential development and resort expansion, particularly associated with Telluride. Proliferation of off-highway vehicles is another major vulnerability because of the vast network of roads associated with previous logging programs. Other threats include water pollution from abandoned mines near Rico and above Telluride, noxious weeds, and logging.

## Upper Colorado-Dolores Watershed Unit List

### *Core Wilderness*

Lizard Head Wilderness Area (Uncompahgre NF, San Juan NF)
Mount Sneffels Wilderness Area (Uncompahgre NF)
Tabeguache Wilderness Area (US BLM)

### *Proposed Core Wilderness (on San Juan NF, Uncompahgre NF, & US BLM)*

Blackhawk Mountain (San Juan NF)
Unaweep (BLM)
Maverick
Fish Creek
Hermosa
McKenna Peak (BLM)
Ryman Creek
San Miguel Peaks
San Miguel River Canyon
Snaggletooth (BLM)
Stoner Mesa
Storm Peak

### *Core Agency*

Dry Creek Basin State Wildlife Area (CO State)
Lone Cone State Wildlife Area (CO State)
Lost Canyon Creek (San Juan NF)

### *Core Private*

San Miguel River Preserves (TNC)

### *Wildlife Movement Linkage*

Glade Blade
Groundhog
Lizard Head Pass

*Compatible Use Low/Medium*

All National Forest Lands not protected as Wilderness (or proposed as such)

All BLM lands not protected as Wilderness (or proposed as such)

*Compatible Use Medium*

Lone Mesa State Park (CO State)

### 3.3 San Juan Watershed

#### *Description and Ecological Values*

The San Juan watershed incorporates the San Juan, Navajo, Piedra, Los Pinos, Animas, La Plata, and Mancos private ranches. The Southern Ute Reservation straddles the lower reaches of the San Juan and tributary rivers.

The existing Weminuche and South San Juan Wilderness Areas are largely dominated by spruce-fir and mixed white fir-Douglas-fir-ponderosa pine forests. The last, large tracts of old-growth ponderosa pine are in the Hermosa and HD Mountains roadless areas. The expanse of federal lands, conserved private lands, and state lands provide a wildlife movement linkage throughout the watershed.

The San Juan watershed encompasses the largest expanse of undeveloped roadless country in the Southern Rockies. The quarter-million-hectare Weminuche Wilderness and contiguous roadless lands straddle the watershed divide between the San Juan and Rio Grande. Coupled with the nearby South San Juan Wilderness, these two areas form the core habitat for restoration of carnivores. Colorado Division of Wildlife initiated lynx recovery beginning in 1999. Habitat studies have identified the upper San Juan watershed as among the most suitable sites for recovery of wolverine and wolf. The largest elk herds in Colorado reside in the San Juans. Healthy populations of mule deer, wild turkey (*Meleagris gallopavo*), and other "game" species also occur here.

The last grizzly known in the Southern Rockies came from the Navajo River drainage, in 1979. The San Juan region may be the best site in Colorado for grizzly bear recovery (see Peterson 1995). Several large private ranches in this drainage are managed with a conservation focus. The region ranked high on the composite SITES analysis.

#### *Status*

Designated Wilderness Areas are the Weminuche, South San Juan, and Piedra, with most of the lands along the Continental Divide included in existing Wilderness designations. The quarter-million-hectare Weminuche/Piedra Wilderness could be expanded by another 40,000 ha. The South San Juan Wilderness straddles the Continental Divide generally between Wolf Creek Pass and the New Mexico border. This 64,000-hectare Wilderness could be significantly expanded to the north, west, and south. The Navajo River is well protected by private ranches managed with a conservation focus. Several rivers have been studied and recommended for Wild and Scenic River designation, including the Piedra, Los Pinos, and Vallecito Creek. No action has occurred on these recommendations. A major Bureau of Reclamation off-river dam project (Animas-LaPlata) threatens to dewater the lower Animas River.

#### *Recommendations*

Develop and campaign for new Wilderness Area and Wild & Scenic River designations on National Forests, including Hermosa, HD Mountains, additions to Weminuche and South San Juans. Utilize existing proposals of such state and local conservation organizations as Colorado Wilderness Network and San Juan Citizens Alliance. Support the ongoing reintroduction of lynx. Advocate for a similar reintroduction program for wolverine in the near future. Longer-term goals include wolf reintroduction. Research the need for supplemental river otter reintroduction. Pursue elimination of domestic sheep grazing in alpine allotments in Wilderness Areas. Restore natural fire regimes to low elevation ponderosa pine forests. Restore heavily clearcut spruce forests through road closures and replanting as necessary.

#### *Justification*

Contains vital large, relatively intact habitat area for focal species. In particular, the Weminuche/South San Juan complex is a core area for wolf reintroduction. The expanse of federal land offers excellent opportunities for wildlife linkages throughout the watershed and among neighboring watersheds.

#### *Further Study*

Study potential for grizzly bear recovery. Prepare for wolverine and wolf reintroductions.

#### *Vulnerability*

The two most serious threats are resort development and oil and gas exploration. Resort development threatens to disrupt key landscape linkages in the headwaters of the San Juan's East Fork and at Durango Mountain Resort (Purgatory Ski Area) in the Animas drainage. Coalbed methane development threatens to overwhelm lower elevation ponderosa pine ecosystems in the HD Mountains.

Other threats include proliferating off-highway-vehicles, water pollution from abandoned mines in the Animas River headwaters, noxious weeds, and logging.

## San Juan Watershed Unit List

*Core Wilderness*

Weminuche Wilderness Area (San Juan NF)
South San Juan Wilderness Area (San Juan NF)
Piedra Special Management Area (San Juan NF)

*Proposed Core Wilderness*

Blackhawk Mountain (San Juan NF)
Hermosa (San Juan NF)
HD Mountains (San Juan NF)
Treasure Mountain (San Juan NF)
San Miguel (San Juan NF)
Weminuche additions (San Juan NF)
South San Juan additions (San Juan NF)
Piedra Additions (San Juan NF)
Menefee Mountain (BLM)

*Core Agency*

West Mancos River (San Juan NF)
East Mancos River (San Juan NF)
Junction Creek (BLM)
Perrins Peak SWA (CO state land)
Florida River/East Animas (San Juan NF)
Ryman Creek (San Juan NF)

*Core Private*

Banded Peaks Ranch (Navajo River watershed)
Other private ranches in the Navajo River watershed

*Wildlife Movement Linkage*

East Fork of San Juan headwaters
Wolf Creek Pass
Molas Pass/Coalbank Summit
Yellowjacket Summit

*Compatible Use Low/Medium*

All National Forest Lands not protected as wilderness (or proposed as such)

All BLM lands not protected as wilderness (or proposed as such)

## 3.4 Upper Rio Grande Watershed

*Description and Ecological Values*

The Upper Rio Grande watershed encompasses the headwaters of the Rio Grande and the San Luis Valley. The San Juan Mountains, La Garitas, and the Continental Divide form the western boundary of the watershed, with the Cochetopa Hills and Poncha Pass to the north. The Sangre de Cristo Range forms the basin's eastern boundary. The Rio Grande National Forest manages the higher elevations of the watershed, while the foothills fall under the jurisdiction of U.S. Bureau of Land Management. In addition, the National Park Service and The Nature Conservancy manage large tracts of land.

The Upper Rio Grande watershed encompasses approximately one-half of the 200,000 ha Weminuche Wilderness and contiguous roadless lands that straddle the watershed divide between the Rio Grande and the San Juan. This watershed also takes in the bulk of the South San Juan Wilderness. The two Wilderness Areas form the core habitat for restoration of large carnivores such as wolves and bears. Colorado Division of Wildlife initiated lynx recovery in 1999, centered in the area near Creede. Habitat studies have identified the upper Rio Grande watershed as among the most suitable sites for recovery of wolverine and wolf. The largest elk herds in Colorado reside in the San Juan Mountains. Healthy populations of mule deer, wild turkey, pronghorn and other "game" species also occur here. Small populations of Rio Grande cutthroat trout occur in isolated streams in the Weminuche Wilderness.

The San Luis Valley and the upper Rio Grande watershed contain a wide diversity of forests. The San Juan Mountains are dominated by spruce-fir and mixed-conifer (white fir-Douglas-fir-ponderosa pine) forests. Farther north, lodgepole pine forests become common at higher elevations in the Cochetopa Hills. Large stands of aspen characterize much of the mid-elevation reaches of the watershed. The fault-block Sangre de Cristo Range creates an impressive barrier east of the San Luis Valley. The San Luis Valley is a closed basin, so most of the water that runs off the surrounding mountains is naturally trapped in the valley and historically created massive wetlands throughout much of the valley. Modern agricultural practices have drained many wetlands and diverted the runoff to irrigation systems.

*Status*

Wilderness Areas are the Weminuche, South San Juan, La Garita, Sangre de Cristo and Great Sand Dunes National Park, with most lands along the Continental Divide included in designated Wilderness. The 200,000 ha Weminuche Wilderness could be expanded by another 40,000 ha. The South San Juan Wilderness straddles the Continental Divide generally between Wolf Creek Pass and the New Mexico border. This 64,000 ha wilderness could be significantly expanded to the north, east, and south. La Garita Wilderness encompasses the Continental Divide north of Creede. The Sangre de Cristo Range includes 90,000 ha of designated Wilderness that runs for 130 km along the eastern boundary of the San Luis Valley. The recently expanded Great Sand Dunes National Park consists of 60,000 ha that spans ecosystems from the valley floor at 2,150 m to the crest of the Sangres at 4,000 m, contiguous with the Sangre de Cristo Wilderness. The Nature Conservancy's Medano-Zapata Ranch covers an adjacent 40,000 ha south and west of the Park. Other key conservation efforts on private lands are taking place along Saguache Creek and Rock Creek. Several state wildlife areas protect remnant wetlands in the heart of the San Luis Valley. The Rio Grande from Alamosa to the New Mexico state line has been evaluated for Wild and Scenic River protection, and local conservation groups and water interests are proposing special status protection. Several large private ranches cover the majority of the Sangre de Cristos south of La Veta Pass, to the New Mexico line.

*Recommendations*

Develop proposals and campaigns for new Wilderness Area and Wild & Scenic River designations on National Forests, including Cochetopa Hills, Pole Creek Mountain, additions to La Garita, Weminuche, South San Juans Wilderness Areas, and U.S. Bureau of Land Management roadless areas. Utilize existing proposals of state and local conservation organizations such as Colorado Wilderness Network and San Luis Valley Ecosystem Council. Support the ongoing reintroduction of lynx. Longer-term goals include wolf reintroduction. Protect remnant habitat of Rio Grande cutthroat and expand to other suitable habitat. Pursue elimination of domestic sheep grazing in alpine allotments, especially in Wilderness Areas. Restore natural fire regimes to low elevation ponderosa pine forests. Restore heavily clearcut spruce forests through road closures and replanting as necessary. Restore wetlands in San Luis Valley.

*Justification*

Contains vital large, relatively intact habitat area for focal species such as bear, wolf, cutthroat trout and pronghorn. In particular, the Weminuche/South San Juan/La Garita complex is a core area for wolf reintroduction. The Great Sand Dunes/Sangre de Cristo/Medano-Zapata Ranch complex encompasses almost 200,000 contiguous hectares of the most diverse habitat in the Southern Rockies, from high desert to alpine tundra.

*Further Study*

Study potential for grizzly bear recovery in South San Juans and prepare for wolf reintroduction. Identify Rio Grande cutthroat trout habitat.

*Vulnerability*

Threats include resort development, logging, and off-road vehicle use. A condominium development with 2,000 units is proposed for Wolf Creek Ski Area on Wolf Creek Pass. Subdivisions sprawl across the Forbes-Trinchera Ranch in the Sangres south of La Veta Pass. The adjacent 30,000 ha (formerly named) Taylor Ranch is ravaged by logging and associated roads. The Rio Grande National Forest exercises relatively little control over off-highway vehicle use. The Summitville Superfund site (an open-pit gold mine) oozes acid mine water at the top of the Alamosa River watershed.

## Upper Rio Grande Watershed Unit List

*Core Wilderness*

Great Sand Dunes National Park (NPS)
La Garita Wilderness Area (Rio Grande NF)
Sangre de Cristo Wilderness Area (Rio Grande NF)
South San Juan Wilderness Area (Rio Grande NF)
Weminuche Wilderness Area (Rio Grande NF)

*Proposed Core Wilderness*

Cochetopa Hills (Rio Grande NF)
Handies Peak (BLM/Rio Grande NF)
La Garita additions (BLM)
Pole Creek Mountain (Rio Grande NF)
Rio Grande (BLM)
San Luis Hills (BLM)
South San Juan additions (Rio Grande NF)
Weminuche additions (Rio Grande NF)

*Core Private*

Medano-Zapata Ranch (TNC)

*Core Agency*

Alamosa NWR

Great Sand Dunes National Park (non-wilderness portion)
Monte Vista NWR
New Baca Ranch NWR
State Parks/Wildlife Areas in heart of San Luis Valley like Mishak Lakes

### *Linkages*

Heart of San Luis Valley through Mishak Lakes
La Veta Pass
Poncha Pass
Southern San Luis Valley dispersal corridor from Vermejo Ranch
Spring Creek Pass
Stony Pass
Wolf Creek Pass

### *Low/Medium Compatible Use*

Remainder of BLM and Forest Service lands

## 3.5 Upper Arkansas Watershed

### *Description and Ecological Values*

The Arkansas watershed headwaters are near Leadville, Colorado, around 3,100 meters in elevation, and the watershed extends south and southeast to near Pueblo, Colorado. Tributaries include the South Arkansas in the upper watershed, and the Cucharas, Huerfano, Apishapa and Purgatoire that drain the Sangre de Cristo and Culebra ranges. The existing Wilderness Areas in the Arkansas watershed include all or parts of Holy Cross, Mount Massive/Hunter-Frying Pan, Buffalo Peaks, Collegiate Peaks, Sangre de Cristo, Greenhorn Mountain, and Spanish Peaks. The San Isabel National Forest, and to a lesser extent the Pike National Forest, manage most of the mountainous areas, and there are significant Bureau of Land Management lands, especially on each side of the Arkansas River from Buena Vista to Fort Carson Military Reserve. The west borders of the drainage are the Continental Divide from Leadville south along the Sawatch Range, crossing over to the crest of the Sangre de Cristo/Culebra mountains south to New Mexico. The Arkansas watershed has the full range of life zones from foothills to alpine tundra, but alpine tundra/barren rock and spruce-fir forests dominate the high mountain slopes on the north and west, with lodgepole pine (*Pinus contorta*) on the lower slopes. The Mosquito Range is similar on the north, but becomes ponderosa pine/Douglas-fir and piñon-juniper as it nears the Arkansas River, while the Wet Mountains reverse that order as they begin near the river corridor and rise to Greenhorn Mountain on the south. The inter-mountain valleys provide significant shrub and grasslands ecotypes, but are primarily in private ranching operations. The Arkansas corridor between Salida and its egress from the foothills is piñon-juniper woodlands intermixed with sage and semi-desert vegetation in the rocky dry canyons. It is unlikely that large amounts of old growth forest remain outside of the Wilderness Areas, but Black Mountain has some of the oldest bristlecone pine in Colorado.

The region has black bear, deer, elk, bighorn sheep, pronghorn, beaver (*Castor canadensis*), turkey, and grouse, to mention only a few. Elk and pronghorn are common in the river valleys, especially in winter. The native greenback cutthroat trout (*Oncorhynchus clarki stomias*), almost wiped out by overharvest and competition from non-native trout, has been recently reintroduced into the Rock Creek drainage, and a pure strain has been newly discovered on Pikes Peak. Mexican spotted owls (*Strix occidentalis lucida*), once more common in Colorado, still are found in the deep canyons northwest and south of Cañon City. The Sangre de Cristos include good wolf habitat and form a dispersal linkage between cores for both black bears and wolves. The region ranked very high on the combined SITES analysis.

### *Status*

Many of Colorado's 14,000 foot peaks are located along the Sawatch, Mosquito, and Sangre de Cristo ranges – picture postcard views of rock and ice bordered by forested slopes above the valleys. The Sangre de Cristo and Collegiate Peaks Wilderness Areas are two of the 10 largest Wilderness Areas in the Southern Rockies region, and the other Wilderness lands provide islands of protection in the watershed. Current Wilderness proposals for Bureau of Land Management lowlands along the Arkansas River between Buena Vista and Cañon City include Browns Canyon, McIntyre Hills, Grape Creek and Beaver Creek. Large tracts of unprotected roadless lands lie along the Wet Mountains north of the Greenhorn Wilderness Area, on the Pikes Peak massif, and in the upper watershed along the Mosquito and Sawatch Ranges. If designated as Wilderness, these lands would provide wild connections along each side of the main river corridor, as well as south along the Wet Mountains, although there would remain significant gaps in the chain of protected lands.

The Nature Conservancy, Trust for Public Land, and San Isabel Foundation are working to acquire conservation easements in the Wet Mountain Valley. Several ranches in the south Wet Mountain Valley participate in the Division of Wildlife's Ranching for Wildlife program.

Mueller State Park and North Lake and Bosque del Oso State Wildlife Areas provide relatively secure wildlife habitat as well as connectivity in the central and southeastern watershed.

The extensive federal and state lands, combined with thoughtful private ranching, provide for wildlife movement throughout the watershed.

*Recommendations*

Use existing proposals of the Colorado Wilderness Network and the Upper Arkansas and South Platte Project for new Wilderness Area designations, which include the unprotected roadless areas high in the watershed, along the eastern front of the Sangre de Cristo range, in the Wet Mountains north of the Greenhorn Wilderness and west of Spanish Peaks. Promote recovery of greenback cutthroat trout and strong protection for Mexican spotted owls and lynx habitat. Restore natural fire regimes to low elevation ponderosa pine-Douglas-fir forests. Restore forests through road closures and replanting as necessary. Retain the roadless areas by closing and rehabilitating old logging and mining roads and opposing expanding motorized recreation proposals. Restore riparian areas by closing motorized routes, especially those that intrude into otherwise roadless areas. Work with land trusts and agencies to protect important valley habitat for pronghorn and ungulate winter range.

*Justification*

Contains vital large, relatively intact habitat area for focal species, including black bear, greenback cutthroat trout, and wolf. The expanse of federal land and unprotected roadless land provides opportunities for wildlife linkages throughout the watershed and among neighboring watersheds. The southeast section of the watershed, the Wet Mountains, the Sangre de Cristo Mountains, and the upper reaches of the watershed scored highly on the combined SITES run.

*Further Study*

Investigate implications of existing good wolf and bear habitat as connecting linkages for future wolf and grizzly recovery. Investigate ways to connect proposed and existing Wilderness Areas across the intervening mountain valleys and uplands, including state/BLM lands east of Buena Vista and Salida, various lands between the Wet Mountains and Sangres, and especially to the adjoining areas in northern New Mexico. Study barriers created by major transportation linkages. Identify roads that could be closed to increase habitat quality. Study the effects of thinning to reduce fuel load on the ecosystems, and how best to restore a natural fire regime.

*Vulnerability*

Serious threats are housing development along the river linkages and in the lower forested areas throughout the watershed; and industrial recreation, including two ski areas, many popular snowmobile routes, and major off-highway vehicle use west of Buena Vista, north of Texas Creek, along the Gold Belt Scenic Byway, and in the Wet Mountains. These pose the most vexing and widespread impact on intact ecosystems, especially in the lower elevations in the upper Arkansas and along the foothills from Colorado Springs to Trinidad. In general, the watershed already is heavily roaded, with only the Sangre de Cristos having much area that is more than 3.2 km from a road. Fuels reduction projects, many an outgrowth of the 2002 fires, pose serious ecological threats unless carefully conducted as a step toward a natural fire regime. Oil and gas development may be a threat near the New Mexico state line. The upper stretches of the Arkansas River drainage are among the state's highest density areas for active and abandoned mines, and water quality suffers as a result.

## Upper Arkansas Watershed Unit List

*Core Wilderness*

Buffalo Peaks Wilderness Area (Pike NF & San Isabel NF)
Collegiate Peaks Wilderness Area (San Isabel NF)
Greenhorn Wilderness Area (San Isabel NF)
Holy Cross Wilderness Area (San Isabel NF)
Mount Massive Wilderness Area (San Isabel NF)
Sangre de Cristo Wilderness Area (San Isabel NF)
Spanish Peaks Wilderness Area (San Isabel NF)

*Core Agency*

Beaver Creek State Wildlife Area (CO state land)
Bosque del Oso State Wildlife Area (CO state land)
Florissant Fossil Beds National Monument (NPS)
Leadville National Fish Hatchery
Mueller State Park and Dome Rock State Wildlife Area (CO state land)
North Lake State Wildlife Area (CO state land)
Spanish Peaks State Wildlife Area (CO state land)

*Proposed Core Wilderness*

Antelope PeakBadger CreekBears HeadBeaver Creek

(BLM, San Isabel NF)
Big Union and Marmot Peak additions to Buffalo Peaks Wilderness (Pike-San Isabel NF)
Browns Canyon (San Isabel NF)
Chipeta (San Isabel NF)
Cisneros Creek, Greenhorn Creek, Apache Creek, Santana Butte, and Badito Cone additions to Greenhorn Mountain Wilderness (San Isabel NF)
Lake Creek, Greenleaf Creek, Horn Creek, Crystal Falls, Upper Grape Creek, Bruff Creek, May Creek, Carbonate Mountain, Blanca Peak, and Slide Mountain additions to Sangre de Cristo Wilderness (San Isabel NF)
Elk Mountains (San Isabel NF)
Grape Creek (San Isabel NF)
Hardscrabble (San Isabel NF)
Highline (San Isabel NF)
Kreutzer-Princeton (San Isabel NF)
La Plata Gulch, Pine Creek, and Frenchmane Creek additions to Collegiate Peaks Wilderness (San Isabel NF)
McIntyre Hills (BLM)
Mt. Antero (San Isabel NF)
Mt. Elbert (San Isabel NF)
Porphyry (San Isabel NF)
Purgatoire (San Isabel NF)
Scraggy Peaks (San Isabel NF)
Starvation Creek (San Isabel NF)
Table Mountain (BLM)
West Pikes Peak (Pike NF)
Williams Creek (San Isabel NF)

### *Core Private Wilderness*

The Nature Conservancy - Aiken Canyon Preserve

### *Private Compatible Use Lands*

Easements on various ranches in the Wet Mountain Valley (TNC, Trust for Public Land, and San Isabel Foundation)
CDOW Ranching for Wildlife ranches

### *Private Compatible Use Study Area (for Dispersal Linkage)*

Area across private lands in the south end of Wet Mountain Valley around the De Weese
Reservoir and towns of Silver Cliff/Westcliffe has a low ecological cost for restoration
Huerfano and Cucharas River drainages to connect south Wet Mountains, Sangre de Cristos and Spanish Peaks.

### *Wildlife Movement Linkage*

Cochetopa Hills
Hagerman Pass
Sangre de Cristos chain for wolves and bears
Unprotected roadless areas along the Arkansas River from Salida to Cañon City
Unprotected roadless areas between Collegiate Peaks and Sangre de Cristo Wilderness areas

### *Wildlife Movement Linkage/Low Compatible Use Study Area*

Heckendorf State Wildlife Areas

Holy Cross eastern front, Chicago Ridge, Mt. Arkansas, N. Cottonwood Creek, Arnold Gulch, Kaufman Ridge, St. Charles Peak, Williams Creek West, and Cuchara – generally roadless areas that are not proposed for Wilderness.

State lands around Black Mountain, Agate Mountain, Long Gulch, Antelope Gulch, and Waugh Mountain

BLM Lands between the north end of the Sangre de Cristos and Wet Mountains

Portions of Ft. Carson Military Reserve

### *Transportation Linkage Study Area*

Crossing Rt. 24 (in particular between Johnson Village and Granite)
Crossing Rt. 285 (Cochetopa Hills, Poncha Pass)
Crossing Rt. 50 between Cañon City and the Continental Divide (in particular along the north side of the Sangre de Cristos
Crossing Rt. 69 in the Wet Mountain Valley

### *Compatible Use Low to Medium*

All National Forest lands not protected as Wilderness or proposed for Wilderness
All undesignated state lands
All BLM lands not protected as Wilderness or proposed for Wilderness
Garden of the Gods, Bear Creek Canyon, North Cheyenne Canyon Park (CO state land)
Royal Gorge Park (CO state land)
Temple Canyon (Cañon City)
Trinidad Lake State Park (CO state land)

## 3.6 Gunnison Watershed

### *Description and Ecological Values*

The Gunnison watershed includes drainages from the Uncompahgre, Cimarron, Little Cimarron, North Fork and Lake Fork of the Gunnison, East, Slate, and Taylor Rivers. The land falls on the Uncompahgre National Forest, Grand Mesa National Forest, Gunnison National Forest, Black Canyon of the Gunnison National Park, and Bellman's. The watershed holds the West Elk mountains, the Ruby Range, the Ragged Mountains, the Elk Mountains, and the Sawatch Range.

The watershed has parts or all of 11 Wilderness Areas: Maroon Bells-Snowmass, West Elk, Collegiate Peaks, La Garita, Uncompahgre, Raggeds, Powderhorn, Fossil Ridge, Black Canyon of the Gunnison National Park, Gunnison Gorge, and Mt. Sneffels. The Gunnison Gorge National Conservation Area (BLM) and Curecanti National Recreation Area (BLM) are also in the watershed.

Plant forms include Douglas-fir (*Pseudotsuga menziesii*), Englemann spruce (*Picea engelmannii*), subalpine fir (*Abies lasiocarpa*), Gambel oak (*Quercus gambelii*), aspen, limber pine (*Pinus flexilis*), lodgepole pine (*Pinus contorta*), piñon pine (*Pinus edulis*), ponderosa pine (*Pinus ponderosa*), bristlecone pine (*Pinus aristata*), cottonwood (*Populus* spp.), willow (*Salix* spp.), juniper (*Juniperus* spp. and *Sabina* spp.), and sage (*Artemisia* spp.). Wildlife includes elk, mule deer, black bear, mountain lion, coyote (*Canis latrans*), bobcat (*Lynx rufus*), badger (*Taxidea taxus*), beaver, bighorn sheep, and many smaller mammals and birds. The Colorado River cutthroat trout (*Oncorhynchus clarki pleuriticus*) is native.

### *Status*

Federal Wilderness Areas are well distributed in the watershed, with the exception of the Grand Mesa NF where there is currently no federally designated Wilderness. Adding unprotected roadless areas to this protection would benefit conservation by enlarging existing Wilderness Areas, establishing Wilderness in underrepresented areas, and reconnecting them. The amount of state and federal lands allows opportunity to reconnect areas for wildlife movement. However, several inventoried roadless areas within the Gunnison and Grand Mesa National Forests are currently compromised with regard to wilderness quality and may therefore be difficult to designate as such (i.e. – south/southwest and northwest portions of West Elk inventoried roadless area [IRA] have extensive user created motorized routes and range "improvements", and the northwest portion is further threatened by potential coal mine expansion). The Clear Creek IRA and Drift Creek IRA north of McClure Pass are threatened by oil and gas leasing; Clear Creek, already has wells and access roads. The Springhouse Park IRA has been leased for oil and gas almost in its entirety, though these leases are currently being challenged at the IBLA, and it also has motorized trails. And finally, the Sheep Flats/Salt Creek IRA has a large timber sale slated for the "non-roadless" surrounding lands, and the IRA may eventually be logged as well – a decision to that effect has been deferred.

### *Recommendations*

Use existing proposals for new Wilderness Areas and Scenic River designations from Colorado Wilderness Network, High Country Citizen's Alliance, and Western Slope Environmental Research Council. The unprotected roadless areas offer needed additions to the wilderness network and could significantly enlarge each existing Wilderness Area. One of the four cores for wolf reintroduction is in the watershed, and several other areas provide good wolf habitat. Investigate wolf and Colorado cutthroat trout recovery. Work to restore natural fire regimes. Restore forests through road closures and replanting as necessary.

### *Justification*

Contains large, relatively intact habitat area for focal species, including cutthroat trout, black bear, and wolf. The expanse of federal land and unprotected roadless land provides opportunities for wildlife linkages throughout the watershed and among neighboring watersheds. Much of the watershed rated highly on the combined SITES run.

### *Further Study*

Study potential for wolf and Colorado cutthroat trout recovery. Identify connections between Wilderness Areas, using in particular the unprotected roadless areas. Study how best to restore a natural fire regime. Identify roads that could be closed to increase habitat quality. Study sensitive areas for wildlife to cross roads.

### *Vulnerability*

Population growth, visitation levels, exurban housing, and industrial recreation compromise the existing Wilderness Areas. Logging, oil and gas exploration, heavy fall hunting pressure (including motorized travel), and livestock grazing threaten unprotected roadless and compatible use areas. Exotic trout affect the native cutthroats.

## Gunnison Watershed Unit List

*Core Wilderness*

Black Canyon of the Gunnison National Park (NPS)
Collegiate Peaks Wilderness Area (Gunnison NF)
Fossil Ridge Wilderness Area (Gunnison NF)
Gunnison Gorge Wilderness (BLM)
La Garita Wilderness Area (Gunnison NF)
Maroon Bells—Snowmass Wilderness Area (Gunnison NF)
Mt. Sneffels Wilderness Area (Uncompahgre NF)
Powderhorn (Gunnison NF & BLM)
Raggeds Wilderness Area, (Gunnison NF)
Uncompahgre Wilderness Area (Uncompahgre NF)
West Elk Wilderness Area (Gunnison NF)

*Core Agency*

Almont Triangle State Wildlife Area (CO state land)
Billy Creek State Wildlife Area (CO state land)
Buckhorn Lakes Park (CO state land)
Cebolla Creek State Wildlife Area (CO state land)
Cimarron State Wildlife Area (CO state land)
Curecanti National Recreation Area (BLM)
Escalante State Wildlife Area (CO state land)
Gunnison Gorge National Conservation Area (BLM)
Gunnison State Wildlife Area (CO state land)
Lake Fork Gunnison State Wildlife Area (CO state land)
Sapinero State Wildlife Area (CO state land)

*Proposed Core Wilderness*

Adobe Badlands (BLM)
Bangs Canyon (BLM)
Cannibal Plateau (Gunnison NF)
Cochetopa Hills (Gunnison NF)
Crystal Peak (Uncompahgre NF)
Dominguez Canyon (BLM)
East Elk Creek (Gunnison NF)
Redcloud Peak (Gunnison NF)
West Elk Addition (BLM)
Powderhorn Addition (BLM)
Roubideau (Uncompahgre NF)
Other unprotected roadless areas on federal Land

*Wildlife Movement Linkage*

Romley
Mt. Antero
Sargents
Tomichi Dome
Seven Creek
Cochetopa Dome
Middle Fork
Cathedral Creek
Slumgullion
Carson Peak
Dry Basin

*Transportation Linkage Study Area*

Cerro Pass on Rt. 50 for bears
Crossing Rt. 133 along North Fork of the Gunnison River
Crossing Rt. 135 along Slate River
Crossing Rt. 149 between Uncompahgre WA and Powderhorn WA (Uncompahgre NF & BLM)
Crossing Rt. 50 along Gunnison River
Crossing Rt. 50 along Quartz Creek
Crossing Rt. 50 along Tomichi Creek
Crossing Rt. 550 along Uncompahgre River

*Compatible Use Low to Medium*

All National Forest lands not protected or proposed for Wilderness
All BLM lands not protected or proposed for Wilderness
All undesignated state lands

*Compatible Use Medium*

Crawford State Park (CO state land)
Swetzer Lake State Road (CO state land)

## 3.7 Colorado Headwaters Watershed

*Description and Ecological Values*

The watershed contains drainages of the Roaring Fork, Frying Pan, Eagle, Blue, and Piney Rivers. The source of the Colorado River is in the western part of Rocky Mountain National Park. The watershed contains all or parts of Rocky Mountain National Park and the Arapaho, White River, and Grand Mesa National Forests, various BLM lands, and the Arapaho National Recreation Area. The Colorado Canyons National Conservation Area (BLM) and Colorado National Monument (NPS) lie just outside the boundary of the Southern Rockies ecoregion, but along the Colorado River.

Mountain ranges in the Colorado Headwaters Watershed include the Gore Range, Never Summer

Mountains, Rabbit Ears Range, Williams Fork Mountains, Williams Mountains, Elk Mountains, and the Sawatch Range and half of the Flat Tops.

There are parts or all of 12 Wilderness Areas in the Upper Colorado watershed, all on National Forest land: Flat Tops, Maroon Bells-Snowmass, Collegiate Peaks, Eagle's Nest, Holy Cross, Mt. Massive/Hunter-Frying Pan, Indian Peaks, Raggeds, Never Summer, Vasquez Peak, Ptarmigan Peak, and Byers Peak.

The portion of Rocky Mountain National Park that is west of the Continental Divide (106,291 ha) lies in the watershed. Eighty-nine percent of Rocky Mountain National Park (94,267 ha) is managed as wilderness. Nearly a third of Rocky Mountain National Park is above 3,385 meters (11,000 feet), so the park contains large areas of alpine tundra. The west slope of Rocky Mountain National Park is moister than the east, and it has more lodgepole pine than the east part (Wuerthner 2001). Other trees and shrubs in the watershed are Douglas-fir, Englemann spruce, subalpine fir, Gambel oak, aspen, limber pine, cottonwood, willow, juniper, and sage. Wildlife includes elk, mule deer, black bear, mountain lion, coyote, bobcat, badger, beaver, bighorn sheep, and many smaller mammals and birds. The Colorado cutthroat trout is native to the Colorado River Basin.

### *Status*

Rocky Mountain National Park contains a large wilderness area, but the high number of visitors and the unnaturally large elk populations are negatively affecting both the flora and fauna of the park (Singer and Zeigenfuss 2002). As mentioned above, federal Wilderness Areas are well distributed in the Upper Colorado River watershed. Adding unprotected roadless areas to this protection would benefit wildlife. The amount of state and federal lands affords opportunities to connect areas for wildlife movement.

### *Recommendations*

Use existing proposals for new Wilderness Area and Scenic River designations identified by the Colorado Citizens Network and the White River Conservation Project. The unprotected roadless areas provide needed additions to the Wilderness network and could significantly enlarge each existing Wilderness Area. One of the four cores for wolf reintroduction is in the watershed, and several other areas provide good wolf habitat. It is important to investigate wolf and Colorado cutthroat trout recovery, how to restore natural fire regimes, and how to restore forests through road closures and replanting as necessary.

### *Justification*

Contains large, relatively intact habitat area for focal species, including Colorado cutthroat trout, black bear, and wolf. The expanse of federal land, including unprotected roadless areas, provides opportunities for wildlife linkages throughout the watershed and among neighboring watersheds. Much of the watershed rated highly on the combined SITES run.

### *Further Study*

Study potential for wolf and Colorado cutthroat trout recovery. Identify connections between Wilderness Areas, using in particular the unprotected roadless areas. Study how best to restore a natural fire regime. Identify roads that could be closed to increase habitat quality and security. Study road barriers to wildlife movement.

### *Vulnerability*

Population growth, exurban housing, and industrial recreation compromise the existing Wilderness Areas. Rocky Mountain National Park is heavily visited, including the part managed as Wilderness. Recreational pressure is thus high, and growing, throughout the watershed. Timber thinning to prevent fire from damaging houses is an impact, particularly if poorly planned. Exotic trout affect the native cutthroats.

## Colorado Headwaters Watershed Unit List

#### *Core Wilderness*

Byers Peak Wilderness Area (Arapaho NF).
Collegiate Peaks Wilderness Area (White River NF)
Eagle's Nest Wilderness Area (Arapaho & White River NF)
Flat Tops Wilderness Area (White River NF)
Holy Cross Wilderness Area (White River NF)
Indian Peaks Wilderness Area (Arapaho NF)
Maroon Bells—Snowmass Wilderness Area (White River NF)
Mt. Massive/Hunter-Frying Pan (White River NF)
Never Summer Wilderness Area (Arapaho NF)
Ptarmigan Peak Wilderness Area (White River NF)
Raggeds Wilderness Area (White River NF)
Vasquez Peak Wilderness Area (Arapaho NF)

#### *Core Agency*

Arapaho Recreation Area (Arapaho NF)
Basalt State Wildlife Area (CO state land)
Coke Oven State Wildlife Area (CO state land)
Garfield Creek State Wildlife Area (CO state land)

Hot Sulfur Springs State Wildlife Area (CO state land)
Junction Butte State Wildlife Area (CO state land)
Piceance Creek State Wildlife Area (CO state land)
Plateau Creek State Wildlife Area (CO state land)
Radium State Wildlife Area (CO state land)
Twin Sisters State Wildlife Area (CO state land)
West Rifle Creek State Wildlife Area (CO state land)

*Proposed Core Wilderness*

Williams Fork (Roosevelt NF)
Pisgah Mountain CWP (BLM)
Castle Peak (BLM)
Bull Gulch (BLM)
Flat Tops Addition (BLM, White River NF)
Deep Creek (BLM)
Other unprotected roadless areas on federal land Grand Hogback (BLM)
Roan Plateau (BLM)
Maroon Bells (BLM)
Thompson Creek (BLM, White River NF)
Rocky Mountain National Park (NP)

*Wildlife Movement Linkage*

Between Eagles Nest WA and Mt. Evans WA (Arapaho NF)
Between Indian Peaks WA, Ptarmigan WA, Vasquez WA, and Byers Peak WA
Between Vasquez Peak WA and Byers Peak WA (Arapaho NF)
Between Eagles Nest WA and Flat Tops WA
Between Flat Tops WA and Roan Plateau
Between Holy Cross WA and Eagles Nest WA
Unprotected areas recommended for Wilderness

*Transportation Linkage Study Area*

Crossing Interstate 70
Crossing Rt. 40 between I 70 and Winter Park
Crossing Rt. 82 along Roaring Fork River
Crossing Rt. 9 along Blue River
Crossing Rt. 9 between Ptarmigan WA and Eagles Nest WA
Crossing Rt. 24 between Ptarmigan Hill RA and Holy Cross WA
Crossing Rt. 131 between Eagles Nest WA and Castle Peak WSA
Crossing Rt. 133 between Maroon Bells-Snowmass WA and Thompson Creek CWP area.
Crossing CR 13 between Flat Tops and Roan Plateau

*Compatible Use Low to Medium*

All National Forest lands not protected as or proposed for wilderness
All BLM lands not protected as or proposed for wilderness
All undesignated state lands

*Compatible Use Medium*

Middle Carter-Gunsight State Lands (CO state land)
Middle Park State Lands (CO state land)
Milk Creek State Lands (CO state land)
Rifle Gap State Park (CO state land)
Troublesome Ranch State Lands (CO state land)
Vega State Park (CO state land)
West Carter Mountains State Lands (CO state land)
Whiskey Creek State Lands (CO state land)
Windy Ridge State Lands (CO state land)

## 3.8 South Platte Watershed

*Description and Ecological Values*

In addition to the North Fork and mainstem of the South Platte, the watershed includes the Clear Creek, Big Thompson and Cache la Poudre drainages. The Laramie Range forms the northern border. The west border is formed by the Continental Divide until it meets the Arkansas Watershed, where the Mosquito Range continues the west boundary. The south border follows the divide along Thirty-nine Mile Mountain and north of the Pikes Peak massif. Included in the watershed are Front Range mountains such as the Mummy Range, Platte River, Kenosha and Tarryall Mountains, and the Rampart Range. The watershed contains all or parts of Rocky Mountain National Park, and Lost Creek, Mt. Evans, Indian Peaks, Commanche Peak, Buffalo Peaks, James Peak, Neota, and Cache la Poudre Wilderness Areas. Rocky Mountain National Park has 94,267 ha (89%) proposed as Wilderness since 1976; although Congress has not approved the proposal yet, the Park Service manages that 89% for wilderness values (Wuerthner 2001). In addition, sections of the Cache la Poudre are Colorado's only designated Wild and Scenic River.

Vegetation in the watershed is highly varied and ranges from piñon pine, juniper, Douglas-fir, and ponderosa pine in the lower elevations to bristlecone pine, limber pine, aspen, blue spruce (*Picea pungens*), Englemann spruce, and subalpine fir, mountain meadows, riparian species, as well as alpine tundra in the higher mountains. The Mt. Evans Wilderness

Area contains small regions of arctic tundra. Unlike typical Colorado alpine tundra, which is dry and brittle once the snow recedes, arctic tundra holds numerous small pools of water. The Mt. Goliath Natural Area has a stand of bristlecone pines that are 2,000 years old. The watershed has extensive wetlands, especially in South Park, which is the highest large intermountain basin in North America. Its fen ecosystems support more rare plant communities than any other ecosystem type in Colorado, and possibly in the Southern Rockies (Baron 2002).

Wildlife includes elk, moose (*Alces alces*), mule deer, black bear, mountain lion, coyote, bobcat, badger, beaver, and many smaller mammals. The Tarryall Mountains have one of the healthiest bighorn sheep herds in the region. Birds include northern goshawks (*Accipiter gentilis*), chickadees (*Poecile* spp.), hummingbirds, gray jays (*Perisoreus canadensis*), red-tailed hawk (*Buteo jamaicensis*), white-tailed ptarmigan, osprey (*Pandion haliaetus*), bald eagle (*Haliaeetus leucocephalus*), flammulated owls (*Otus flammeolus*), and golden eagle (*Aquila chrysaetos*). Mountain goats (*Oreamnos americanus*) were introduced a half-century ago for hunting opportunities on Mt. Evans, although it is unlikely they are native to the area. A few streams, particularly in Rocky Mountain National Park, have the rare and endangered greenback cutthroat trout, but native cutthroat trout are threatened by interbreeding with brown (*Salmo trutta*), rainbow (*Oncorhynchus mykiss*), and brook trout (*Salvelinus fontinalis*).

### *Status*

In addition to the significant Wilderness Areas listed above, Rocky Mountain National Park provides a large wilderness area, but visitation in the park and over Trail Ridge Road is very high. In addition, elk populations are negatively affecting both the flora and fauna (Singer and Zeigenfuss 2002). Because the South Platte watershed is close to Colorado Springs, Denver, and Fort Collins, recreation pressure is enormous. For example, the South Platte Ranger District (184,000 ha) of the Pike National Forest has 2,500,000 visits a year, a number that exceeds the total use on 47 National Forests across the Nation. Indian Peaks Wilderness is one of most heavily used Wilderness Areas in the United States.

The checkerboard pattern of federal and state ownership, which covers much of the northeast section of the Arapaho-Roosevelt Forest, is a problem for habitat connectivity and consistent management.

The South Platte River downstream from Elevenmile Reservoir and on the North Fork is eligible for Wild and Scenic River designation. In a multi-year planning process, numerous stakeholders crafted the South Platte River Protection Plan which will protect the outstandingly remarkable values in lieu of designation and is generally supported by all parties. State lands are scattered throughout the watershed, and several are sizeable (Antero, Reinecker, Elevenmile, Spinney, Golden Gate Canyon, Mt. Evans and Cherokee Park). Roxborough State Park, though small, is managed for foot-powered recreation and wildlife. Large tracts of unprotected roadless land provide good options for connecting fragmented areas of protection.

### *Recommendations*

Use existing proposals for new Wilderness Areas of the Upper Arkansas and South Platte Project, the Colorado Wilderness Network, and the South Platte River Protection Plan. The unprotected roadless areas provide needed additions to the wilderness network and could significantly enlarge each existing Wilderness Area. While none of the four wolf reintroduction areas is in the watershed, several areas, including Mt. Evans and Lost Creek Wildernesses, would provide good wolf habitat. Investigate wolf and greenback cutthroat trout recovery. Restore natural fire regimes in the backcountry and pursue ecologically sound fuels reduction in the wild-urban interface. Restore forests through road closures and replanting as necessary.

Explore protection options for other non-federal and private lands in South Park. Encourage conservation easements and Park County's interest in protecting large ranches from future subdivision. Promote wildlife land bridges and underpasses on Highway 285 and similar transportation corridors. Support efforts to maintain the backcountry character of the Guanella Pass Road. Explore land trades and consolidation of ownership in the Arapaho-Roosevelt National Forest.

### *Justification*

Contains large, relatively intact habitat area for focal species (cutthroat trout, pronghorn and wolf). The expanse of federal land, both protected and unprotected, provides opportunities for wildlife linkages throughout the watershed and among neighboring watersheds. Much of the watershed rated highly on the combined SITES run.

### *Further Study*

Study potential for wolf and greenback cutthroat trout recovery. Identify connections between Wilderness Areas, using in particular the unprotected roadless areas. Study how best to restore a natural fire regime. Identify roads that could be closed and obliterated to increase habitat quality and security. Identify potential conservation easements in South Park. Study barriers to wildlife movement, especially along the highway 285 corridor.

### *Vulnerability*

Population growth along the Front Range compromises the existing Wilderness, roadless areas and Rocky Mountain National Park, both from encroaching development and from recreation demands. Indian Peaks is the most heavily visited Wilderness Area in the United States, and the Rampart Range Recreation Area is devoted to intensive all-terrain vehicle (ATV) and motorcycle recreation. Ski resorts have major impacts at Loveland Basin and Winter Park. Recreational pressure is high, and growing, throughout the watershed.

Timber thinning, ostensibly to prevent fire from damaging houses, is a threat, particularly if poorly planned and located in the backcountry, rather that in the wild-urban interface. A few large fuels reduction projects, notably the Upper South Platte Restoration Project and the Trout West Project, could possibly reduce catastrophic fire, but at the expense of Wilderness and ecological values if roads are constructed. The 2002 Hayman fire of 55,200 ha highlighted the fire potential of fire-suppressed ponderosa pine-Douglas-fir forests, as well as the magnitude of fire exacerbated by drought, low humidity and high winds. The heightened fears of fire have brought a spate of salvage logging, fuels reduction and logging-for-water proposals from both federal and state agencies. Unfortunately, these proposals are generally not informed by fire ecology and would perpetuate flawed fire policies.

Exotic trout affect the native cutthroats, and encroaching non-native weeds are an increasing problem. Blister rust (*Cronartium* spp.) has been documented recently in the Red Feather Lakes area; and outbreaks of various insects and other pathogens, although natural, have increased in occurrence and severity with fire suppression.

Water storage and use radically affects the aquatic and riparian habitats of the watershed. The mainstem of the South Platte River, which provides 70% of Denver's drinking water (Foster-Wheeler, 1999), has five reservoirs before it reaches the foothills, including major augmentation with west slope water through the Roberts Tunnel. Horsetooth Reservoir, on a tributary of the Cache la Poudre, and Carter Lake on the Big Thompson are the 7th and 13th largest reservoirs in the Southern Rockies (Shinneman et al. 2000). Inter-mountain diversions, both existing and proposed, increased capacity of existing reservoirs, and new dam proposals are all high threats in light of the current drought conditions. Water quality continues to be affected by domestic, agricultural and industrial uses. There are many active and abandoned mines in the watershed whose acid and heavy metal depositions affect humans and wildlife alike. The eastern part of the watershed has the highest density of EPA regulated facilities, as well as the largest cluster of Superfund/CERCLA sites in the ecoregion (Shinneman et al. 2000).

## South Platte Watershed Unit List

#### *Core Wilderness*

Bowen Gulch Wilderness Area (Arapaho NF)
Buffalo Peaks Wilderness Area (Pike NF)
Cache La Poudre Wilderness Area (Roosevelt NF)
Comanche Peak Wilderness Area (Roosevelt NF)
Indian Peaks Wilderness Area (Roosevelt NF)
James Peak Wilderness Area (Arapaho NF)
Lost Creek Wilderness Area (Pike NF)
Mt. Evans Wilderness Area (Pike & Arapaho NF)
Neota Wilderness Area (Roosevelt NF)
Never Summer Wilderness Area (Arapaho & Routt NF)
Rawah Wilderness Area (Roosevelt NF)
Vasquez Peak Wilderness Area (Arapaho NF)

#### *Core Agency*

Antero State Lands (CO state land)
Cherokee Park State Wildlife Units (CO state land)
Elevenmile Canyon State Park (CO state land)
Golden Gate Canyon State Park (CO state land)
Green Ridge (Roosevelt NF)
Lion Gulch (Roosevelt NF)
Mount Goliath Research Natural Area (Arapaho NF)
Mt. Evans State Wildlife Area (CO state land)
North Lone Pine (Roosevelt NF)
Reinecker Ridge State Wildlife Area (CO state land)
Roxborough State Park (CO state land)
Spinney Mountain State Park (CO state land)
St. Louis Peak (Arapaho NF)
Tarryall Reservoir State Wildlife Area (CO state land)

#### *Proposed Core Wilderness*

Big Union, Lynch Creek, and Salt Creek additions to Buffalo Peaks Wilderness (Pike & San Isabel NF)
Boreas (Pike NF)
Burning Bear (Pike NF)
Cache La Poudre additions (Roosevelt NF)
Cherokee Park (Roosevelt NF)
Comanche Peak additions (Roosevelt NF)
Elk Creek and other additions to Mt. Evans Wilderness (Arapaho & Pike NF)
Farnum (Pike NF)

Front Range (Pike NF)
Green Mountain (Pike NF)
Green Ridge East (Roosevelt NF)
Gunbarrel (Pike NF)
Hoosier Ridge (Pike & Arapaho NF)
Indian Peaks additions (Roosevelt NF)
Jefferson (Pike NF)
Neota additions (Roosevelt NF)
Puma Hills (Pike NF)
Rocky Mountain National Park (NPS)
Sheep Rock (Pike NF)
Silverheels (Pike NF)
Square Top (Arapaho & Pike NF)
Thirty-Nine Mile (Pike NF)
Thunder Butte (Pike NF)
Vedauwoo (Medicine Bow NF)
Weston Peak (Pike & San Isabel NF)
White Pine Mountain (Roosevelt NF)
Wildcat Canyon (Pike NF)
Williams Fork (Arapaho NF)

*Core Private Wilderness*

High Creek Fen TNC Preserve
Phantom Canyon Ranch TNC Preserve

*Private Compatible Use Study Area (for Dispersal Linkage)*

South Park

*Wildlife Movement Linkage*

Public lands, especially in South Park
Unprotected areas recommended for Wilderness

*Wildlife Movement Linkage/Low Compatible Use Study Area*

South Park State/BLM lands, including Reinecker Ridge and Red Ridge
Lost Creek Arm, Northrup/Longwater Gulches, Indian Creek, Noddle Heads, Jenny Gulch, Trout Creek, Jackson Creek, Limbaugh Canyon, and Stanley Canyon - generally roadless areas that are not proposed for Wilderness
South Platte River corridor

*Transportation Linkage Study Area*

Crossing Highway 9
Crossing Interstate 70
Crossing Route 285
Crossing Rt. 14
Crossing Rt. 34
Crossing Rt. 36
East end of Highway 24

*Compatible Use Low to Medium*

All National Forest lands not protected as or proposed for Wilderness
All BLM lands not protected as or proposed for Wilderness
All undesignated state lands

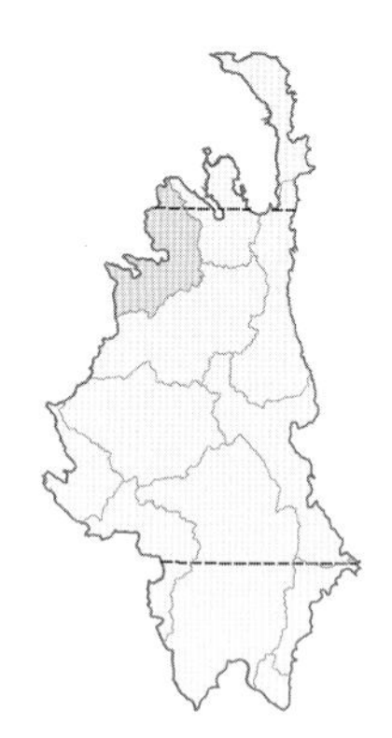

## 3.9 Yampa-White River Watershed

### *Description and Ecological Value*

Three National Forests are represented in the watershed: the Medicine Bow in Wyoming (Brush Creek/Hayden District), and the Routt and White River National Forests in Colorado. The Yampa-White River Watershed is separated from the North Platte Watershed by the Continental Divide. It borders the Upper Colorado Watershed to the south.

Huston Park Wilderness Area in the Medicine Bow National Forest straddles the Continental Divide in Wyoming. The Mount Zirkel Wilderness Area of the Routt National Forest straddles the continental divide in Colorado between the Yampa-White Watershed and the North Platte Watershed. In the southwestern part of the watershed, the Flat Tops Wilderness Area is part of the Routt National Forest and the White River National Forest. Large roadless areas that are still unprotected exist around the Mt. Zirkel Wilderness Area and link the Mt Zirkel Wilderness to the Sarvis Creek Wilderness Area of the Routt National Forest. The Sarvis Creek Wilderness Area is located in the southeastern part of the watershed. A large block of unprotected roadless areas also border the Flat Tops Wilderness Area and the Huston Park Wilderness Area, and numerous blocks of isolated unprotected roadless areas are found throughout the Sierra Madres.

Elevations are high, ranging from 2,100 meters in the irrigated valleys to 4,000 meters along the Divide. Heavy winter snows are common and frost may occur anytime during the short summer season. Habitat varies across the watershed from alpine meadows to high desert of sagebrush. Forests include stands of lodgepole pine, Englemann spruce, subalpine fir, and aspen (sometimes interspersed with open parks and brushy meadows). In the fall of 1997, a storm blew down 3,200 ha of trees within the Mt. Zirkel Wilderness Area and 5,200 ha in total along the west side of the Continental Divide to the north of Steamboat Springs.

Elk and deer are common residents, using the higher elevations during the summer months, but moving to lower elevations for the winter. The region has bighorn sheep, mountain lion, coyote, northern goshawks, pronghorn, American marten, beaver, yellow-bellied marmot (*Marmota flaviventris*), white-tailed ptarmigan, osprey, eagles, and Colorado River cutthroat trout. The Colorado pikeminnow (*Ptychocheilus lucius*) was recently re-discovered in the Little Snake River on the slope of the Sierra Madre Mountains. Black bear have core habitat in the western part of the watershed from the Wyoming/Colorado border through the southern part of the watershed, although the bears of this region are smaller in size than those found farther south in Colorado. The watershed is north of the large Gambel oak stands and south of the whitebark pine (*Pinus albicaulis*) range—both rich sources of food for bears. Much of the region is good wolf habitat, and the watershed includes the northern part of the 2,500 $km^2$ wolf core found in northwestern Colorado.

### *Status*

Two of the 10 largest Wilderness Areas in the Southern Rockies region lie partially in this watershed: the Flat Tops and Mount Zirkel. Two smaller Wilderness Areas, Sarvis Creek and Huston Park, are also found in the Yampa-White watershed. In Wyoming, there are state lands, including several school sections close to Huston Park. State and BLM lands lie at the edge of the Southern Rockies ecoregion boundaries in Wyoming. In Colorado, state lands are scattered throughout the watershed, and there are significant BLM lands along the Yampa River. There are several state parks and recreation areas. Large tracts of unprotected roadless land provide good options for connecting fragmented areas of protection.

### *Recommendations*

Use existing proposals for new Wilderness Area and Scenic River designations on the Medicine Bow, Routt, and White River National Forests identified by Biodiversity Conservation Alliance, Colorado Wilderness Network, the White River Conservation Project, and the Routt Citizens Alternative. The unprotected roadless areas provide needed additions to the wilderness network and could significantly enlarge each existing Wilderness Area. Indeed, Sarvis Creek Wilderness Area could be connected to Mount Zirkel Wilderness Area by officially protecting the roadless area that lies between them. The Flat Tops Wilderness Area is one of the four cores identified for wolf reintroduction, and the watershed provides a great deal of secondary wolf habitat.

Identify ways to connect the Flat Tops Wilderness Area to the Sarvis Creek Wilderness Area. Investigate wolf and cutthroat trout recovery, how to restore natural fire regimes, and the ecological impact of the blowdown in the Routt National Forest. Restore heavily clearcut forests through road closures and replanting as necessary. Explore options for easements with ranchers.

### *Justification*

Contains large, relatively intact habitat area for focal species, including Colorado River cutthroat trout and wolf. The expanse of federal land and unprotected roadless land provides opportunities for wildlife linkages throughout the watershed and among neighboring watersheds. Most of the watershed found in the Southern Rockies rated high on the combined SITES run. The region holds species that are not found elsewhere in the Southern Rockies—*e.g.* Pacific trillium (*Trillium ovatum*).

### *Further Study*

Study the potential for wolf and cutthroat trout recovery. Identify connections between Wilderness Areas, using in particular the unprotected roadless areas. Study how best to restore a natural fire regime. Identify roads that could be closed to increase habitat quality and security. Study barriers to wildlife movement. Study the possibility of adding the large block of unprotected roadless areas around Vermillion Bluffs/Sevenmile Ridge (Bureau of Land Management lands) to the Southern Rockies boundary area as a federally protected Wilderness Area (at present it borders the Southern Rockies ecoregion boundary in northern Colorado).

### *Vulnerability*

Logging is a threat on unprotected National Forest lands, and the U.S. Forest Service once proposed to log up to the border of the Mount Zirkel Wilderness Area. Off-road vehicles need to be eliminated. Roads and associated edge effects in the Medicine Bow National Forest cover as much as 20% of the land surface.

## Yampa-White River Watershed Unit List

#### *Core Wilderness*

Flat Tops Wilderness Area (Routt NF, White River NF)

Huston Park Wilderness Area (Medicine Bow NF)
Mount Zirkel Wilderness Area (Routt NF)
Sarvis Creek Wilderness Area (Routt NF)

*Proposed Core Wilderness*

Scenic River proposed for North Fork of the Elk River (USFS)
Yampa River CWP (BLM)
Huston Park adjacent (Medicine Bow NF)

*Core Wilderness Study*

Unprotected roadless areas between Mt Zirkel Wilderness Area and Sarvis Creek Wilderness Area (Routt NF)
Unprotected roadless areas around Vermillion Bluffs and Sevenmile Ridge (BLM)
All unprotected roadless areas in the Sierra Madres range of the Medicine Bow NF

*Core Agency*

Indian Run State Wildlife Area (CO state land)
Jensen State Wildlife Area (CO state land)
Oak Ridge State Wildlife Area (CO state land)
Radium State Wildlife Area (CO state land)
Yampa River State Wildlife Area and state lands (CO state land)

*Wildlife Movement Linkage*

Buffalo Pass
Dunckley Pass
Gore Pass
Rabbit Ears Pass
Ripple Creek Pass
Unprotected roadless areas between Sarvis Creek and Mount Zirkel

*Wildlife Movement Study Area*

BLM and CO state lands that could link Sarvis Creek and Flat Tops

*Transportation Linkage Study Area*

Crossing CO Rt. 131 between Flat Tops and Sarvis Creek
Crossing CO Rt. 40 from Rabbit Ears Pass to Steamboat Springs
Crossing CO Rt. 13 between the Elkhead Mts. and Pole Gulch State and BLM Lands
Crossing WY Rt. 230 from Colorado line to Riverside

*Compatible Use Low/Medium*

All National Forest lands not protected or proposed as Wilderness
All BLM lands not protected or proposed as Wilderness

*Compatible Use Medium*

Grassy Creek State Lands (CO state land)
Pearl Lake State Park (CO state land)
Pole Gulch State Lands (CO state land)
Stagecoach State Recreation Area (CO state land)
Steamboat Lake State Recreation Area (CO state land)

## 3.10 North Platte Watershed

### *Description and Ecological Values*

The west boundary of the North Platte Watershed includes the east slope of the Continental Divide through the Park Range of the Mount Zirkel Wilderness Area (Routt National Forest) and the Sierra Madres of the Medicine Bow National Forest. The east slope of the Mount Zirkel Wilderness Area holds the headwaters for the North Platte and Encampment Rivers. Other drainages that begin in the North Platte Watershed form the Laramie, Medicine Bow, and Michigan Rivers. The east boundary of the watershed runs along the Mummy and Laramie Ranges. The Laramie Range is the northern extension of the Colorado Front Range.

Wilderness Areas include parts or all of the Mount Zirkel, Rawah, Huston Park, Platte River, Never Summer, Savage Run, Encampment River, and Neota. All are located on Forest Service lands (Medicine Bow, Routt, and Roosevelt Forests).

The watershed contains a variety of habitats from alpine to sagebrush steppe. Engelmann spruce, limber pine, Douglas-fir, subalpine fir, lodgepole pine, and quaking aspen are the dominant trees at higher altitudes. The North Platte River moves through a large riparian floodplain, and native cottonwood stands support Wyoming's second largest population of nesting bald eagles. Many other birds also use the riparian areas, such as black-crowned night heron (*Nycticorax nycticorax*), western grebe (*Aechmophorus occidentalis*), common loon (*Gavia immer*), Wilson's phalarope (*Phalaropus tricolor*), ducks, and sandpipers. Bighorn sheep, deer, elk, mountain lion, porcupine (*Erethizon dorsatum*), and pure strains of Colorado River cutthroat trout live in the region. There are some black bears, but they are small in size and number. The region is north of good Gambel oak areas and south of the whitebark pine, both rich sources of food for bears.

Some species that the Medicine Bow-Routt National Forests consider "sensitive" are American marten, boreal owl, tiger salamander (*Ambystoma tigrinum*), northern goshawk, and clustered-lady's slipper (*Cypripedium fasciculatum*). Moose were introduced to North Park in 1978 and now about 1,000 live in northern Colorado. They are exotic to Colorado, but the species has a great deal of public support.

Recreational use by humans includes camping, hiking, cross-country skiing, mountain bike riding, horseback riding, and off-road vehicles (including driving snowmobiles). Rabbit Ears Pass has recently seen conflict between snowmobile use and cross-country skiing, as some snowmobile users have not respected the boundaries separating the two uses.

### *Status*

Parts or all of eight Wilderness Areas are in the region, but six are smaller than 15,000 ha. State lands extend throughout the region, and the Wyoming Game and Fish Department has special Wildlife Management Areas for bighorn sheep, deer, and elk (e.g. Sheep Mountain and in the Laramie Range). The U.S. Bureau of Land Management has extensive areas bordering the Southern Rockies ecoregion boundary in Wyoming, and they have holdings along the Platte River. The Colorado State Forest State Park borders the Rawah Wilderness Area. Unprotected roadless areas offer opportunities to expand the designated Wilderness Areas, particularly in the Snowy Range and in the northern Laramie Range (around Laramie Peak).

### *Recommendations*

Use existing proposals for new Wilderness Area and Scenic River designations on the Medicine Bow, Routt, and Roosevelt National Forests, as in the Biodiversity Conservation Alliance Citizen's Proposal for the Medicine Bow National Forest. The unprotected roadless areas provide needed additions to the wilderness network and could significantly enlarge some Wilderness Areas. Investigate wolf potential, cutthroat trout recovery, and how to restore natural fire regimes. Restore heavily clearcut forests through road closures and replanting as necessary.

### *Justification*

The watershed contains large, relatively intact habitat area for focal species (pronghorn and wolf). The expanse of federal land (including unprotected but roadless land) provides opportunities for wildlife linkages throughout the watershed and among neighboring watersheds. The Laramie Range rated high on the combined SITES run.

### *Further Study*

Study potential for wolf and lynx recovery. Identify connections between Wilderness Areas, using the unprotected roadless areas. In particular, study how road closure could be combined with existing unprotected roadless areas to produce a large Wilderness Area around Laramie Peak in the Laramie Range. The same type of study could produce a Wilderness Areas proposal in the Snowy Range.

Study how best to restore a natural fire regime. Identify roads that could be closed to increase habitat quality and security, particularly in the Medicine Bow National Forest. Study sensitive areas for wildlife to cross roads. In particular, study how to improve the ability of elk and deer to cross Interstate 80 at Elk Mountain.

In the Laramie Range, study the possibility of meshing the Southern Rockies ecoregion boundaries to the west with the Heart of the West plan. From the Medicine Bow National Forest of the Laramie Range, the region to the west includes the Chalk Hills, Shirley Basin, Seminoe Mountains, and Ferris Mountains. The northern part of this extension is nearly solid U.S. Bureau of Land Management Land. From the Seminoe Reservoir to the south, the land is checkerboard between U.S. Bureau of Land Management and private holdings. The area is rich in pronghorn and raptors. Shirley Basin has a small number of black-footed ferrets, recently reintroduced.

Additionally, the land along the North Platte River from Saratoga to the Colorado state line could be added to the ecoregion boundary. That would include many of the important riparian areas and a significant Bureau of Land Management holding.

### *Vulnerability*

Over the last 40 years, approximately 15% of all forested acres on the Medicine Bow Forest have been logged. Over the last 20 years, approximately 7% of the Routt National forest has been logged. The Forest Service considers 31% of the Routt and 37% of the Medicine Bow as suitable for timber harvest. They now consider timber harvest in roadless areas on a case-by-case basis (via internal analysis and public involvement). Approximately 20% of the Medicine Bow forest is affected by roads or by edge effects from those roads. The Snowy Range alone has nearly 350 km of groomed and ungroomed snowmobile trails. Livestock grazing on these public lands is another problem.

## North Platte Watershed Unit List

### *Core Wilderness*

Encampment River Wilderness Area (Medicine Bow NF)
Huston Park Wilderness Area (Medicine Bow NF)
Mount Zirkel Wilderness Area (Routt NF)
Neota Wilderness Area (Roosevelt NF)
Never Summer Wilderness Area (Routt NF)
Platte River Wilderness Area (Medicine Bow NF)
Rawah Wilderness Area (Roosevelt NF)
Savage Run Wilderness Area (Medicine Bow NF)

### *Proposed Core Wilderness*

Elk Mountain area (Medicine Bow NF)
Pennock Mountain Roadless Area (Medicine Bow NF)
Unprotected roadless areas around Huston Park and Encampment River Wilderness Area (Medicine Bow NF)
Unprotected roadless areas in the Laramie Range (Medicine Bow NF)
Unprotected roadless areas in the Snowy Range (Medicine Bow NF)
Unprotected roadless around Savage Run and Platte River Wilderness Area (Medicine Bow NF)

### *Core Agency*

Arapahoe National Wildlife Refuge (US FWS)
East Delany Butte State Wildlife Area (CO state land)
Lake Johns State Wildlife Area (CO state land)
Odd Fellows State Wildlife Area (CO state land)
Owl Mountain State Wildlife Area (CO state land)

### *Private Compatible Use Lands Study Area*

Private lands along the North Platte from Saratoga WY to the Colorado border are important.

### *Wildlife Movement Linkage*

Laramie Range
Michigan River, Roaring Fork, and North Platte between Mt. Zirkel and Rawah WA
Forest Service land around existing and proposed Wilderness Areas

### *Transportation Linkage Study Area*

Crossing CO 125 along the North Platte River
Crossing CO 14 from Muddy Pass through Colorado State Forest State Park
Crossing Interstate 80 along the Laramie Range
Crossing Interstate 80 near Elk Mountain
Crossing Wyoming 230 along the North Platte River
Crossing Wyoming 34 through Sybille Canyon
Crossing Wyoming 130 from Brush Creek to Centennial

### *Compatible Use Low/Medium*

All National Forest lands not protected or proposed as Wilderness
All BLM lands not protected or proposed as Wilderness

### *Compatible Use Medium*

Colorado State Forest State Park (CO state land)
Curt Gowdy State Park (WY state land)
Walden Reservoir State Wildlife Area (CO state land)

*One can either curse the darkness or light a candle to find the way out.*

-Adlai Stevenson

*We cannot solve the problems we have created with the same thinking that created the problem.*

-Albert Einstein

# 10. IMPLEMENTATION AND CONSERVATION ACTION

Jean Smith, Jen Clanahan, Margaret DeMarco, Robert Howard, Dave Foreman, Rich Reading, Brian Miller

*(This chapter draws from the Sky Islands Wildlands Network Vision and the New Mexico Highlands Wildlands Network Vision, but was extensively rewritten to reflect Southern Rockies implementation opportunities.)*

## 1. Introduction

The Southern Rockies Wildlands Network Vision provides the bold, visionary framework for the ecoregion, but ultimately the vision must be embodied in actual protection and restoration on-the-ground. We offer this chapter to stimulate discussion about implementation strategies for the ecoregion. The Wildlands Network Vision and implementation through conservation action are two sides of a single process. Work on the Wildlands Network Vision precedes detailed consideration of implementation, and that work should not be limited by implementation considerations, but the two cannot be considered in isolation. This chapter provides some practical examples for those working toward the Vision's goals, for it will take many people working on many different aspects to make the Southern Rockies Wildlands Network Vision a reality.

The jigsaw puzzle is a useful metaphor for implementing the Wildlands Network Vision. The completed wildlands network is the picture on the cover of the puzzle box. Inside the box are the pieces of the puzzle (implementation steps) that, when fitted together, will make the complete picture (the Wildlands Network Vision). The implementation steps include all the different campaigns and action items necessary to realize the conservation vision goals and objectives. The whole puzzle will not be put together in one fell swoop as conservationists did, most spectacularly, with the Alaska National Interest Lands Conservation Act. Rather, different cooperating groups will place separate pieces down on the table from time to time.

**GOALS OF THE SOUTHERN ROCKIES WILDLANDS NETWORK VISION**

1. Protect and recover native species.
2. Protect and restore native habitats.
3. Protect, restore and maintain ecological and evolutionary processes.
4. Protect and restore landscape connectivity.
5. Control and remove exotic species.
6. Reduce pollution and restore areas degraded by pollution.

How do we place the pieces on the table? Conservationists have a well-equipped toolbox, including organizing, public relations, working with government agencies, fundraising, crafting Wilderness proposals and other protective legislation, advocacy and lobbying, legal action, writing management plans, monitoring agency projects, applying science in the defense or promotion of lands, doing scientific research, working with private landowners, facilitating land purchases and exchanges, doing ecological restoration, fostering recovery and protection for native species, engaging volunteers in on-the-ground activities and advocacy, to name only a few of our tools. Different conservationists have expertise in using different tools, and circumstances often dictate which are appropriate to reach certain goals. And there are always new tools that we need to learn and use.

This chapter is not an implementation plan. Its purpose is to suggest some key activities and illustrate successful conservation action with selected regional and national examples. The conservation action plan specific to this Wildlands Network Vision will evolve, formally and informally, as we work together to identify existing initiatives, convene watershed and regional meetings, and build concerted action toward fulfilling the Southern Rockies Wildlands Network Vision goals and objectives.

## 2. Values, implementation, and policy

In this section, we provide a description of values, attitudes, and beliefs that can affect a conservation plan, followed by a discussion of the policy process.

### *How values affect the implementation of a conservation plan*

Conservation planning is typically high profile, contentious, and complex. As a result, conservationists must address polarized attitudes and controversy. The resistance to restoring biodiversity rests with human perceptions and values. These perceptions and values in turn are enacted through societal choices which change through education and clarification of societal values over time.

#### *Values, attitudes, and beliefs*

Values, attitudes, and beliefs help define who we are as individuals and what we do. They impose order and consistency to our complex and chaotic world (Tessler and Shaffer 1990, Olson and Zanna 1993). Many people use the terms "values", "attitudes", and "beliefs" almost interchangeably. Although these terms all interact, they do differ (Bright and Barro 2000). So we offer some definitions for clarification (Bem 1970, Rokeach 1972).

A *value* is a preferred mode of behaving (*e.g.*, honesty) or existing (*e.g.*, equality), and values are affected by perception, context, and knowledge of the situation (Rokeach 1972, Williams 1979, Brown 1984, Brown and Manfredo 1987). Because the relative strengths of values are not equal, people arrange them in a hierarchical fashion. Thus, when a person is faced with a situation in which two or more values clash, that individual usually relies upon more strongly held values (core values) over less strongly held ones (peripheral values, Williams 1979).

*Attitudes* are affinities or aversions toward something (*e.g.*, wilderness), and that affinity or aversion is based on beliefs (Bem 1970, Rokeach 1972). A *belief*, in turn, is based on our perception of how an event or entity affects a given situation. So, context is important. An example might be a livestock owner's belief about the predatory nature of wolves (*Canis lupus*) after discovering a fresh livestock kill. Extreme attitudes tend to be based on more simple belief systems than moderate attitudes (Bright and Barro 2000).

*Perceptions* are formed by what a person senses and understands about an issue. Perceptions evolve from information, cultural values, and personal experiences (Brown 1984). *Context* describes a person's situation; for example, how frequently someone is exposed to an issue. The present social setting is also important. Traditional customs, peer pressure, level of socialization by institutions, and other social factors interact to determine the social setting (Brown and Manfredo 1987, Chaiken and Stangor 1987). Finally, a person's mood (*e.g.*, level of satisfaction), and physical state (*e.g.*, physical or economic health) can also be important. Thus, a message will be received more effectively when it is presented frequently, presented by people who are similar to the target audience, and presented in the appropriate context.

*Knowledge* is the acquisition, comprehension, and retention of information, and it depends on exposure, receptivity, perception, interpretation, and memory (Petty et al. 1997). While knowledge is an important determinant of values, attitudes, and beliefs, its importance is often over-estimated, especially among people who value knowledge greatly, such as biologists and conservationists (Reading 1993, Kellert 1996, Kellert et al. 1996). Knowledge is only one of several factors influencing values, attitudes, and beliefs, and its influence is often relatively weak. As an example, ranchers and members of conservation organizations scored more highly than all other groups when tested about knowledge of black-footed ferrets (*Mustela nigripes*, Reading 1993, Miller et al. 1996). Yet, ranchers and members of conservation groups were diametrically opposed on the issue of reintroducing black-footed ferrets.

When values, attitudes, and beliefs are strongly held, new knowledge is often selectively received (accepting only the parts of the information with which one already agrees) and selectively interpreted (Tessler and Shaffer 1990, Olsen and Zanna 1993). In other words, people often focus on facts that support their existing attitudes. Indeed, values, attitudes, and beliefs can even affect memory, with information supporting a pre-existing opinion memorized and remembered more easily than information contradicting such an opinion (Tessler and Shaffer 1990, Olsen and Zanna 1993). These interactions are strengthened if information is poor, ambiguous, or too complex to be easily understood (Tessler and Shaffer 1990, Olson and Zanna 1993).

Thus, simply supplying knowledge in an implementation campaign will have little effect on people who strongly hold negative attitudes. Indeed, there is little chance of changing their minds in time for effective action. Implementation on a timely scale must somehow neutralize the negative effects of strong opponents. A strategy of supplying additional knowledge is most effective at gaining support for conservation when it is used on people who are undecided or who do not hold strong opinions on an issue (*i.e.* their core values are not negative to conservation). As mentioned above, context and frequency of presentation are important.

#### *Myths*

The strongly entrenched attitudes and beliefs toward conservation, whether positive or negative, can be viewed as part of the larger worldview that people hold. Social scientists refer to such worldviews as *myths*. Myths are nothing more than the political perspectives most firmly accepted by a community (Lasswell and Kaplan 1950). Thus, myths are based on a number of fundamental assumptions, regardless of their truth (they can be true). Over time, a given community no longer questions these assumptions. For example, in American society most people never question the assumption that economic growth is "good," they simply operate in a manner that takes this assumption as a given.

Myths are therefore powerful belief systems based on unquestioned assumptions, and they are supported by powerful symbols (*e.g.* flags). Myths help people understand, relate to, and operate in a complex world. In a sense, they are like blinders on a horse that is walking down a busy street. Myths promote solidarity among people who share them. Problems arise when people refuse to accept that there are different perspectives, thus their own subjective experience becomes a substitute for reality (Arendt 1958).

Myths, and their symbols, are typically not defended with logic, but often elicit emotional responses. Many people who can hold an intellectual conversation about the pros and cons of a given technique become quite irrational when their fundamental myth is challenged. Emotional responses can become violent. For this reason, leaders use and manipulate myths to further their political agendas, usually by associating loyalty to their agenda with loyalty to the dominant myth (and its symbols). Thus pro-development forces present energy exploitation as sustaining the American way of life.

In sum, it is important to understand the values of people who oppose and people who favor a given conservation plan. That will help neutralize opponents and garner support from the undecided. Simultaneously, people working to implement a plan must not let such strategies erode the existing support for the plan.

### *Forming Policy*

If we are ever to achieve more wilderness, then we must come to understand and improve the policy process (Clark 2000). In general terms, the policy process is viewed as a human social dynamic that determines who gets what, how, and why. In conservation policy, however, it is important to remember that social science is a human construct, and as such can too easily define "common interest" only in human terms. Any definition of common interest should include Nature and all affected species as equal partners. Social science can be of great help for moving conservation plans through societal organizations. But to benefit Nature, such plans must be strongly grounded in biology. That includes ecology as well as the biological reasons humans act (not just the cultural reasons).

All the chapters in this conservation Vision address different parts of the overall policy process in one way or another. It is thus important to integrate those parts into comprehensive action that maps, understands, evaluates, and improves the path needed to implement a Wildlands Network Vision. We tend to think only of legal actions as policy, but there is much more. Legal actions are formal policy, but policy-making is also a sequence of actions by many people and organizations in the decision process (informal policy). Indeed, formal policy can also be changed greatly during implementation (by poor decisions made in the field, sand-bagging, etc.).

A policy is a commitment and a process toward a preferred outcome (Clark 2000). In response to a problem, it represents an alternative to the processes that produced the trouble in the first place. In other words, it is a difference between goals and trends (Clark 2000, Clark et al. 2001). In the rush to solve problems, activists, scientists, politicians, and the general public are typically solution-oriented rather than problem-oriented. Yet, effective solutions to problems cannot be constructed unless the root of the problem is fully analyzed, understood, and defined. Being solution-oriented, also called "problem-blind," leads people to promote solutions before they fully understand the problem. Thus being solution-oriented can itself be a major problem. Five interrelated tasks will help address any problem. They are (Clark 2000):

- Clarify goals: What events or processes are preferred outcomes?
- Describe trends: What are the historic and recent events that can affect goals?
- Analyze conditions: What factors shape trends?
- Make projections: What future developments are likely under various circumstances?
- Invent, evaluate, and select alternatives: What course of action is likely to help realize your goals?

If these five tasks are fully carried out, rational choices can be established. The rational thing to do is use trends, conditions, and projections to "choose the alternative that you expect... to be the best means of realizing your goals" (Brunner 1995:3). By best we mean the most effective, efficient, and equitable alternative. The process is procedural. Each iteration should reconsider previous findings in light of new information and changing circumstances.

When one or more of the five tasks is omitted or poorly treated, a gap exists in the policy argument. Sometimes a gap is a sign of propaganda or censorship designed to manipulate viewpoints on controversial issues (Brunner 1995). In addition, decisions can be misrepresented to cover motives and increase credibility. While good analysis thrives on alternative choices, politics often depends on restricting potential alternatives to control policy outcomes (Clark 2000). Thus, these five steps can also be used to understand the arguments of opponents to conservation and identify their attempts to manipulate the process.

## 3. Working together

### *Conservation Action*

Many organizations, agencies, landowners, decision makers, and scientists are already working toward a wilder Southern Rockies with programs that complement and help implement the Wildlands Network Vision. Thus, an early step in developing an implementation plan is to catalog all the compatible conservation initiatives in the region. It is beyond the scope of this chapter to describe the thousands of regional, state, and local activities that work toward the Southern Rockies Wildlands Network Vision, but we will highlight a few examples later.

### SKY ISLANDS IMPLEMENTATION
#### NEW MEXICO

The Sky Islands implementation plan and action steps were developed in consultation with conservation groups, land users, academic experts, and government agencies. In 1999, the Wildlands Project and the Sky Islands Alliance hosted a three-day workshop at Rex Ranch near Tucson to discuss implementation tactics. Thirty participants including conservation campaigners, economists, media consultants, biologists, ranchers, outdoor recreationists, hunters and fishers, federal and state agency staff, and social scientists discussed implementation and conservation action in detail. Since then, various pieces of the Sky Islands puzzle have been put in place.

In October 2002, a day-long workshop brought over 250 individuals from broadly different areas of interest, including ranchers, government agencies, conservation groups (over 100 different groups), land trusts, universities, private researchers, scientists, private land owners, foundations, and interested members of the public. Some were new to the Wildlands Network concept and many are already working on their own piece of the puzzle. Now all the players realize how successful their cumulative effort has been. This gathering inspired and motivated both new and old participants, and demonstrated that working together toward a common vision will produce results.

Conservation action will continue in its vigorous and myriad forms, but the overarching vision embodied in this document provides a unifying force, which can strengthen existing efforts, create new initiatives, and build synergy. The Wildlands Project and Southern Rockies Ecosystem Project intend to convene watershed and regional meetings to facilitate flexible and creative methods of working together toward wildlands protection and restoration. The Wildlands Project and its partners conducted implementation planning of the Sky Islands Network through such meetings, making significant progress over the past few years in directly addressing the objectives of their vision.

A Wildlands Network Vision implementation is not something entirely new or conceptually difficult. Most of the steps necessary to implement the Southern Rockies Wildlands Network Vision are well known among the skilled conservation community practitioners. Of course, identifying the goals is simpler than achieving them. The scope and boldness of the goals may require a shift in perspective of how humans view Nature, and their place in it, before the goals can be reached.

Given the fluid nature of public policy and the dynamics of building common strategies among varied interests, there certainly will never be a lock-step action plan. It is possible that the cat will leap onto the table and scatter the puzzle pieces momentarily, or that some pieces are lost for a long time. There will be genuine differences in philosophy, strategy, and tactics that will require open dialogue, a willingness to consider different strategies, and a commitment to resolve the inevitable conflicts that arise when passionate people from different backgrounds seek a common goal.

The practical form of this Vision will change as new science and techniques are developed and as public policy opportunities arise. It may take decades or even generations to realize this Vision; however, we hope similar projects will band together for more effective action and innovative solutions will be shared more broadly, unleashing a powerful force for change.

### *Outreach*

The Wildlands Project and cooperators such as the Southern Rockies Ecosystem Project function through networks of people protecting networks of land (Soulé 1995). An outreach plan for the Southern Rockies Wildlands Network Vision will build connections within and between the networks of people, and develop and distribute important information among various constituencies.

Communication is critical to the plan's success. It is important that organizations and agencies know how their ongoing initiatives and programs and those of other organizations contribute to the implementation of the Southern Rockies Wildlands Network Vision. In order to understand how these efforts will affect the plan, it will be helpful for partner organization to understand that:

- The Southern Rockies Wildlands Network Vision has its foundation in conservation biology.
- Strategy and conservation action are crafted within the context of the Wildlands Network goals.
- The vocabulary (core wild areas, compatible use areas, linkages, connectivity, rewilding, *etc.*) is drawn both from science and from our vision for the future.
- Compatible programs and efforts promote and complement the overall Wildlands Network.

In addition to partnerships and active participation of many organizations and government agencies, successful implementation will require considerable public support and understanding of the Southern Rockies Wildlands Network Vision. Outreach and education must, at some point, focus on the larger public, media, and policy makers. Working with other groups will be vital, as each group will have different constituencies, connections, and abilities to reach different audiences. Distribution of the Southern Rockies Wildlands Network Vision through various media to interested parties will begin the process of public outreach, and partner organizations hopefully will incorporate concepts relevant to network implementation into their media and education plans. Through a network of people and organizations conducting education and outreach programs to a variety of audiences, we can all successfully protect and restore a network of wildlands in the Southern Rockies.

In the rest of this chapter we will look at opportunities for conservation action, highlight some compatible conservation initiatives, discuss social and economic facets of a Wildlands Network Vision and suggest monitoring methods to judge the success of implementation.

## 4. Conservation Action and the Wildlands Network Goals

In this section we suggest a few activities to illustrate the variety of actions one might do as part of an implementation plan. Some ecoregions are already into the implementation phase of their Wildlands Network Design/Conservation Vision; so we will share some of their success stories. Other regions, including the Southern Rockies, can look to ongoing efforts as initial "pieces of the puzzle" which can be important to a strong implementation plan.

Conservation action should be responsive to the objectives and goals of the Wildlands Network Vision, so we use that framework for this discussion. In many cases, a step will affect more than one goal, and such a synergy is important for effective action, but we will not attempt to describe all the possible relationships.

### *Goal One: Protect and Recover Native Species*

Efforts on behalf of native species should promote healthy ecosystems that provide for the needs of selected focal species and the many others species at risk, recovering and protecting them from further endangerment. See the species accounts in the Appendix for more details on the needs of Southern Rockies focal species.

Key activities in achieving this goal include the following:

- Advocate for wolf restoration and support existing wolf reintroductions in adjacent regions.
- Ensure successful lynx (*Lynx canadensis*) restoration.

- Explore restoration of the grizzly bear (*Ursus arctos*) to the large Wilderness Areas, especially in the San Juan Mountains.
- Protect existing native cutthroat trout (*Oncorhynchus clarki*) populations and expand restoration.
- Explore habitat protection for pronghorn (*Antilocapra americana*), bighorn sheep (*Ovis canadensis*), American marten (*Martes americana*), and beaver (*Castor canadensis*).
- Oppose unwarranted sport hunting such as prairie dog (*Cynomys* spp.) shoots and spring bear hunts.
- Encourage the Forest Service and the Bureau of Land Management to monitor Management Indicator Species and Sensitive Species and take necessary action to prevent decline of those species that are native.

## SUCCESS STORY FROM THE SKY ISLANDS WILDLANDS NETWORK SONORAN DESERT CONSERVATION PLAN

### PIMA COUNTY ARIZONA

The biggest threat to wildlife is loss of habitat; so many conservation efforts proactively focus on habitat protection and restoration. Nevertheless, some conservation plans are initiated as a response to species loss resulting from habitat destruction that has already occurred. In Pima County, the federal listing of the cactus ferruginous pygmy owl (*Glaucidium brasilianum cactorum*) caused officials to seek broader habitat protection measures through a multi-species habitat conservation plan.

An area of habitat for the owl, as well as other species of concern, was located in Tucson's primary new home construction zone. Facing legal battles with developers and the Fish and Wildlife Service over construction in this owl habitat, Pima County officials realized the time had come to develop a multi-species habitat conservation plan. The Sonoran Desert Conservation Plan (SDCP), scheduled for official adoption in 2003, identifies a core biological zone in which development will be reduced or eliminated, depending on protection requirements for included species. The SDCP also identifies less ecologically valuable lands where development can continue to occur, giving builders an opportunity to plan for the future without facing legal and permit challenges. The SDCP process includes species identification, mapping, and linkage determination, which led Pima County to seek input from the Wildlands Project and the Sky Island Alliance. The science from the Sky Islands Wildlands Network Conservation Vision was used to help justify and shape the SDCP and to reaffirm the county's selection of areas identified for protection of biodiversity and connectivity.

## BRINGING BACK THE LYNX

### COLORADO

The lynx's historic range in the Southern Rockies, based on historic sighting and specimen records, is thought to include the mountainous areas between south-central Wyoming and north-central New Mexico. Records indicate that the lynx was much more abundant at the turn of the century than at present. For example, 210 lynx were trapped on the Routt National Forest in 1916 alone (LCAS 2000). The last confirmed native lynx sightings in Colorado were in the mid-1970s in the Vail ski area (Halfpenny 1989).

Lynx are a focal species for the Southern Rockies and New Mexico Highlands Wildlands Networks, so the reintroduction program of Colorado Division of Wildlife (CDOW) is of great importance.

CDOW and FWS have released 129 lynx in Colorado during the past three years. Of these, 45 are known to have died and 20 are unaccounted for. Starvation, vehicle collision and gunshot wounds were the top known causes of death (Colorado Division of Wildlife 2003).

Survival rates have improved considerably and 16 lynx kittens were found in the spring of 2003, the first evidence of reproduction and a major milestone for the program. A public outreach campaign in the fall of 2002 led by Sinapu and Center for Native Ecosystems resulted in approval by the Colorado Wildlife Commission of a plan to release up to 130 more lynx in the state over the next four years.

## THE I-70 CORRIDOR: BREACHING THE BARRIER

### COLORADO

Animals, by nature, are characterized by the capacity for locomotion. They move across landscapes to meet their daily, seasonal, and lifetime needs. Because animals were created to be mobile, barriers in the landscape that restrict their movement pose hazards to both individuals and populations of animals. Human activities and developments are the leading threat to animal movement. Highways and development fragment the natural landscape, reduce the dispersal ability of animals, and impact ecosystem processes.

The effect of roads can be lessened. Closing roads, careful design and planning, and a variety of construction options can facilitate wildlife movement, such as fencing, underpasses, culverts, and overpasses.

In Colorado, underpasses and deer-proof fencing were constructed on I-70 for mule deer (*Odocoileus hemionus*) movement. They have proved fairly successful. About five wildlife crossings have been constructed in Colorado. In addition, the Colorado Department of Transportation (CDOT) has conducted research on animal movement patterns which they hope will guide future wildlife crossing structures.

*I-70 Corridor*

I-70 spans the state of Colorado and is known to be a major barrier to wildlife movement for many species of animals. To address this issue, CDOT analyzed I-70 for the most critical wildlife crossing areas using data and expert opinion. This map will be released with CDOT's Programmatic Environmental Impact Statement.

The Southern Rockies Ecosystem Project (SREP) also analyzed I-70 for likely crossing zones for several species of mammals. Species considered in this study were black bear, gray wolf, Canada lynx, mountain lion, elk, and mule deer. Locations of probable crossing by wildlife were derived from three types of data: areas where the highway bisects core suitable habitat, areas where highway crossings were recorded, and areas where wildlife dispersal modeling indicates an individual would be likely to cross the highway on its way between two areas of core suitable habitat.

The results are a combination of areas where the highway crosses suitable core habitat and areas of probable crossing predicted by the least cost path analysis (Figure 10.1).

SREP, CDOT, the Federal Highway Administration, the Wildlands Project, and The Nature Conservancy will be co-hosting a *Missing Linkages* conference in 2004 to prioritize the remaining wildlife linkages in the Southern Rockies and Colorado.

## Goal Two: Protect and Restore Native Habitats

Strictly protected, core wild areas are essential elements of a natural landscape with ecological integrity. These cores, along with connecting linkages, should afford permanent protection and opportunities for restoration of representative amounts of all habitat types. Wilderness designation, the most protective land designation available, is critical to expanding the amount of permanently protected land. National Conservation Areas, Forest Service regulations protecting inventoried roadless areas, Outstanding Natural Water Resource designation, and Wild and Scenic Rivers designation are among the many other tools that can be used.

Key activities in achieving this goal include the following:

- Keep existing roadless areas intact by completing roadless area inventories on US. Forest Service and BLM Lands.
- Defend areas from projects that would degrade habitat or create new travel routes.
- Design and pass Wilderness legislation.
- Work to close non-system routes, reduce overall road density on public lands, and stop expansion of illegal travel ways.
- Incorporate the recommendations of the Southern Rockies Wildlands Network Vision into future Forest and BLM management plan revisions.
- Ensure the best application of the Wildlands Network Vision recommendations for Forests and BLM Resource Areas with recently completed management plan revisions.
- Engage in restoration projects for degraded riparian and terrestrial areas.
- Ensure that logging activities are ecologically sustainable and lead to improvement of habitat in compatible-use areas.
- Promote responsible recreation policies that prevent further loss or degradation of habitat.

### WINNING WILDERNESS STRATEGIES
### SOUTHERN ROCKIES ECOREGION

*From forests to canyons*

Following the 1993 designation of 260,000 ha of Wilderness in the National Forests of Colorado, the Colorado Wilderness Network took on the job of inventorying and identifying 640,000 ha of BLM potential Wilderness. Approximately half of this proposal is in the Southern Rockies Wildlands Network Vision, including Red Cloud-Handies Peak and Troublesome, each more than 40,000 ha. Representatives Diana DeGette and Mark Udall introduced the citizens' proposal as legislation in 1999 with the support of more than 350 businesses, organizations, and local governments.

The Biodiversity Conservation Alliance is inventorying Wyoming BLM lands, which are especially threatened by new oil and gas wells, coal mining, and coalbed methane wells. The New Mexico Wilderness Alliance has 48 proposed BLM Wilderness Areas in their draft plan, several of them in the Southern Rockies ecoregion.

*Stand-alone designations*

In the current political climate, statewide proposals are difficult to pass. At the risk of picking off the crown jewels, conservation groups supported a number of stand-alone Wilderness bills. Spanish Peaks, a "further study area," was the first Wilderness designated since 1993. The Colorado Canyons National Conservation Area encompasses 48,896 ha along the Colorado River, including the 30,574 ha Black Ridge Canyons Wilderness with its spectacular arches and wildlife habitat. Black Ridge Canyons Wilderness is also contiguous to the Colorado National Monument, much of which is wild because of rugged topography. In a similar fashion, Gunnison Gorge National Conservation Area is just downstream from the Black Canyon of the Gunnison National Park, and includes the Gunnison Gorge Wilderness. Recent legislation designated The Great Sand Dunes as a National Park and included authorization to purchase adjacent land to add to the Park. This creates a contiguous protected complex of the Sangre de Cristo Wilderness, Great Sand Dunes Wilderness, the expanded National Park, and The Nature Conservancy's 100,000-acre Medano-Zapata Ranch. Finally, the James Peak legislation, after vigorous debate and many compromises, added 5,666 ha of Wilderness in 2002, bringing the Colorado total to 41 Wilderness Areas which cover 1,369,600 ha, about 5% of the state.

*Subregional proposals*

As the Southern Rockies Conservation Alliance groups complete wilderness evaluations for some 2,240,000 ha of roadless land in Colorado and southern Wyoming, many subregional proposals will likely emerge. For example, the Central Colorado Wilderness Coalition is looking at roadless areas inventoried by Upper Arkansas and South Platte Project in Custer, Fremont, Teller, Park, El Paso, Douglas and Jefferson counties as the basis for a Wilderness proposal for their Congressional District.

## PREDATOR-FRIENDLY RANCHING
## GRAZING STRATEGIES FROM AROUND THE WEST

Innovative ranchers in the West are establishing predator friendly herd management and creating markets for their products. Will and Jan Holder, who ranch in the Apache NF (AZ), are producing and marketing Ervin's Natural Beef. Jim Winder, of Heritage Ranches in Deming, New Mexico, produces and markets Wolf Country Beef. Both have publicly supported recovery of the Mexican wolf (*Canis lupus baileyi*). Predator Friendly, Inc., based in Montana, is a coalition of sheep producers, environmentalists, scientists, and entrepreneurs who are creating markets for wool products produced without killing predators. These efforts expand the "green market," allowing environmentally concerned consumers to put their dollars where their ideologies are, and increased income helps the ranches stay in production.

Thoughtful modification of traditional grazing practices can greatly reduce predation on cattle. For example, Jim Winder has no problem with coyote predation on calves. He manages his cows so they calve in April when coyotes have abundant natural food. Ranchers who claim problems with coyotes often manage their herds so that their cows calve in January when coyotes have very little natural food.

*Depredation Compensation* -Defenders of Wildlife has established a fund to pay for any livestock killed by wolves. Between August 1987 and October 2002, Defenders paid 210 ranchers a total of $242,097 for livestock depredations by wolves in Montana, Idaho, Wyoming, Arizona, and New Mexico. Their goal is to shift economic responsibility for wolf recovery away from the individual rancher and toward the millions of people who want to see wolf populations restored.

*Grass Bank-* The original Grass Bank is located in southern New Mexico/Arizona, largely on private lands, but affects portions of nearby National Forest, BLM, and state trust lands. This Grass Bank was started on the 119,891 ha Gray Ranch. Participating ranches pool their herds and move them from ranch to ranch. Long periods of rest after intense grazing allows grasses to recover and provide the fine fuels needed to restore fire. Fire reinvigorates the fire-adapted grasses and the species that depend upon them, and kills the invasive shrubs which out-compete perennial grasses. Participating ranchers also donate a conservation easement on their ranch to prevent development.

Grass Banks are not a panacea for all ecological ills on the western rangelands, and should not be viewed as an alternative to buy-out campaigns like the National Public Lands Grazing Campaign. Grass banks on public lands are problematic in that they perpetuate livestock grazing on public lands. The aim of Grass Banks and similar initiatives should always be ecological restoration and sustainability, and the result of sustainability can often mean greater economic stability for the rancher.

*Voluntary Retirement Option* -Andy Kerr (1998a and b) has proposed the Voluntary Retirement Option. He argues: "It would be easier—and more just—for the federal government to fairly compensate the permit holders as it reduces cattle numbers. Since the government spends substantially more than it receives for grazing, in a few years the savings realized by reducing livestock numbers can pay for the compensation." He proposes changing federal law to allow permit holders to choose not to graze their allotment, to sell or donate their allotment back to the agency which would retire it, to allow an environmentalist, state fish and wildlife agency or private conservation organization to compensate a permittee for retiring his or her allotment, and various other mechanisms that would provide fair compensation to the rancher.

*Buyout*- In some cases, the best solution for conservationists is to simply buy a ranch or base property with a federal or state grazing lease.

*State Lease Bidding*- Several conservation groups, including Forest Guardians in New Mexico, have outbid ranchers for state land grazing permits. As of August 2000, Forest Guardians had 1,400 ha of New Mexico State Land under lease, along four different rivers. These lands are no longer being grazed, and volunteers are doing ecological restoration, such as planting of willow and cottonwood, to heal past grazing wounds.

### Goal Three: Protect, Restore, and Maintain Ecological and Evolutionary Processes

Some of the steps necessary to permanently protect and restore ecological and evolutionary processes have been discussed in other sections. Fully functional natural processes are necessary for healthy ecosystems and native species.

Key activities in achieving this goal include the following:

- Identify eligible free-flowing river segments in the Southern Rockies and secure Wild and Scenic River designation or other protective legislation.
- Advocate for water conservation and increased capacity of current reservoirs over new dam building and for management of reservoirs to simulate natural seasonal flows, gradients, and temperatures.
- Support carefully designed ecological restoration of fire-dependent forest types and a return to natural fire regimes. Steps in that direction may include thinning of small diameter trees followed by prescribed fire in the wild-urban interface.
- Encourage the Forest Service, BLM, and state agencies to address the root causes of extreme outbreaks of diseases or insects, rather than the symptoms.
- Improve in-stream flow protection.

#### TOWARD SUCCESSFUL RESTORATION OF HYDROLOGICAL CYCLES ON THE SOUTH PLATTE RIVER
#### DENVER, COLORADO

Of the 18,080 km of rivers in the National Wild and Scenic Rivers System, only 344 km are in the Southern Rockies ecoregion. Wild and Scenic designation gives permanent protection from impoundments, but this often brings fierce opposition.

For the South Platte River, a multi-year consensus process involving water utilities, conservation groups, the Forest Service, and county agencies created the South Platte River Protection Plan (SPRPP). In lieu of Wild and Scenic designation the SPRPP provides:

- A commitment to not build any water works facilities in Cheesman and Eleven Mile Canyons;
- Flow management that includes temperature goals, ramping outflows, minimum stream flows and other major improvements to the stream flows;
- Protection for outstanding scenic, recreation and wildlife values;
- Water quality improvement initiatives;
- A 1 million dollar endowment from local governments and water providers;
- Withdrawal of applications for conditional storage rights for the Two Forks reservoir site and seeking alternatives to development of Denver's right-of-way;
- Public management and oversight mechanisms.

This proposal is now awaiting a Forest Service decision. While not perfect, the SPRPP likely is a win-win proposal that helps restore some of the natural hydrological cycles, benefits the adjacent wildlife habitat and provides flexibility for management of a river that carries 60% of metro Denver's water supply.

### Goal Four: Protect and Restore Landscape Connectivity

Protection of the land from further fragmentation, and restoration of functional connectivity for all species native to the region, are essential. Much of the land critical for large-scale connectivity is in private hands, ranging from the large intermountain parks to the bottomland in nearly every river. Therefore finding common ground and working with private landowners and land trusts will be important. Many ways a private landowner can contribute to biodiversity preservation are discussed in the economic section at the end of this chapter.

Key activities in achieving this goal also include the following:

- Hold a Southern Rockies *Missing Linkages* conference to identify key linkages and prioritize them so that other groups can mobilize on their behalf.
- Address the barriers posed by major transportation cor-

ridors such as Interstates 70 and 80, and advocate for wildlife crossings, including land bridges, underpasses, and similar mitigation.

- Develop legislative or administrative protection standards for linkages and compatible-use areas on federal lands. A workshop should be held with representatives from conservation groups, science, land users, decision makers and agencies to develop preliminary guidelines and standards.
- Cooperate with conservation-minded land owners on management plans compatible with the Southern Rockies Wildlands Network Vision.
- Work with county and state planners to support and guide "open space" acquisition and protection efforts so that they prioritize key linkages and meet the management guidelines for Southern Rockies Wildlands Network Vision.
- Work with tribes and pueblos to identify lands important for preservation and restoration in ways responsive to biological values and their cultural context.
- Protect all National Forest and BLM roadless areas of 400 ha or more. Encourage the Forest Service to enforce the existing Roadless Initiative and to oppose efforts to rescind it. Declare a moratorium on road-building and logging in National Forest roadless areas of 400 ha or more.
- Work with land management agencies on travel management plans that incorporate road density standards, designated routes, adequate signs and enforcement, and that emphasize protection of the land, especially riparian zones.
- Advocate for responsible recreation that respects the land, forgoes entering sensitive or roadless areas, and provides for quiet backcountry use.
- Suggest land exchanges based on the Southern Rockies Wildlands Network Vision.
- Assess land exchanges that aggregate land within one agency, ensuring that the biotic community benefits and fair market value is obtained for the exchange.

## MISSING LINKAGES
### CALIFORNIA

In California, regional work on wildlands protection and habitat linkages is proving highly successful. The California Wildlands Project, a project of the California Wilderness Coalition (CWC), has completed Wildlands Reserve Designs for three ecoregions in the state including the south coast. In November 2000, more than 200 land managers and conservation ecologists working in the South Coast region participated in the Missing Linkages Conference where they identified and mapped critical wildlife linkages throughout the state.

Since then, the California Wildlands Project, The Nature Conservancy, partner organizations, and State and Federal agencies, have been integrating a Wildlands vision into conservation work around the state. The South Coast Wildlands Project (SCWP) transformed a wildlands vision into an organization, thanks to the dedication and hard work of people in that region.

Sixty critical linkages were identified within the South Coast Ecoregion at the Missing Linkages Conference, plus seven major linkages to other ecoregions, and two cross-border linkages to Baja California, Mexico. SCWP conducted a formal evaluation of these 69 linkages based on biological irreplaceability and vulnerability to urbanization. This process identified 15 linkages of crucial biological value that are likely to be irretrievably compromised by development projects in the next decade unless immediate conservation action occurs.

SCWP is currently developing on-the-ground conservation designs for these 15 critical linkages via two one-day workshops on each linkage. The outcome will be a detailed comprehensive report describing threats and conservation opportunities for each of the 15 critical yet unprotected linkage areas. This report will guide the newly formed coalition's efforts to protect each critical linkage.

Eleven project partners have participated in this endeavor, including county, state, and federal agencies and non-profit organizations. Four of these partners have provided funding.

## SNOWMOBILES BANNED ON TRAIL RIDGE ROAD
### ROCKY MOUNTAIN NATIONAL PARK, COLORADO

Snowmobile use interrupts connectivity, crushes subnivean habitat, and stresses animals at a time when they are particularly vulnerable. Rocky Mountain National Park is the third busiest park for snowmobile use, exceeded only by Yellowstone and Voyageurs. Park Service staff worked with snowmobilers, the local community, and environmental groups to develop the plan. Closing Trail Ridge Road will protect important winter habitat for bighorn sheep and elk, as well as potential habitat for lynx and wolverine. As part of the plan, the North Supply Access Road, which crosses the southwest corner of the Park, will remain open to snowmobiling. Grand Lake, the Colorado Snowmobile Association, the Colorado Mountain Club, and the Wilderness Society, among others, support this decision.

## *Goal Five: Prevent the Spread of Exotic Species*

Exotic species are now the second greatest threat to North American biodiversity, behind habitat loss, and thus it is necessary to combat bioinvasion.

Key activities in achieving this goal include the following:

- Eliminate or control exotic species, including bullfrog (*Rana catesbeiana*) and non-native fish; non-native grasses, especially those that change fire regime, and other invasive plants such as spotted knapweed (*Centaurea biebersteinii*) and purple loosestrife (*Lythrum salicaria*); species that alter soil chemistry; and exotic pathogens, such as white pine blister rust (*Cronartium* spp.).
- Develop a strategy that prevents the planting/release of known invasive exotics on state and federal lands.
- Lobby USDA to adopt effective controls on imports of foreign wood including a ban on raw log imports and establish a monitoring system to detect introductions of forest pests.
- Restrict activities that cause soil disruption and that introduce exotics in sensitive areas, which activities include off-road vehicle driving, livestock grazing, and road and powerline construction.

### NON-NATIVES VS. CUTTHROAT TROUT
#### REMOVAL OF EXOTIC SPECIES IN COLORADO

In Colorado, a litany of introduced non-native sport fish exist: rainbow trout (*Oncorhynchus mykiss,* in the 1880s), brown trout (*Salmo trutta,* in the 1890s) and brook trout (*Salvelinus fontinalis,* in the late 1800s). These non-native trout interbreed with and out-compete the native greenback (*O. c. stomias*), Colorado (*O. c. pleuriticus*) and Rio Grande (*O. c. virginalis*) cutthroats. These introductions and over-harvesting have decimated native populations. In the presence of non-native trout, reintroductions of natives are expensive and complicated.

One project for Turkey Creek south of Colorado Springs involved killing all brook trout above Monkey Falls, a natural barrier that will prevent later upstream migration of the brookies, prior to reintroducing greenback cutthroats. This one step alone involved engaging certified pesticide operators, determining days of optimum water temperature, placing a series of drip applicators, monitoring toxicant levels throughout the stream segment, treating the outflow at a detox station and establishing an emergency detox station further downstream, monitoring sentinel fish in live cages below the detox station, and many person-hours of planning and implementation. One year later, if all goes well, a batch of greenbacks will be released into the clean stream.

## *Goal Six: Prevent or Reduce the Further Introduction of Pollutants*

Clean up polluted lands, waters, and air that harm native biodiversity.

Key activities in achieving this goal include the following:

- Prohibit oil and gas well development in ecologically sensitive areas.
- Clean up and eliminate the release of toxic, long-lasting chemicals like dioxin, mercury, and PCBs to the air, water, and land.
- Create mass transit systems that can adequately substitute for cars, thereby reducing traffic congestion, global warming, and air pollution.
- Promote the development of alternative energy sources, such as wind and solar power, both of which have a promising future in the Southern Rockies, particularly in New Mexico.
- Promote the development of more efficient, cleaner fuels and fuel cell technology.
- Advocate against new nuclear facilities and ensure current nuclear waste is safely and securely stored.

### SIERRA CLUB'S ENERGY CAMPAIGN
#### ROCKY MOUNTAIN CHAPTER

The Rocky Mountain Chapter Sierra Club is working throughout Colorado to promote energy generation and efficiency that minimize harmful impacts on public health and the environment while improving the state's economy and security. The Energy Campaign's goals are to meet 20% of our electricity needs through clean, renewable sources and become 20% more efficient in energy consumption by 2020. Currently, Colorado gets almost all its electricity from coal, which is associated with air pollution, global warming, and health problems. To keep up with growth, utilities want to meet future electricity demands with coal and natural gas. A few counties in the state have embraced renewable energy and energy efficiency, and Denver will soon make long-term decisions regarding the renewal of its franchise agreements with Xcel Energy, the largest electrical provider in Colorado. If significant renewable energy management provisions are included in the final agreement, this will have implications far beyond Denver. We will call on Xcel Energy to become a leader in 21st century energy technologies by providing Denver residents with electricity from clean, Colorado-based energy sources.

## 5. Compatible Conservation Initiatives

As mentioned at the beginning of this chapter, implementation of the Wildlands Network Vision through concerted conservation action is only possible through the joint efforts of hundreds, perhaps thousands, of people, organizations, and agencies. The Southern Rockies Ecosystem Project and the Wildlands Project will identify complementary efforts, facilitate coordination within the Vision framework, and initiate some direct efforts.

We offer a few activities from the non-governmental and private sectors below to illustrate the scope of possible action. However, this is only the tip of the iceberg. Given the vigorous conservation community in the Southern Rockies region, the growing interest in private lands conservation, and the plethora of town, county, state and federal programs, one might imagine a reference volume of hundreds of pages. We apologize to groups and projects that we did not include, and we invite readers to help create a comprehensive list by sending a brief description and contact information, especially on more local projects. We also note that these projects are not necessarily associated with the Southern Rockies Wildlands Network Vision, and many pre-date it. These groups are pursuing their own goals under their own direction, but we recognize their valuable contributions to realizing the overall network goals and objectives.

### *Large Landscape Conservation Planning*

#### *New Mexico Highlands Wildlands Network and Heart of the West Wildlands Network*

These are the adjacent Wildlands Network Designs whose conservation visions for northwestern New Mexico and central Wyoming/northeastern Utah will protect and restore biologically critical wildlands and reconnect them with wildlands in adjacent regions. These use the same principles of conservation biology as applied to network design and follow the three track approach of representation, special elements, and focal species planning to creating specific Wildlands Network Designs for their planning areas. Both plans overlap the Southern Rockies Wildlands Network, and we regularly coordinate our efforts (Foreman et al. 2003, Jones et al. 2003).

#### *The Nature Conservancy*

TNC recently released *The Southern Rocky Mountains: An Ecoregional Assessment and Conservation Blueprint v. 1.0* (Neely et al. 2001). The Executive Summary states, "A proactive approach to conservation is needed to prevent future federal listings, extinctions and extirpations of species, and further losses of communities and systems." This ecoregional assessment is a timely first step toward addressing conservation needs of the ecoregion's biodiversity. They used 148 carefully defined terrestrial and aquatic ecological systems, as well as 79 rare plant communities, 177 plants and 206 animals as their conservation targets. The assessment presents a "portfolio" of conservation areas – 140 in Colorado, 23 in Wyoming, and 13 in New Mexico – which are priority for protection. TNC's approach is complementary to the Southern Rockies Wildlands Network Vision which focuses more on large wild cores, wide-ranging mammals, carnivores, and other major linkages across the landscape.

#### *Predator Conservation Alliance*

This group is facilitating the Northern Plains Conservation Network, which is partially contiguous with the Wyoming part of the Southern Rockies. Their vision includes restoration of some areas of the northern Great Plains to an ecosystem dominated by large native mammals, and transected by free-flowing rivers with healthy populations of native fish species. These areas are large enough to restore wildlife populations, traditional migration patterns, and other natural processes. Given the declining agricultural base, existing land ownership patterns, and a shrinking human population, wildlife restoration efforts for species such as prairie dogs, black-footed ferrets, and mountain plovers (*Charadrius montanus*) will help rebuild and diversify regional communities and economies.

### *Public Lands and Open Space*

Hundreds of conservation organizations are at work in the region to protect public lands and open space. Many are very local, taking responsibility for a particular place, such as Friends of the Eagles Nest, Amigos Bravos, or the Ridgeway Ouray Community Council. There are forest watch groups for each National Forest (see Citizens Alternatives below);

and others, such as Colorado Environmental Coalition, Colorado Wild, Western Colorado Congress, New Mexico Wilderness Alliance, and Biodiversity Conservation Alliance in Wyoming, focus on statewide activities. Some focus on special constituencies, such as the Quiet Use Coalition or the Backcountry Skiers Alliance. In addition, national conservation organizations, such as Audubon Society, the Sierra Club, the National Wildlife Federation, and The Wilderness Society have thousands of local members, regional offices, and active groups in all three states. All defend important areas from further degradation and work proactively toward permanent protection for wildlands and wild rivers, both through their organizations' programs and through joint campaigns with other groups. We highlight just a few below.

***Southern Rockies Conservation Alliance (SRCA)***

In 2003, the Colorado Wilderness Network (CWN) and the Southern Rockies Forest Network (SRFN) merged to form a powerful force for public lands protection. This coalition of local, regional and national conservation groups protects wilderness quality lands, conserves and restores biodiversity, and ensures "responsible management" of motorized recreation. Each network had significant accomplishments prior to the merger, including successfully getting diverse constituencies working together toward common goals.

CWN has long been the driving force for citizen support of major Wilderness legislation, including the 1993 forest legislation and the current BLM legislation. Inventories of more than 640,000 ha of BLM land documented wilderness qualities and engaged activists. Concerted outreach to businesses, organizations and local governments has been successful. For example, the campaign to protect Roan Plateau in west-central Colorado secured resolutions from all the town councils in Garfield County — Rifle, Silt, Parachute, Glenwood Springs, New Castle and Carbondale — asking the BLM to allow drilling at the Roan's base, but preserve 13,200 ha on its top and sides for wildlife and "primitive recreation."

SRFN, an alliance of 26 organizations in southern Wyoming and Colorado, was formed in 1999 to bring together groups interested in forest lands protection. Three campaigns – roadless area protection, biodiversity conservation, and responsible motorized recreation have produced significant practical results. In particular, member groups are inventorying approximately 2,240,000 ha of roadless lands on the Arapaho-Roosevelt, Medicine Bow-Routt, White River, Grand Mesa-Uncompahgre-Gunnison, San Juan, Rio Grande, and Pike-San Isabel National Forests. SRFN provided some funding for inventory staff, and digital cameras and GPS units; and many groups held volunteer mapping weekends or deployed volunteer teams. The Southern Rockies Ecosystem Project designed and maintains the regional roadless area database and provided maps and GIS services. By the end of 2002, 76% of the inventory was complete, and there are nearly 3,000 route and 12,5000 photo records in the database, with additional data input to be completed. These roadless areas are evaluated for their wilderness qualities and should become the basis of the next Wilderness designations.

With the merger, SRCA will build on the strengths of both coalitions and mount even more effective protection campaigns across the region.

***New Mexico Wilderness Alliance (NMWA)***

NMWA is dedicated to the protection, restoration, and continued enjoyment of New Mexico's wildlands and Wilderness Areas. NMWA has a major fieldwork effort underway to update citizens' Wilderness Area proposals for BLM lands in New Mexico, many of which are within the New Mexico Highlands planning area. NMWA is also beginning fieldwork for National Forests in New Mexico within the NM Highlands Wildlands Network. NMWA incorporates conservation biology principles in the creation of Wilderness Area proposals, and supports the vision of a connected network of people working to protect networks of wildlands in New Mexico and beyond.

***Citizens' Alternatives for Forest Plan Revisions***

Various groups in the Southern Rockies participated in public lands management through the forest plan revision process. Several years ago citizens' alternatives were presented to the Rio Grande, Routt and Arapaho-Roosevelt National Forests. They were incorporated by the agency into the range of alternatives in the Draft Environmental Impact Statement. They broadened the scope of the final selected alternative, although the Forest Service fell far short of the strong protection called for by citizen groups.

Biodiversity Conservation Alliance's *Keep the Med-Bow Wild* (Medicine Bow NF), the White River Conservation Project's *Bring Wilderness Home* (White River NF), Upper Arkansas and South Platte's *Wild Connections* (Pike-San Isabel NF), and the San Juan Citizens Alliance's *Wild San Juans* (San Juan NF) address current National Forest plan revisions. Groups such as the San Luis Valley Ecosystem Project (Rio Grande NF), Citizens for Arapaho-Roosevelt, Western Slope Environmental Resource Council/High Country Citizen's Alliance (GMUG), and New Mexico Wilderness Alliance (Carson and Santa Fe NFs) are crafting updated and future citizens' conservation alternatives for their Forests. All of these are conservation biology-based proposals that call on the Forest Service to protect wild roadless areas, usually with permanent protection as Wilderness, and to fulfill their responsibility to protect biodiversity and native species. The organizations listed above are merely the coordinators, equally important are the many local groups and individual citizens involved in each of these.

*Local conservation groups*

At an even more local level, Sierra Clubs, Audubon Societies, and hundreds of conservation councils contribute pieces to the puzzle of wildlands protection. We highlight just two examples:

Aiken Audubon Society in Colorado Springs created the Great Pikes Peak Birding Trail to guide birders around the nearby mountain locations and onto the plains. A color brochure with map and a web site with detailed descriptions for each location are easily available to the public. They participated in Audubon's Important Bird Area (IBA) program, which is a national, voluntary, non-regulatory method of protecting habitat vital to bird migration, breeding, and wintering, and included the IBAs in the Birding Trail.

Sierra Club, with Trapper's Lake Group leading the way, successfully litigated the clean-up of the Hayden power plant in the Yampa valley, a coal-fired operation that accounted for virtually all of the area's emissions of sulfur and nitrogen oxides — the components of acid rain. The acidity of the snow pack in nearby Mt. Zirkel Wilderness tested higher than any of the other 200 federally monitored sites west of the Mississippi. Sierra Club sued the owners of the Hayden plant for more than 17,000 violations of its Clean Air Act permit. The Federal Court ordered Hayden's owners to spend $130 million to upgrade the power plant, and pay $2 million in fines to the U.S. Treasury and another $2 million to environmental projects in the Yampa Valley. The Club directed the money to purchase a conservation easement at the base of a proposed major ski resort complex; and $250,000 was directed to the Routt County Board of Health to provide rebates to residents converting from wood-stove heat to natural gas or propane.

## Species Recovery and Protection

Turner Endangered Species Fund (TESF) is undertaking several projects designed to ecologically reconnect landscapes and habitats that extend far beyond the borders of Turner's three ranches located in New Mexico. TESF operates the Arrmendaris and Vermejo Ranches in the New Mexico Highlands region. Vermejo has a captive rearing facility for black-footed ferrets, and a black-tailed prairie dog (*Cynomys ludovicianus*) colony restoration project, and is developing a Mexican wolf experience center to acclimate wolves prior to reintroduction to wild. Vermejo is also restoring greenback trout and fire regimes.

Southern Rockies Wolf Restoration Project is a coalition of regional and national conservation organizations whose mission is to ensure the restoration of wolves to their ecological role in the Southern Rocky Mountains. One of the best locations to restore wolves in the Southern Rockies, as shown by focal species, habitat, and prey analyses, is in the four core wolf habitat areas identified by Carroll et al. 2003 in western Colorado and northern New Mexico (see the focal species account for wolves in Appendix 1).

Sinapu works to restore wolves, grizzly bears, wolverines, lynx, and river otter (*Lontra canadensis*) to their rightful place in the wild. While recognizing that loss of habitat threatens these species, Sinapu believes that the greatest threat of all continues to be human intolerance of predators and misunderstanding about the vital ecological role these animals play in Nature. In addition to reintroduction and direct protection, Sinapu seeks a proper balance between the expansive tracts of unroaded wildlands needed for grizzly bears and the wild places for species that can thrive in areas with some intrusion.

The Center for Native Ecosystems (CNE) is an environmental advocacy group dedicated to conserving and recovering native and naturally functioning ecosystems in the Greater Southern Rockies and Plains. CNE uses the best available science to forward its mission through participation in policy and administrative processes, legal action, public outreach and education. Their current projects include working with other groups to list the Colorado River cutthroat trout under the Endangered Species Act and seeking protection for Utah (*Cynomys parvidens*), white-tailed (*C. leucurus*), and black-tailed prairie dogs.

The Colorado Grizzly Project was established with the

goal of reintroduction of the grizzly bear to Colorado's wild areas. They educate the public about the vital role that grizzlies fulfill in healthy ecosystems, conduct field research to determine if a relict population of grizzly bears still exists in the San Juan Mountains, and intervene on behalf of the grizzly whenever its well-being is threatened.

Grizzly Bear *(Ursus arctos)*

The Colorado, Utah and Wyoming Councils of Trout Unlimited participated with fish and wildlife agencies from the three states and the Ute Indian Tribe, the U.S. Fish and Wildlife Service, the U.S. Forest Service, the Bureau of Land Management and the National Park Service in a long-term strategy to restore imperiled Colorado River cutthroat trout in the central Rockies. "It was important to us that the plan looks at range-wide recovery and not just museum piece populations in a few watersheds," said Tom Krol, Chairman of TU's Colorado Council.

### *Private and tribal lands*

Taos Pueblo-Blue Lake Tribal Wilderness: The Carson National Forest, created in 1906, carved away thousands of hectares of what was originally Taos Pueblo land. This land, viewed as sacred by the native Indians, included Blue Lake, a vital religious shrine in the Taos Pueblo religion. For many years, the Taos Pueblo lobbied and organized to have their claim to these lands recognized, as this sacred site became increasingly degraded due to overuse and vandalism. In 1970, President Nixon signed a bill to return 19,200 ha of the Carson National Forest, including Blue Lake, in trust for the sole use of the Taos Pueblo. Since that time, the Taos Pueblo has shown exemplary land stewardship and continues to conserve the area as a tribal wilderness, allowing restricted access to the tribe for subsistence and ceremonial uses.

We hope this brief introduction stimulates the imagination and encourages us to look at all the pieces, from the largest landscapes to the smallest creatures.

## 6. Socioeconomic Benefits and Opportunities

While we believe that Nature has its own intrinsic value and that humans have a responsibility to protect biodiversity, we can also describe the economic and social values of Nature and Nature's services. This is certainly not an exhaustive report on the topic; in fact, an in-depth study of the economic issues surrounding wildlands network design would be helpful. There are other efforts not cited here, such as the extensive body of work by Tom Powers at the University of Montana, The Wilderness Society, and others.

For human inhabitants of a wildlands network planning area, the socioeconomic implications of wildlands protection represent both new opportunities and new challenges. These include rethinking the ways in which the land is used and managed, making commitments for a better quality of life, recognizing the importance of "ecological services" provided by healthy ecosystems, creating a foundation for long-term economic stability by addressing new markets, products, and services, and taking advantage of programmatic incentives. There is a critical role for information sharing and education about the many socioeconomic benefits and opportunities made available through implementation of the Wildlands Network Vision.

### *Rethinking Land Use and Management*

A necessary first step to realize the benefits and opportunities of nearby wildlands networks is for every individual, family, and community to rethink their approaches to land use and land management. Historically, our approach to western lands has been one of resource extraction — logging old-growth trees, mining valuable minerals, trapping and hunting wildlife, and often exporting those resources outside the region. Even farming and ranching, through inappropriate plowing or overgrazing, have often led to erosion and topsoil loss or to permanent loss of desirable vegetative cover.

A new approach requires viewing the entire landscape and natural world as a living resource where the social and economic health of the individual and the community is inextricably tied to the health of the land. Uses of land by individuals will need to shift away from unsustainable extractive uses, and both private and public land management will need to shift toward rewilding, restoration, and sustainable, holistic landscape health.

## *Social and economic benefits*

### *Quality of Life Benefits*

Quality of life issues are a concern of every segment of society, particularly in rural areas where economic pressures and impending residential development may lead to major lifestyle changes. Thousands of families in the region are affected by uncertainty due to "boom and bust" economic cycles, increasing traffic congestion, subdivision encroachment on open space, air and water pollution, and fear that we are destroying the natural world for future generations.

Wildlands conservation can be an effective, ethical response to many of the threats that reduce quality of life for residents of the Southern Rockies states. For many years, protecting native ecosystems has been a recognized tool for improving the well being of sensitive species, but now it is also recognized as an effective way to ensure the quality of life of current and future generations of humans.

Social benefits and opportunities provided by nearby wildlands networks include aesthetic benefits, stabilization of local economies, and educational, cultural, and recreational opportunities. Although, we must guard that our use does not destroy those very things that foster the benefits.

### *Ecological Services Benefits*

Healthy ecosystems provide a complete life support system for all species, including humans. This life support system provides "ecological services" that make life possible and give it meaning. Research sponsored by the National Science Foundation lists 17 categories of services with an estimated value worldwide of between $16 and $54 trillion per year (Constanza et al. 1997). These benefits are provided to all individuals and communities within and adjacent to a wildlands network:

- Air we breathe is filtered and oxygenated by plants.
- Water we drink is purified by wetlands.
- Food we eat is grown in soils fertilized and renewed by ecosystem processes.
- Climates are tempered and made livable by the effects of regional ecosystems.
- Floods, droughts, and fires are mitigated by intact ecosystems.
- Pollination of beneficial plants and dispersal of seeds is assured.
- Beneficial byproducts are provided, such as antitoxins and nutrients.
- A wild, healthy genetic pool of plants is maintained in healthy ecosystems.
- Spiritual and intellectual stimulation is provided by undisturbed natural beauty.
- Carbon is stored in healthy forests, which is increasingly important as rising atmospheric carbon dioxide levels, due largely to fossil fuel burning and deforestation, cause global overheating.

### *Economic Benefits and Opportunities for Communities and Families*

Ecological services provide not only our life support system, but also economic benefits to communities (Constanza et al. 1997). As the dollar value of ecological services becomes better documented, protection of ecosystems will play an increasingly important role in economic decisions. Economists have developed a total valuation framework to assess the extent of all goods and services in wildlands. This compares wilderness values to opportunity costs. In contrast, a financial analysis only considers costs and benefits at market values, and often just short-term. This is inadequate, as many wilderness benefits do not have a market value. Such short-term practices have led corporations to externalize costs by passing them to society as a whole (*e.g.* the cost of polluted water from sedimentation that follows clear-cutting forests). An economic analysis should account for the non-priced benefits and costs as well as the externalized costs that are passed to society. For more detail, see The Wilderness Society website and the work by Pete Morton included therein.

**YELLOWSTONE WOLVES AND THE ECONOMY**

The Yellowstone wolf recovery program provides an example of economic benefits from wild Nature. "... wolf restoration has benefited local economies by bringing in more tourist dollars. Since wolves returned to Yellowstone National Park in 1995, the region has a seen a $10 million increase in economic activity, indicating that wolves are clearly having a positive impact on the economy of the greater Yellowstone area. Moreover, US Fish and Wildlife Service studies project that the wolf reintroduction program will continue to attract more park visitors, eventually bringing an additional $23 million annually to Yellowstone. Wolf recovery in Yellowstone has introduced a new source of revenue into the area. Visitors to the park now rank the wolf as the number one animal they come to see, thereby creating new demand for near-by lodging, guided wolf-watching tours and a variety of wolf-related merchandise" (McNamee 1997).

Direct economic benefits are possible because communities near protected conservation areas offer a clean, healthy, highly desirable lifestyle that invites long-term economic investment from those who value natural areas. Protection of wildlands and surrounding compatible-use areas reduces threats to quality of life and creates an attractive foundation for sustainable economic growth.

While economics plays a role in conservation, we offer a caveat. Our culture has traditionally viewed the idea of constant economic growth as necessary for improvement. Much of this economic growth is based on a culture of consumption. This is an idea that cannot persist if we are to live within the limits of our finite system. As Hannah Arendt (1957: 253) stated, "Under modern conditions, not destruction but conservation spells ruin because the very durability of conserved objects is the greatest impediment to the turnover process, whose constant gain in speed is the only constancy left wherever it has taken hold."

### *Economic Incentives for Private Landowners*

Approximately 70% of the land in the United States is privately owned. Even in the Southern Rockies, where a large percentage of the land is in federal ownership, 37.6% of the region is in private ownership (Shinneman et al. 2000). Conservation on public lands has been discussed throughout this document, so we would like to highlight private lands conservation here. These private lands are critical to biodiversity because they often encompass riparian and wetland habitats and lower elevation areas. In addition, private lands are often strategically located between federal land holdings and serve as movement linkages for wildlife.

#### *Land Protection Tools*

Many tools are available for protection of natural areas or features on private land. An excellent reference to these techniques is the Rincon Institute's publication *Conservation Options for Landowners* (Vint et al. 1998), which lists 23 land protection tools for achieving various landowner goals. Goals include enhancement of income, tax savings of various kinds, and life style options for retaining ownership and continuing to live and work on the land. Land protection goals may be for personal land or for neighboring lands, and income goals may be for one-time income or for continuing income. Tax savings goals include reducing income taxes, reducing estate taxes, reducing gift taxes, reducing property taxes, or offsetting or avoiding capital gains tax. Lifestyle goals may be to retain land ownership and/or to continue to live and/or work on the land.

**LAND PROTECTION TOOLS**

- Bargain sale
- Charitable gift annuity trust
- Conservation buyer
- Conservation easement donation
- Conservation easement sale
- Deed restrictions
- Donating conservation land
- Donating trade land
- Donating land by will
- Donating a remainder interest
- Donations of undivided partial interests
- Installment sale
- Leasing land to conservation groups
- Like-kind exchange
- Limited development
- Management agreement
- Mutual covenant
- Purchase option
- Reverter or conditional transfer
- Right of first refusal
- Sale/leaseback
- Sale with reserved life estate
- Special use valuation

(from Vint et al. 1998, with permission)

#### *Conservation Easements*

Conservation easements are one of the most useful tools. Many private landowners want to maintain a legacy for their children and future generations. If they can receive some financial benefits or compensation and continue to live on their land, they frequently are willing and even eager to place a conservation easement on their land, or to donate or sell the land to a land trust. Conservation easements typically are a restriction on a warranty deed or title to real property that legally limits the use of the land, for conservation purposes, in perpetuity and prevents residential or commercial development of the property. A conservation easement can permanently protect open space while leaving the land in private ownership. It can also permit continuation of agricultural uses, including crop and livestock production, and forestry. Furthermore, donation of a conservation easement to a qualified non-profit organization or public agency carries a variety of tax benefits for the landowner.

#### *Land Trusts*

Land trusts are private, non-profit organizations that work with landowners who want to voluntarily protect land with important natural, scenic, archeological, recreational, agricultural, or historic value for the public benefit. Land trusts acquire land directly through donation and purchase, hold conservation easements on properties, and often provide stewardship of protected lands (Vint et al. 1998).

Private landowners who want to conserve their land may donate or sell the land itself or a conservation easement to a land trust. Land trusts will be most effective in conserving biodiversity if they incorporate the principles of conservation biology into easements and examine the larger landscape context of individual parcels. The nation's private, non-profit land trusts have been tremendously successful at protecting land from housing development and urban sprawl. More than 1,880,000 ha have been at least partially protected by local and regional land trusts, including approximately 560,000 ha by conservation easements alone, according to the Land Trust Alliance's National Land Trust Census (Nudel 2001).

#### *Federal, state, and local programs*

There are many existing programs specifically providing economic incentives to encourage activities that are beneficial for the ecosystem. These programs can serve as tools to protect or restore ecologically important areas. Most of these have to do with private land management, and they should be promoted as aspects of Wildlands Network implementation. Examples of programs that provide economic incentives for private landowners include the Forest Legacy Program, Forest Resource Management Program, Forestry Incentives Program, Grassland Reserve Program, Conservation Reserve Program, Environmental Quality Incentives Program, Wetlands Reserve Program, Wildlife Habitat Incentive Program, and Partners for Wildlife Program.

#### *Tax system changes*

Another economic strategy for implementation is to change the way the tax system works. Currently millions of dollars in tax breaks and other subsidies are given to resource extractors and polluters. For example, the Forest Service loses millions of dollars each year by paying for the construction and maintenance of logging roads (Talberth and Moskowitz 2000). Instead the government should subsidize activities that repair or protect the natural environment and tax those that destroy it. *Tax Shift*, a book published by Northwest Environment Watch, provides information on this subject (Durning and Bauman 1998). One organization working to put this idea into policy is Taxpayers for Common Sense. Through their Green Scissors program, they work to cut wasteful and environmentally harmful spending, subsidies, and tax breaks. This program brings together fiscal conservatives and conservationists into a successful partnership.

#### *Implications for Implementation*

Although the new social and economic opportunities associated with wildlands protection have proven merit, people are often resistant to adopting new lifestyles and jobs. Achieving full implementation of wildlands network designs will require working with residents to address their concerns and listen to their ideas. There is a critical role for information sharing and education about the many socioeconomic benefits and opportunities made available through implementation of the Wildlands Network Vision. This implies that socioeconomic change will come about only with significant, sustained, and cooperative education and outreach efforts.

## 7. Basic Elements to an Implementation Campaign

Although giving detailed components of an implementation campaign is beyond the scope of this plan, we would like to address some basic campaign steps for this Vision.

Working with many kindred groups, agencies, and individuals, we intend to implement this Vision through the use of science, education, advocacy, and policy. Realizing the time constraints most groups have, we intend for this Vision to fit within the existing framework, goals, and workplan of interested conservation organizations within the Southern Rockies.

The first step in an implementation campaign is identifying these interested parties. We will sit down, face-to-face, with conservation groups, agencies, and individuals to assess how this Vision can best support and strengthen the group's work and goals. We will educate the general public by using credible and influential spokespeople, relevant materials, and media. We recognize the importance of developing a detailed media component and we will evaluate our campaign so that the next effort is well informed.

Conservation biologists should make available the best scientific information to state and federal wildlife managers, policy makers, and all interested parties. Information should be shared in easily understood presentations, as well as printed material, tailored in easy-to-understand language given to substantiate oral presentations.

## 8. Monitoring

Implementing and protecting a region-wide wildlands network is an ambitious and long-term undertaking. Many of the actions will take years to implement; some will take decades. How will success be measured? Only through continual and rigorous spot-checking and evaluation can one be sure that the goals are being achieved. All regional conservation plans must include a commitment to continual and long-term ecological monitoring and evaluation as well as to adjusting the Vision and associated management according to new information (Noss et al. 1997).

There are many different kinds of monitoring, but the focus here will be primarily on effectiveness monitoring. Monitoring can and should be done on many different scales, e.g. by watershed, region, species abundance, etc. The monitoring program itself needs to be developed, planned, and initiated, with clearly defined organizational responsibility for data collection, documentation, and archiving. See Noss and Cooperrider (1994) for more detail.

## 9. Conclusion

Numerous organizations and individuals are already working toward achieving the vision laid out in this Network Design. Many more will be needed. This section provides a starting place, a guideline, for achieving a healthy functioning ecosystem. We hope this document will prove useful for all who share this vision, and will inform, inspire, and support your individual conservation work. While it can sometimes seem like an overwhelming task to preserve biodiversity, it can be accomplished one step at a time, as we've seen from the success stories from other Wildlands Network Visions. In regions across North America, and beyond, conservationists are laying down those puzzle pieces that will create the big picture. Together we are all stronger than the sum of our parts.

*You must be the change you want to see in the world.*

-Mohandas Ghandi

# CONCLUSION

The Southern Rockies Wildlands Network Vision is a launching pad into on-the-ground action and implementation. Those of us working on behalf of the Southern Rockies intend to disseminate this Vision into the hands of those who can implement it. We will work on its behalf, and continue working as our Vision continues to evolve.

Of particular importance is our future work with The Nature Conservancy in Boulder, CO. As noted throughout this document, this is the first iteration of our Vision. It complements the work done by The Nature Conservancy, yet stands alone as an important and useful document for the conservation community. We hope to work with The Nature Conservancy on the next iteration of this Vision or other comparable work, so that we can join our dedication and knowledge of the Southern Rockies in full coordination. Most likely this will happen in steps, as both organizations compare and contrast their conservation plans.

Work with the local and national Wildlands Project office will also be vital to the success of this Vision. We look forward to working with our colleagues at the Wildlands Project on implementation of this Vision for the Southern Rockies.

If you are interested in becoming more involved with implementing the Southern Rockies Wildlands Network Vision, please contact the Southern Rockies Ecosystem Project or the Wildlands Project:


**Southern Rockies Ecosystem Project**
**Margaret DeMarco, Executive Director**
**4990 Pearl East Circle, Suite 301**
**Boulder, CO 80301**
**www.RestoreTheRockies.org**
**303.258.0433**
*Vision@RestoreTheRockies.org*

**Wildlands Project**
**Jen Clanahan, Rocky Mountain Director**
**2260 Baseline Rd., #205C**
**Boulder, CO 80302**
**www.twp.org**
*jenc@wildlandsproject.org*

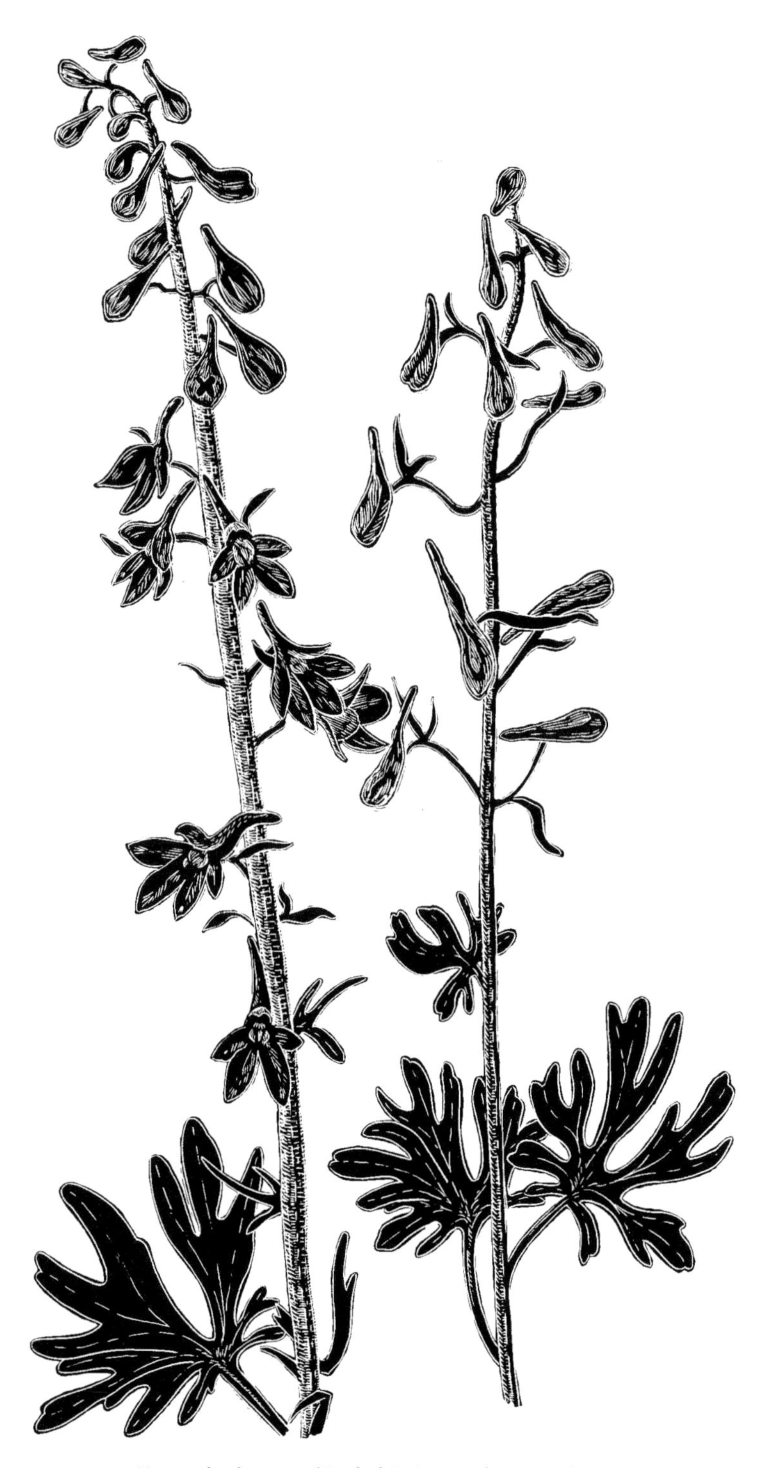

Sierra larkspur (*Delphinium glaucum*)

# LITERATURE CITED


Abbott, C., S.J. Leonard, and D. McComb. 1982. Colorado: A history of the Centennial state. Revised edition, Boulder, CO: University Press of Colorado.

Adams, L.G, F.J. Singer, and B.W. Dale. 1995. Caribou calf mortality in Denali National Park, Alaska. Journal of Wildlife Management 59: 584-594.

Albert, S. and T. Trimble. 2000. Beavers are partners in riparian restoration on the Zuni Indian Reservation. Ecological Restoration 18(2): 87-92.

Allen, A.W. 1983. Habitat suitability index models: Beaver. U.S. Fish and Wildlife Service. FWS/OBS-82/10.30 Revised.

Allen, C.D. 1994. Ecological perspective: Linking ecology, GIS, and remote sensing to ecosystem management. Pages 111-139 in V.A. Sample (ed.). Remote sensing and GIS in ecosystem management. Washington, D.C.: Island Press.

Allen, C.D., J.L. Betancourt, and T. W. Swetnam. 1998. Landscape Changes in the Southwestern United States: Techniques, Long-term Data Sets, and Trends. Pp. 71-84 in Sisk, T.D. (ed.). Perspectives on the land-use history of North America: a context for understanding our changing environment. U.S. Geological Service, Biological Resources Division, Biological Science Report USGS/BRD/BSR-1998-0003.

Allendorf, F.W. and R.F. Leary. 1988. Conservation and distribution of genetic variation in a polytypic species, the cutthroat trout. Conservation Biology 2: 170-184.

Allendorf, F.W., R.B. Harris, and L.H. Metzgar. 1991. Estimation of effective population size of grizzly bears by computer simulation. Pp. 650-654, Proceedings of the Fourth International Congress of Systematics and Evolutionary Biology. Portland, OR: Fourth Doscorides Press.

Alverson, W.S., D.M. Waller and S.L. Solheim. 1988. Forests too deer: edge effects in northern Wisconsin. Conservation Biology 2: 348-358.

Alverson, W.S., W. Kuhlmann and D.M. Waller. 1994. Wild forests: Conservation Biology and Public Policy. Washington DC: Island Press.

Andelman, S.J., I. Ball, F. Davis, and D. Stoms. 1999. SITES V 1.0: An analytical toolbox for designing ecosystem conservation portfolios. Report for The Nature Conservancy. Arlington, VA.

Anderson, C.R., M.A Ternent, D.S. Moody, M.T. Bruscino, and D.F. Miller. 1997. Grizzly bear-cattle interactions on two cattle allotments in northwest Wyoming. Wyoming Game and Fish Department, Lander. 78pp.

Andrews, R.R., and R.R. Righter. 1992. Colorado birds. Denver Museum of Natural History, Denver, CO.

Andromidas, J. 2001. The conservation biology proposal for grizzly bear reintroduction to the San Juan Ecosystem of southwestern Colorado and northern New Mexico. Colorado Grizzly Project, P.O. Box 1511, Boulder CO 80306.

Arendt, H. 1958. The Human Condition. Chicago, IL: The University of Chicago Press.

Arizona Wilderness Coalition 1987. Arizona Wilderness. Alpine, AZ.

Aubry, K.B., G.M. Koehler, and J.R. Squires. 1999a. Ecology of Canada lynx in southern boreal forests. Pages 373- 396 in L.R. Ruggiero, K.B. Aubry, S.W. Buskirk, G.M. Koehler, C.J. Krebs, K.S. McKelvey, and J.R. Squires. Ecology and conservation of lynx in the United States. Department of Agriculture, Forest Service, Rocky Mountain Research Station. General Technical Report RMRS-GTR-30WWW.

Aubry, K.B., L.F. Ruggiero, J.R. Squires, K.S. McKelvey, G.M. Koehler, S.W. Buskirk, and C.J. Krebs. 1999b. Conservation of lynx in the United States: a systematic approach to closing critical knowledge gaps. Pages 455-470 in L.R. Ruggiero, K.B. Aubry, S.W. Buskirk, G.M. Koehler, C. . Krebs, K.S. McKelvey, and J.R. Squires. Ecology and conservation of lynx in the United States. Department of Agriculture, Forest Service, Rocky Mountain Research Station. General Technical Report RMRS-GTR-30WWW.

Bailey, V. 1931. Mammals of New Mexico. North American Fauna 53. Washington, D.C. Department of Agriculture, Bureau of Biological Survey.

Baker, W.L. and Veblen, T.T. 1990. Spruce beetles in the nineteenth-century subalpine forests of western Colorado. Arctic Alpine Research 22:65-80.

Baker, W.L. and G.M. Walford. 1995. Multiple stable states and models of riparian vegetation succession on the Animas River, CO. Annals of the Association of American Geographers 85:320-338.

Baker, W.L., J.A. Munroe, and A.E. Hessel. 1997. The effects of elk on aspen in the winter range in Rocky Mountain National Park. Ecography 20: 155-165.

Ballard, W.B., L.A. Ayres, P.R. Krausman, D.J. Reed, and S.G. Fancy. 1997. Ecology of wolves in relation to a migratory caribou herd in northwest Alaska. Wildlife Monographs 135: 1-47.

Ballard, W.B., J.S. Whitman, and C.L. Gardner. 1987. Ecology of an exploited wolf population in south-central Alaska. Wildlife Monographs 98: 1-54.

Ballard, W.B., H.A. Whitlaw, B.F. Wakeling, R.L. Brown, J.C. DeVos, and M.C. Wallace. 2000. Survival of female elk in northern Arizona. Journal of Wildlife Management 64: 500-504.

Ballard, W.B., D. Lutz, T.W. Keegan, L.H. Carpenter, and J.C. deVos, Jr. 2001. Deer-predator relationships: A review of recent North American studies with emphasis on mule and black-tailed deer. Wildlife Society Bulletin 29: 99-115.

Bangs, E.E. 1998. Evaluation of the interim wolf control plan. U.S. Fish and Wildlife Service, Helena, MT.

Bangs, E.E., T.N. Bailey, and M.F. Portner. 1989. Survival rates of adult female moose on the Kenai Peninsula, Alaska. Journal of Wildlife Management 53: 557-563.

Baron, J.S. ed. 2002. Rocky Mountain Futures: An Ecological Perspective. Washington: Island Press.

Bartmann R.M., G.C. White, L.H. Carpenter. 1992. Compensatory mortality in a Colorado mule deer population. Wildlife Monograph 121: 1-39.

Bauer, S.B. 1985. Evaluation of nonpoint source impacts on water quality from forest practices in Idaho: relation to water quality standards. Pp. 455-458, Perspectives in Nonpoint Source Pollution, Proceedings of a National Conference. Kansas City, MO.

Baxter, G.T., and J.R. Simon. 1970. Wyoming Fishes. Wyoming Game and Fish Department, Cheyenne. p. 167.

Beale, D.M. and A.D. Smith. 1970. Forage use, water consumption, and productivity of pronghorn antelope in western Utah. Journal of Wildlife Management 34: 570-582.

Bean, M.J., and M.J. Rowland. 1997. The evolution of national wildlife law. Westport, CT: Praeger Publishing

Beck, T.D.I. 1991. Black bears of west-central Colorado. Colorado Division of Wildlife, Technical Publication 39.

Beever, E.A., P. F. Brussard, and J. Berger. 2003. Patterns of apparent exptirpations among ilsolated populations of pikas. Journal of Mammology 84 (1): 37-54.

Behnke, R.J. 1992. Native trout of western North America. Bethesda, Maryland: American Fisheries Society Monograph 6.

Belsky, A.J., A. Matzke, and S. Uselman. 1999. Survey of live-stock influences on stream and riparian ecosystems in the western United States. Journal of Soils and Water Conservation 54: 419-431.

Bem, D.J. 1970. Beliefs, Attitudes, and Human Affairs. Belmont, CA: Brooks/Cole Publishing Company.

Benedict, A.D. 1991. A Sierra Club Naturalist's Guide to the Southern Rockies: The Rocky Mountain Regions of Southern Wyoming, Colorado, and Northern New Mexico. Sierra Club Books, San Francisco, CA.

Bennett, L.E. 1994. Colorado gray wolf recovery: A biological feasibility study. Final Report. U.S. Fish and Wildlife Service and University of Wyoming Fish and Wildlife Cooperative research unit, Laramie, WY.

Berg, W.E. and D.W. Kuehn. 1994. Demography and range of fishers and American martens in a changing landscape. Pages 262-271 in S. W. Buskirk, A. S. Harestad, M. G. Raphael, and R. A. Powell, editors. Martens, Sables, and Fishers: Biology and Conservation. Ithaca, New York: Comstock Publishing Associates.

Berger, J. 1990. Persistence of different-sized populations: An empirical assessment of rapid extinctions in bighorn sheep. Conservation Biology 4: 91-98.

Berger, J., J.E. Swenson, I. Persson. 2001a. Recolonizing carnivores and naïve prey: Conservation lessons from the Pleistocene extinctions. Science 291: 1036-1039.

Berger, J., P.B. Stacey, L. Bellis, and M.P. Johnson. 2001b. A mammalian predator-prey imbalance: Grizzly bear and wolf extinction affect avian neotropical migrants. Ecological Applications 11: 947-960.

Bergerud, A.T. and J.P. Elliot. 1998. Wolf predation in a multiple-ungulate system in northern British Columbia. Canadian Journal of Zoology 76: 1551-1569.

Best, A. 2002. Colorado's lynx are feeding, but not breeding. Colorado Central Magazine 96: 16.

Binns, N.A. and R. Remmick. 1994. Response of Bonneville cutthroat trout and their habitat to drainage-wide habitat management at Huff Creek, Wyoming. North American Journal of Fisheries Management 14: 669-680.

Biodiversity Conservation Alliance. 2003 http://www.biodiversityassociates.org./mbnf/mbnfprogram.html

Bishop, N.A. 1998. Child lifting by a wolf in India. International Wolf 8: 17-18.

Blair, B. 1996. Origin of landscapes. Pp. 3-17 in B. Blair (ed.). The Western San Juan Mountains: Their Geology, Ecology, and History. Fort Lewis College Foundation, Niwot, CO: University Press of Colorado.

Bleich, V.C., J.D. Wehausen, and S.A. Holt. 1990. Desert-dwelling mountain sheep: conservation implications of a naturally fragmented distribution. Conservation Biology 4(4): 383-390.

Bogan, M.A., C.D. Allen, E.H. Muldavin, S. P. Platania, J.N. Stuart, G.H. Farley, P. Mehlhop, and J. Belnap. 1998. Southwest. Pp. 543-592 in M.J. Mac, P.A. Opler, C.E. Puckett Haecker, and P.D. Doran (eds.). Status and Trends of the Nation's Biological Resources. Department of the Interior, U.S. Geological Survey, Reston, VA.

Boileau, R., M. Crete, and J. Huot. 1994. Food-habits of the black bear, Ursus americanus, and habitat use in Gaspesie Park, eastern Quebec. Canadian Field-Naturalist 108(2): 162-169.

Bolger, D.T., A.C. Alberts, and M.E. Soulé. 1991. Occurrence patterns of bird species in habitat fragments: sampling, extinction, and nested species subsets. American Naturalist 105: 467-478.

Bonham, C.D. and A. Lerwick. 1976. Vegetation changes induced by prairie dogs on short-grass range. Journal of Range Management 29: 221-225.

Bowker, J.M., D. English and H.K. Cordell. 1999. Projections of outdoor recreation participation to 2050. In Outdoor Recreation in the United States: Results from the National Survey on Recreation and the Environment.

Boydstun, C., P. Fuller, and J.D. Williams. 1995. Nonindigenous fish. Pp. 431-433 in E.T. LaRoe, G.S. Farris, C.E. Puckett, P.D. Doran, and M.J. Mac (eds.). Our Living Resources: A Report to the Nation on the Distribution, Abundance, and Health of U.S. Plants, Animals, and Ecosystems. U.S. Department of the Interior. National Biological Service, Washington, D.C.

Braun, C.E. 1995. Distribution and status of sage grouse in Colorado. Prairie Naturalist 27: 1-9.

Breck, S.W., K.R. Wilson, and D.C. Anderson. 2001. The demographic response of bank-dwelling beavers to flow regulation: A comparison on the Green and Yampa Rivers. Canadian Journal of Zoology 79: 1957-1964.

Breck, S.W., K.R. Wilson, and D.C. Anderson. 2003. Beaver herbivory and its effect on cottonwood trees: Influence of flooding along matched regulated and unregulated rivers. River Research and Applications 19: 43-58

Brewer, G.D. and T.W. Clark. 1994. A policy sciences perspective: improving implementation. Pp. 391-413, in T.W. Clark, R.P. Reading, and A.C. Clarke (eds.). Endangered Species Recovery: Finding the Lessons, Improving the Process. Covelo, CA: Island Press.

Bright, A.D. and S.C. Barro. 2000. Integrative complexity and attitudes: A case study of plant and wildlife species protection. Human Dimensions of Wildlife 5: 30-47.

Brody A.J. and M.R. Pelton. 1989. Effects of roads on black bear movements in western North Carolina. Wildlife Society Bulletin 17(1): 5-10.

Brown, J.S., B.P. Kotlier, and T.J. Valone. 1994. Foraging under predation: A comparison of energetic and predation costs in rodent communities of the Negev and Sonoran Deserts. Australian Journal of Zoology 42: 435-448.

Brown, P.J. and M.J. Manfredo. 1987. Social values defined. Pp. 12-23 in D. J . Decker and G. R. Goff (eds.). Valuing Wildlife: Economic and Social Perspectives. Boulder, CO: Westview Press.

Brown, P.M., M.R. Kaufmann, and W.D. Shepperd. 1999. Long-term, landscape patterns of past fire events in a montane ponderosa pine forest of central Colorado. Landscape Ecology 14: 513-532.

Brown, T.C. 1984. The concept of value in resource allocation. Land Economics 60: 231-246.

Brunner, R.D. 1995. Notes on basic concepts of the policy sciences. Unpublished course notes. Department of Political Science, University of Colorado, Boulder, CO.

Bull, E.L. 2000. Seasonal and sexual differences in American marten diet in northeastern Oregon. Northwest Science 74(3): 186-191.

Bull, E.L. and A.K. Blumton. 1999. Effect of fuels reduction on American martens and their prey. USDA Forest Service, Pacific Northwest Research Station. Research Note PNW-RN-539.

Bull, E.L. and T.W. Heater. 1995. Intraspecific predation on American marten. Northwest Naturalist 76: 132-134.

Bull, E.L. and T.W. Heater. 2000. Resting and denning sites of American martens in northeastern Oregon. Northwest Science 74(3): 179-185.

Bull, E.L. and T.W. Heater. 2001. Survival, causes of mortality, and reproduction in the American marten in northeastern Oregon. Northwestern Naturalist 82: 1-6.

Bull, E.L., R.S. Holthausen, and L. R. Bright. 1992. Comparison of 3 techniques to monitor marten. Wildlife Society Bulletin 20: 406-410.

Busch, D.E., and M.L. Scott. 1995. Western riparian ecosystems. Pages 286-290 in E.T. LaRoe, G.S. Farris, C.E. Puckett, P.D. Doran, and M.J. Mac (eds.). Our Living Resources: A Report to the Nation on the Distribution, Abundance, and Health of U.S. Plants, Animals, and Ecosystems. U.S. Department of the Interior. National Biological Service, Washington, D.C.

Buskirk, S.W. 1984. Seasonal use of resting sites by marten in south-central Alaska. Journal of Wildlife Management 48(3): 950-953.

Buskirk, S.W. and R.A. Powell. 1994. Habitat ecology of fishers and American martens. Pages 283-296 in S.W. Buskirk, A.S. Harestad, M.G. Raphael, and R.A. Powell, editors. Martens, Sables, and Fishers: Biology and Conservation. Ithaca, N.Y.: Comstock Publishing Associates.

Buskirk, S.W. and L.F. Ruggiero. 1994. American marten. Pages 7-37 in L.F. Ruggiero, K.B. Aubry, S.W. Buskirk, L.J. Lyon, and W.J. Zielinski, technical editors. The Scientific Basis for Conserving Forest Carnivores: American Marten, Fisher, Lynx, and Wolverine in the Western United States. USDA Forest Service, Rocky Mountain Forest and Range Experiment Station, Fort Collins, CO. General Technical Report RM-254.

Buskirk, S.W., S.C. Forrest, M.G. Raphael, and H.J. Harlow. 1989. Winter resting site ecology of marten in the central Rocky Mountains. Journal of Wildlife Management 53(1): 191-196.

Buskirk, S.W., L.F. Ruggiero, K.B. Aubry, D.E. Pearson, J.R. Squires, and K.S. McKelvey. 1999. Comparative ecology of lynx in North America. Pages 397-418 in L.R. Ruggiero, K.B. Aubry, S.W. Buskirk, G.M. Koehler, C.J. Krebs, K.S. McKelvey, and J.R. Squires. Ecology and Conservation of Lynx in the United States. Department of Agriculture, Forest Service, Rocky Mountain Research Station. General Technical Report RMRS-GTR-30WWW.

Byers, J.A. 1997. American Pronghorn: Social Adaptations & the Ghosts of Predators Past. Chicago, IL: The University of Chicago Press.

Carbyn, L.N. 1982. Incidence of disease and its potential role in the population dynamics of wolves in Riding Mountain National Park, Manitoba. Pp. 106-116 in F.H. Harrington and P.C. Paquet (eds.). Wolves of the World: Perspectives of Behavior, Ecology, and Conservation. Park Ridge, N.J.: Noyes.

Caldwell, M. M.; L. O. Bjorn, J. F. Bornman; S. D. Flint; G. Kulandaivelu, A. H. Teramura and M. Tevini. 1998. Effects of increased solar ultraviolet radiation on terrestrial ecosystems. Journal of Photochemistry and Photobiology B: Biology 46: 40-52.

California Wild Heritage Campaign. 2002. Retrieved from http://www.californiawild.org/.

Carroll, C. P. Paquet, R. Noss, and J. Strittholt. 1998. Modeling carnivore habitat in the Rocky Mountain region: A literature review and suggested strategy. Unpublished report for World Wildlife Fund.

Carroll, C., R.F. Noss, N.H. Schumaker, and P.C. Paquet. 2001. Is the restoration of the wolf, wolverine, and grizzly bear to Oregon and California biologically feasible? Pages 24-46 in D. Maehr, R. Noss, and J. Larkin (eds.). Large mammal restoration: ecological and social implications. Covelo, CA: Island Press.

Carroll, C., M.K. Phillips, N.H. Schumaker, and D.W. Smith. 2003. Impacts of landscape change on wolf restoration success: Planning a reintroduction program using static and dynamic spatial models. Conservation Biology 17(2): 536-548.

Carroll, C., M.K. Phillips, N.H. Schumaker, B. Martin, K. Kunkle, P.C. Paquet, B. Miller, and D.W. Smith. In press a. Potential for wolves in the Southern Rockies. In The Feasibility of Reintroducing Wolves into the Southern Rockies. Turner Endangered Species Fund and the Denver Zoological Foundation. Bozeman, MT and Denver, CO.

Chaiken, S. and C. Stangor. 1987. Attitudes and attitude change. Annual Review of Psychology 38: 575-630.

Chamberlin, T.W., R.D. Harr, and F.H. Everest. 1991. Timber harvesting, silviculture, and watershed processes in W.R. Meehan, ed. Influence of Forest and Rangeland Management on Salmonid Fishes and Their Habitats. Pp. 181-205. Special Publication 19. American Fisheries Society, Bethesda, MD.

Cheney, E., W. Elmore, and W.S. Platts. 1990. Livestock grazing on western riparian areas. Northwest Resource Information Center, Inc. Eagle, ID.

Church, F. 1973. The Wilderness Act Applies to the East. Congressional Record Senate. January 16, 1973.

Cid, M.S., J.K. Detling, A.D. Whicker, and M.A. Brizela. 1991. Vegetational response of a mixed-grass prairie site following exclusion of prairie dogs and bison. Journal of Range Management 44: 100-104.

City of Boulder. 1999. City of Boulder forest ecosystem management plan. Unpublished report, City of Boulder Open Space/Real Estate Department and Boulder Mountain Parks Division, Boulder, CO.

Clark, T.W. 2000. Interdisciplinary problem-solving in endangered species conservation: The Yellowstone grizzly bear case. Pp. 285-301 in R.P. Reading and B. Miller (eds.). Endangered Animals. Westport, CT: Greenwood Press.

Clark, T.W., T.M. Campbell III, and T.N. Hauptman. 1989. Demographic characteristics of American marten populations in Jackson Hole, Wyoming. The Great Basin Naturalist 49(4): 587-596.

Clark, T.W., D.J. Mattson, R.P. Reading, B.J. Miller. 2001. Interdisciplinary problem-solving in carnivore conservation: An introduction. Pp. 223-240 in J.L. Gittleman, S.M. Funk, D. Macdonald, and R.K. Wayne (eds.). Carnivore Conservation. Cambridge, UK: Cambridge University Press.

Cluff, H.D. and D.L. Murray. 1995. Review of wolf control methods in North America. Pp. 491 - 504 in L.N. Carbyn, S.H. Fritts, and D.R. Seip (eds.). Ecology and Conservation of Wolves in a Changing World. Canadian Circumpolar Institute, Occasional Publication 35. Edmonton, Alberta.

Collen, P. and R. J. Gibson. 2001. The general ecology of beavers (*Castor* spp.), as related to their influence on stream ecosystems and riparian habitats, and the subsequent effects on fish: a review. Reviews in Fish Biology and Fisheries 10(4): 439-461.

Colorado Division of Wildlife. 2001. Harvest/Population Estimate - Elk & Deer 1949-2000. Available: http://wildlife.state.co.us/huntrecap/.

Colorado Division of Wildlife. 2003. Lynx update. Prepared by Tanya Shenk on May 24, 2003. http://wildlife.state.co.us/species_cons/lynx.asp.

Colorado River Cutthroat Trout Task Force. 2001. Conservation agreement and strategy for Colorado River cutthroat trout (*Oncorhynchus clarki pleuriticus*) in the States of Colorado, Utah, and Wyoming. Colorado Division of Wildlife, Fort Collins, CO, 87pp.

Colorado Trout Unlimited. 2002. Logging for Water: Not a Solution for Colorado Logging for Water.htm

Colorado Water Resources Research Institute. 2001. Available: http://water-knowledge. colostate.edu.

Colorado Wilderness Network. 2001. Citizens' WIlderness Proposal. Denver, CO.

Cook, J.R. 1989. The border and the buffalo: All gone. State House Press. Austin, Texas, USA.

Cook, R.R., C.H. Flather, and K.R. Wilson. 2000. Faunal characteristics of the Southern Rocky Mountains of New Mexico: implications for biodiversity analysis and assessment. U.S. Department of Agriculture, Forest Service, Rocky Mountain Research Station. General Technical Report RMRS-GTR-58.

Cooper, D.J. 1993. Sustaining and restoring western wetland and riparian ecosystems threatened by or affected by water development projects. Pp. 27-33 in W.W. Covington and L.F. DeBano (technical coordinators). Sustainable Ecological Systems: Implementing an Ecological Approach to Land Management, July 12-15, 1993. General Technical Report RM-247. Mountain Forest and Range Experiment Station, Fort Collins, CO.

Costanza, R., R. d'Arge, R. de Groot, S. Farber, M. Grasso, B. Hannon, K. Limburg, S. Naeem, R.V. O'Neill, J. Paruelo, R.G. Raskin, P. Sutton, and M. van den Belt. 1997. The value of the world's ecosystem services and natural capital. Nature 387: 253-260.

Costello, D.F. 1970. The World of the Prairie Dog. J.B. lippincott. New York, N.Y.

Costello, C.M., D.E. Jones, K.A. Green Hammond, R.M. Inman, K.H. Inman, B.C. Thompson, R.A. Deitner, and H.B. Quigley. 2001. A study of black bear ecology in New Mexico with models for population dynamics and habitat suitability. Final Report, Federal Aid in Wildlife Restoraton Project W-131-R. New Mexico Department of Game and Fish, Santa Fe.

Covington, W.W., and M.M. Moore. 1994. Southwestern ponderosa forest structure. Journal of Forestry 92: 39-47.

Cowardin, L.M., V. Carter, F.C. Golet, and E.T. LaRoe. 1979. Classification of wetlands and deepwater habitats of the United States: U.S. Fish and Wildlife Service Report FWS/OBS-79/31. Washington D.C.

Cox, G. 2001. Increasing invasion of alien plants. Santa Fe Botanical Garden Newsletter 9 (1).

Crabtree, R.L. and J.W. Shelton. 1999. Coyotes and canid coexistence in Yellowstone. Pp. 127-164 in T.W. Clark, A.P. Curlee, S.C. Minta, and P.M. Karieva (eds.). Carnivores in Ecosystems: The Yellowstone Experience. New Haven, CT: Yale University Press.

Craighead, F.C. and J.J. Craighead. 1972. Grizzly bear prehibernation and denning activities as determined by radiotracking. Wildlife Monographs 32: 1-32.

Craighead, J.J. and J.A. Mitchell. 1982. Grizzly bear. Pp.515-556 in J.A. Chapman and G.A. Feldhamer, editors. Wild Mammals of North America. Baltimore, MD: John Hopkins University Press.

Crawford, C. S., A. C. Cully, R. Leutheuser, M. S. Sifuentes, L. H. White, and J. P. Wither. 1993. Middle Rio Grande ecosystem; bosque biological management plan. Middle Rio Grande Biological Interagency Team, U.S. Fish and Wildlife Service, Albuquerque, N.M.

Crête, M. 1999. The distribution of deer biomass in North America supports the hypothesis of exploitation of ecosystems. Ecology Letters 2: 223-227.

Crête, M. and C. Daigle. 1999. Management of indigenous North American deer at the end of the 20th century in relation to large predators and primary production. Acta Veterinaria Hungarica 47: 1-16.

Crête, M. and M. Manseau. 1996. Natural regulation of cervidae along a 1000 km latitudinal gradient: change in trophic dominance. Evolutionary Ecology 10: 51-62.

Crooks, K.R., and M.E. Soulé. 1999. Mesopredator release and avifaunal extinctions in a fragmented system. Nature. 400:563-566.

Cutler, R. 1977. Memo: Policy for evaluating wilderness potential for National Forest Roadless and Undeveloped Areas. From: Assistant Secretary of Agriculture to Chief of Forest Service, November 2, 1977.

D'Antonio, C.M. and P.M. Vitousek. 1992. Biological invasions of exotic grasses, the grass/fire cycle, and global change. Annual Review of Ecology and Systematics 23: 63-87.

Dahl, T.E. 1990. Wetland losses in the United States: 1780s to 1980s. U.S. Fish and Wildlife Service. Washington, D.C.

DeByle, N.V. 1985. Wildlife. Pp. 135 -152 in N.V. DeByle, and R.P. Winokur (eds.). Aspen: Ecology and management in the western United States, USDA Forest Service General Technical Report RM-119. Washington D.C.

Decker, D., W. Bradford, and J.R. Gunson. 1995. Elk and wolves in Jasper National Park, Alberta, from historical times to 1992. Pp. 85-94 in L.N. Carbyn, S.H. Fritts and D.R. Seip (eds.). Ecology and Conservation of Wolves in a Changing World. Canadian Circumpolar Institute, Occasional Publication 35. Edmonton, Alberta.

Detling, J.K. 1998. Mammalian herbivores: Ecosystem-level effects in two grassland national parks. Wildlife Society Bulletin 26: 438-448.

DeVelice, R.L. and J.R. Martin. 2001. Assessing the extent to which roadless areas complement the conservation of biological diversity. Ecological Applications 11: 1008-1018.

DeVelice, R.L., J.A. Ludwig, W.H. Moir, and F. Ronco, Jr. 1986. A classification of forest habitat types of northern New Mexico and southern Colorado. USDA Forest Service General technical Report RM-131. USDA Forest Service Rocky Mountain and Range Experiment Station, Fort Collins, CO.

Diamond, J. 1992. The Third Chimpanzee. HarperCollins, N.Y.

Dick-Peddie, W.A. 1993. New Mexico Vegetation: Past, Present, and Future. Albuquerque, N.M.: University of New Mexico Press.

Dobson, A., K. Ralls, M. Foster, M. E. Soulé, D. Simberloff, D. Doak, J. A. Estes, L. S.Mills, D. Mattson, R. Dirzo, H. Arita, S. Ryan, E. A. Norse, R. F. Noss, and D. Johns. 1999. Regional and Continental Restoration. Pp. 129-170 in M.E. Soulé and J. Terborgh (eds.). Continental Conservation: Scientific Foundations of Regional Reserve Networks. Washington, D.C.: Island Press.

Dobson, F. S. 1982. Competition for mates and predominant juvenile male dispersal in mammals. Animal Behaviour 30: 1183-1192.

Douglas, C. L. and D. M. Leslie Jr. 1999. Management of bighorn sheep. Pages 238-262 in R. Valdez and P. R. Krausman, editors. Mountain Sheep of North America. Tucson, AZ: University of Arizona Press.

Duff, D. 1996. Bonneville cutthroat trout, *Oncorhynchus clarki utah*. In: D.A. Duff (ed.). Conservation Assessment For Inland Cutthroat Trout: Distribution, Status, and Habitat Management Implications. U.S. Department of Agriculture, Forest Service, Intermountain Region, Ogden, UT.

Dunlap, T.R. 1988. Saving America's Wildlife. Princeton, N.J.: Princeton University Press.

Durning, A. T. and Y. Bauman. 1998. Tax Shift: How to help the economy, improve the environment, and get the tax man off our backs. Seattle, WA: Northwest Environment Watch. Can be downloaded for free at www.north-westwatch.org.

Dusek, G.L. A.K. Wood, and S.T. Stewart. 1992. Spatial and temporal patterns of mortality among female white-tailed deer. Journal of Wildlife Management 56: 645-650.

Elias, S.A. 2002. Rocky Mountains. Washington D.C.: Smithsonian Institution Press.

Ellingson, J.A. 1996. Volcanic Rocks. Pp. 68-79 in B. Blair (ed.). The Western San Juan Mountains: Their Geology, Ecology, and History. Fort Lewis College Foundation, Niwot, CO: University Press of Colorado.

Ellis, J. 1970. A computer analysis of fawn survival in the pronghorn antelope. Ph.D. thesis, University of California, Davis. Davis, CA.

Estes, J.A., N.S. Smith, and J.F. Palmisano. 1978. Sea otter predation and community organization in the western Aleutian Islands, Alaska. Ecology 59: 822-833.

Estes, J.A., D.O. Duggins, and G.B. Rathbun. 1989. The ecology of extinctions in kelp forest communities. Conservation Biology 3: 252-264.

Estes, J.A. and D.O. Duggins. 1995. Sea otters and kelp forests in Alaska:generality and variation in a community ecology paradigm. Ecological Monographs 65:75-100.

Estes, J.A. 1996. Carnivores and ecosystem management. Wildlife Society Bulletin 24: 390-396.

Estes, J.A., M.T. Tinker, T.M. Williams, and D.F. Doak. 1998. Killer whale predation on sea otters linking oceanic and nearshore ecosystems. Science 282: 473-476.

Estes, J.A., K. Crooks, and R. Holt. 2001. Predation and diversity. Pp. 857-878 in S. Levin (ed.). Encyclopedia of Biodiversity. San Diego, CA: Academic Press.

Fahrig, L., J.H. Pedlar, S.E. Pope, P.D. Taylor, and J.F. Wegner. 1995. Effect of road traffic on amphibian density. Biological Conservation 73: 177-182.

Farley, G. H., L. M. Ellis, J. N. Stuart, and N. J. Scott, Jr. 1994. Avian species richness in different-aged stands of riparian forest along the Middle Rio Grande, New Mexico. Conservation Biology 8: 1098-1108.

Fielder, John. 1997. Along the Colorado Continental Divide Trail. Englewood, CO: Westcliffe Publishers.

Finch, D.M., and Ruggiero, L.F. 1993. Wildlife habitats and biodiversity in the Rocky Mountains and northern Great Plains. Natural Areas Journal 13: 191-203.

Findley, J.S. and S. Anderson. 1956. Zoogeography of the montane mammals of Colorado. Journal of Mammalogy 37: 80-82.

Findley, J.S., A.H. Harris, D.E. Wilson, C. Jones. 1975. Mammals of New Mexico. Albuquerque, N.M.: University of New Mexico Press.

Finley, B. 1999. High-tech vs. high altitude: man and machines imperil timberline. Denver Post, February 28, 1999.

Fischer, H. 1989. Restoring the wolf: Defenders launches a compensation fund. Defenders 64:9,36.

Fischer, H., B. Snape, and W. Hudson. 1994. Building economic incentives into the Endangered Species Act. Endangered Species Technical Bulletin 19: 4-5.

Fish, Jim D., ed. 1987. Wildlands: New Mexico Wilderness Coalition Statewide Proposal. New Mexico BLM Wilderness Coalition, Placitas, NM.

Fitzgerald, J.P., C.A. Meaney, and D.M. Armstrong. 1994. Mammals of Colorado. Boulder, CO: Denver Museum of Natural History/University Press of Colorado.

FitzGibbon, C.D. and J. Lazarus. 1995. Antipredator behavior of Serengeti ungulates: Individual differences and population consequences. Pp. 274-296 in A.R.E. Sinclair and P. Arcese (eds.). Serengeti II: Dynamics, Management, and Conservation of an Ecosystem. Chicago, IL: University of Chicago Press.

Fitzsimmons, N.N., S.W. Buskirk, M.W. Smith. 1997. Genetic changes in reintroduced rocky mountain bighorn sheep populations. Journal of Wildlife Management 61: 863-872.

Flannery, T. 2001. The Eternal Frontier. New York, N.Y.: Atlantic Monthly Press.

Fleischner, T.L. 1994. Ecological costs of livestock grazing in western North America. Conservation Biology 8:629-644.

Flores, D.L. 1996. A long love affair with an uncommon country: Environmental history and the Great Plains. Pp. 3-18 in F.B. Sampson and F.L. Knopf (eds.). Prairie Conservation: Preserving North America's Most Endangered Ecosystem. Covelo, CA: Island Press.

Floyd, M.L., W.H. Romme, and D.D. Hanna. 2000. Fire history and vegetation pattern in Mesa Verde National Park, Colorado. Ecological Applications 10: 1666-1680.

Floyd-Hannah, L., A.W. Spencer, and W.A. Romme. 1996. Biotic communities of the semiarid foothills and valleys. Pp. 143-158 in B. Blair (ed.). The Western San Juan Mountains: Their Geology, Ecology, and History. Niwot, CO: University Press of Colorado, Fort Lewis College Foundation.

Foreman, D. 1995. Wilderness areas and National Parks: The foundation for an ecological nature reserve network. Wild Earth 5 No. 4: 60-63.

Foreman, D. 1998. Around the campfire: The ever-robust wilderness idea and Ernie Dickerman. Wild Earth 8: 1.

Foreman, D. 1999a. The Pleistocene-Holocene Event: 40,000 years of extinction. Wild Earth 9: 1-5.

Foreman, D. 1999b. The Wildlands Project And The Rewilding Of North America. Denver University Law Review 76 (2): 535-553.

Foreman, D., K. Daly, B. Dugelby, R. Hanson, R. E. Howard, J. Humphrey, L. Klyza Linck, R. List, and K. Vacariu. 2000. Sky Islands Wildlands Network Conservation Plan. The Wildlands Project, Tucson, AZ.

Foreman, D. and H. Wolke 1992. The Big Outside. New York: Crown Publishers.

Forman, R.T.T. 1995. Land mosaics: the ecology of landscape and regions. Cambridge University Press, Cambridge.

Forman, R.T.T. 2000. Estimate of the area affected ecologically by the road system in the United States. 2000. Conservation Biology 14: 31-35.

Forman, R.T.T. and R.D. Deblinger. 2000. The ecological road-effect zone of a Massachusetts (U.S.A.) suburban highway. Conservation Biology 14: 36-46.

Forman, R.T.T. and M. Godron. 1986. Landscape Ecology. New York, N.Y.: John Wiley and Sons

Forsey, E.S. and E.M. Baggs. 2001. Winter activity of mammals in riparian zones and adjacent forests prior to and following clear-cutting at Copper Lake, Newfoundland, Canada. Forest Ecology and Management 145(3): 163-171.

Foster, M.L. and S.R. Humphrey. 1995. Use of highway underpasses by Florida panthers and other wildlife. Wildlife Society Bulletin 23(1): 95-100.

Frankel, O.H. and M.E. Soulé. 1981. Conservation and Evolution. Cambridge, U.K.: Cambridge University Press.

Franklin, I.R. 1980. Evolutionary change in small populations. Pages 135-149 in M. E. Soulé and B. A. Wilcox, editors. Conservation Biology: An Evolutionary-Ecological Perspective. Sunderland, MA: Sinauer Associates, Inc.

Franklin, J.F. and R.T.T. Forman. 1987. Creating landscape patterns by forest cutting: Ecological consequences and principles. Landscape Ecology 1: 5-18.

Freddy, D.J. 1987. The White River elk herd: a perspective, 1960-1985. Technical Publication Number 37, Colorado Division of Wildlife, Ft. Collins, CO.

Fredrickson, R.J. 1990. The effects of disease, prey fluctuation, and clearcutting on American marten in Newfoundland, Canada. Utah State University, Logan. M.S. Thesis.

Frey, J. K. and T. L. Yates. 1996. Mammalian diversity in New Mexico. Pages 4-37 in E. A. Herrera and L. F. Huenneke, editors. New Mexico's natural heritage: biological diversity in the Land of Enchantment. New Mexico Journal of Science 36.

Fritts, S.H., E.E. Bangs, J.A. Fontaine, W.G. Brewster, J.F. Gore. 1995. Restoring wolves to the Northern Rocky Mountains of the United States. Pp. 107-125 in L.N. Carbyn, S.H. Fritts, and D.R. Seip (eds.). Ecology and Conservation of Wolves in a Changing World. Canadian Circumpolar Institute, Occasional Publication No. 35, Edmonton, Alberta.

Fryxell, J. M. 2001. Habitat suitability and source-sink dynamics of beavers. Journal of Animal Ecology 70: 310-316.

Fryxell, J.M., J.B. Falls, E.A. Falls, R.J. Brooks, L. Dix, and M.A. Strickland. 1999. Density dependence, prey dependence, and population dynamics of martens in Ontario. Ecology 80(4): 1311-1321.

Fuller, T. 1989. Population dynamics of wolves in north-central Minnesota. Wildlife Monographs 105: 1-1.

Fuller, T. 1990. Dynamics of a declining white-tailed deer population in north-central Minnesota. Wildlife Monograph 110: 1-37.

Furniss, M.J., T.D. Roelofs, and C.S. Yee. 1991. Road construction and Maintenance in W.R. Meehan, ed. Influence of Forest and Rangeland Management on Salmonid Fishes and Their Habitats. Pp. 297-323. Bethesda, Maryland: Spec. Publ. 19, American Fisheries Society.

Gasaway, W.C., R.O. Stevenson, J.L. Davis, P.E.K. Shepherd, and O.E. Burris. 1983. Interrelationships of wolves, prey, and man in interior Alaska. Wildlife Monographs 84: 1-54.

Gasaway, W.C., R.D. Boertje, D.V. Grangaard, D.G. Kellyhouse, R.O. Stevenson, and D.G. Larson. 1992. The role of predation in limiting moose at low densities in Alaska and Yukon and implications for conservation. Wildlife Monographs 120: 1-59.

Gauthier, D.A. and J.B. Theberge. 1986. Wolf predation in the Burwash caribou herd, southwest Yukon. Proceedings of the International Reindeer/Caribou Symposium 4: 137-144.

Gibilisco, C.J. 1994. Distributional dynamics of modern Martes in North America. Pages 59-71 in S.W. Buskirk, A.S. Harestad, M.G. Raphael, R.A. Powell, editors. Martens, Sables, and Fishers: Biology and Conservation. Ithaca, N.Y.: Comstock Publishing Associates.

Gilpin, M E. and Soulé, M.E. 1986. Minimum viable populations: the processes of species extinction. Pp. 19-34 in M.E. Soulé (ed.). Conservation Biology. Sinauer Associates, Sunderland, MA.

Glick, D. 2001. Powder burn: Arson, money, and mystery on Vail Mountain. New York, NY. Public Affairs.

Goldblum, D., and T.T. Veblen. 1992. Fire history in ponderosa pine/Douglas fir forest in the Colorado Front Range. Physical Geography 13: 133-148.

Goodrich, J.M. and J. Berger. 1994. Winter recreation and hibernating black bears, *Ursus americanus*. Biological Conservation 67(2): 105-110.

Gordon, C.C. 1986. Winter food habits of the pine marten (*Martes americana*) in Colorado (USA). Great Basin Naturalist 46(1): 166-168.

Gosnell, H. and D. Shinneman. In press. The Human Landscape. In M.K. Phillips, B. Miller, and R. Reading (eds.). The Feasibility of reintroducing wolves into the Southern Rockies. Turner Endangered Species Fund and the Denver Zoological Foundation. Bozeman, MT and Denver, CO.

Greenwood, P.J. 1980. Mating systems, philopatry, and dispersal in birds and mammals. Animal Behavior 28: 1140-1162

Gresswell, R.E. 1995. Yellowstone cutthroat trout. In M.K. Young (ed.). Conservation assessment for inland cutthroat trout. U.S. Department of Agriculture, Forest Service, Rocky Mountain Forest and Range Experiment Station. General Technical RM-GTR-256. pp. 36-54.

Gresswell, R.E., and J.D. Varley. 1989. Yellowstone cutthroat trout: *Oncorhynchus* (formerly *Salmo*) *clarki bouvieri* in T.W. Clark, A.H. Harvey, R.D. Dorn, D.L. Genter, and C. Groves (eds.). Rare, Sensitive, and Threatened Species of the Greater Yellowstone Ecosystem. Northern Rockies Conservation Cooperative, Montana Natural Heritage Program, The Nature Conservancy, and Mountain West Environmental Services. pp. 35-36.

Groom, M., D.B. Jensen, R.L. Knight, S. Gatewood, L. Mills, D. Boyd-Heger,

L.S. Mills, and M.E. Soulé. 1999. Buffer zones: Benefits and dangers of compatible stewardship. Pp. 171-197 in M.E. Soulé and J. Terborgh (eds.). Continental Conservation, Washington, D.C.: Island Press.

Gross, J. E., F. J. Singer, and M. E. Moses. 2000. Effects of disease, dispersal, and area on bighorn sheep restoration. Restoration Ecology 8(4S): 25-37.

Grossman, D.H., D. Faber-Langendoen, A.S. Weakley, M. Anderson, P. Bourgeron, R. Crawford, K.Goodin, S. Landaal, K. Metzler, K. Patterson, M. Pyne, M. Reid, and L. Sneddon. 1998. International classification of ecological communities: Terrestrial vegetation of the United States. Volume I: The National Vegetation Classification Standard. The Nature Conservancy, Arlington, VA.

Gurd, D.B., T.D. Nudds, and D.H. Rivard. 2001. Conservation of mammals in eastern North American wildlife reserves: how small is too small? Conservation Biology 15: 1355-1363.

Gurnell, A. M. 1998. The hydrogeomorphological effects of beaver dam-building activity. Progress in Physical Geography 22(2): 167-189.

Gutzwiller, K.J., and S.H. Anderson. 1987. Multiscale associations between cavity nesting birds and features of Wyoming streamside woodlands. Condor 89: 534-548.

Halaj, J. and D.H. Wise. 2001. Terrestrial trophic cascades: How much do they trickle? American Naturalist 157: 262-281.

Hall, E. R. 1981. The Mammals of North America. Second Edition. New York, N.Y.: John Wiley and Sons.

Harding, L. 2000. Differential habitat use, behavior, and movements of black bears on the East Tavaputs Plateau, UT. M.S. Thesis, Brigham Young University, Provo, UT.

Harrington, M.G., and S.S. Sackett. 1992. Past and present influences on southwestern ponderosa pine old growth. Pp. 44-50 in M.R. Kaufmann, W.H. Moir, and R.L. Bassett (eds.). Old-Growth Forests in the Southwest and Rocky Mountain Regions: Proceedings of a Workshop. USDA Forest Service General Technical Report RM-213, Rocky Mountain Forest and Range Experiment Station, Fort Collins, CO.

Harris, L.D. 1984. The Fragmented Forest: Island Biogeography Theory and the Preservation of Biotic Diversity. Chicago, IL: University of Chicago Press.

Hayes, R.D., and A.S. Harestad. 2000. Demography of a recovering wolf population in the Yukon. Canadian Journal of Zoology 78:36-48.

Hays, S.P. 1996. The trouble with Bill Cronon's wilderness. Environmental History 1: 30.

Hayward, G.D. 1994. Review of technical knowledge: Boreal owls. Pp. 92-127 in G.D. Hayward and J. Verner (eds.). Flammulated, Boreal, and Great Gray Owls in the United States: A Technical Conservation Assessment. USDA Forest Service General Technical report RM-217, Rocky Mountain Forest and Range Experiment Station, Fort Collins, CO.

Hebblewhite, M. 2000. Wolf and elk predator-prey dynamics in Banff National Park. MS thesis, University of Montana, Missoula, MT.

Hejl, S.J. 1994. Human-induced changes in bird populations in coniferous forests in western North America during the last 100 years. Studies in Avian Biology 15: 232-246.

Hendee, J.C., G.H. Stankey, and R.C. Lucas 1990. Wilderness Management. Golden, CO: Fulcrum Publishing.

Hendee, J.C. and D.J. Mattson. 2002. Wildlife in wilderness: a North American and international perspective. Pp. 321-349 in J.C. Hendee and C.P. Dawson. Wilderness Management: Stewardship and Protection of Resources and Values. Third edition. Golden, CO: Fulcrum Press.

Henke, S.E. and F.C. Bryant. 1999. Effects of coyote removal on the faunal community in western Texas. Journal of Wildlife Management 63:1066-1081.

Hepworth, D.K., M.J. Ottenbacher, and C.B. Chamberlain. 2001. Occurrence of native Colorado River cutthroat trout (*Oncorhynchus clarki pleuriticus*) in the Escalante River drainage, Utah. Western North American Naturalist 61(2): 129-138.

Herrero, S.W., W. McCrory, and B. Pelchat. 1983. Using grizzly bear habitat evaluations to locate trails and campsites in Kananaskis Provincial Park. International Conference on Bear Research and Management 6: 187-193.

Hickey, J.R., R.W. Flynn, S.W. Buskirk, K.G. Gerow, and M.F. Wilson. 1999. An evaluation of a mammalian predator, *Martes americana*, as a disperser of seeds. Oikos 87(5): 499-508.

Hickman, T.J. and D.A. Duff. 1978. Current status of cutthroat trout subspecies in the western Bonneville Basin. Great Basin Naturalist 38: 193-202.

Hickman, T. and R.F. Raleigh. 1982. Habitat suitability index models; cutthroat trout. U.S. Fish and Wildlife Service. FWS/OBS -82-/10.5.

Hilderbrand, R.H. 1998. Movements and conservation of cutthroat trout. Ph.D. dissertation, Utah State University, Logan, UT.

Hilderbrand, R.H. and J.L. Kershner. 2000. Conserving inland cutthroat trout in small streams: how much stream is enough? North American Journal of Fisheries Management 20: 513-520.

Hinchman, S. and B. Noreen. 1993. Colorado mining industry strikes again. High Country News, January 25, 1993.

Hitt, N.P., and C.A. Frissell. 1999. Wilderness in a landscape context: a quantitative approach to ranking aquatic diversity areas in western Montana. Presented at the Wilderness Science Conference.

Hollings, C.S. 1978. Adaptive environmental assessment and management. New York, NY. John Wiley and Sons.

Hoogland, J.L. 1995. The black-tailed prairie dog: Social life of a burrowing mammal. Chicago IL: University of Chicago Press.

Hoover, R.L., C.E. Till, and S. Ogilvie. 1959. The antelope of Colorado. Colorado Department of Fish and Game Technical Bulletin 4: 1-110.

Horejsi, B.L., B.K. Gilbert, E.L. Craighead. 1998. Grizzly bear conservation strategy. Calgary, Canada. Western Wildlife Consulting, Ltd.

Hornbeck, G.E. and B.L. Horejsi. 1986. Grizzly bear, *Ursus arctos*, usurps wolf, *Canis lupus*, kill. Canadian Field Naturalist 100: 259-260.

Howard, R.J. and J.S. Larson. 1985. A stream habitat classification system for beaver. Journal of Wildlife Management 49(1): 19-25.

Huggard, D.J. 1993. Prey selectivity of wolves in Banff National Park. I. Prey species. Canadian Journal of Zoology 71: 130-139.

Jacobs, J.F. 1985. Removal of the brown pelican in the southeastern United States from the list of endangered and threatened wildlife. Federal Register 50: 4938-4945.

Jakober, M.J., T.E. McMahon, R.F. Thurow, and C.G. Clancy. 1998. Role of stream ice on fall and winter movements and habitat use by bull trout and cutthroat trout in Montana headwater streams. Transactions of the American Fisheries Society 127: 223-235.

Jamieson, D.W., W.H. Romme, and P. Somers. 1996. Biotic communities of the cool mountains. Pp. 159-173 in B. Blair (ed.). The Western San Juan Mountains: Their geology, ecology, and history. Niwot, CO: University Press of Colorado, Fort Lewis College Foundation.

Janzen, D.H. 1986. The external threat. in M.E. Soulé (ed.). Conservation Biology: The Science of Scarcity and Diversity. Sunderland, MA: Sinauer Associates.

Jedrzejewska, B. and W. Jedrzejewski. 1998. Predation in vertebrate communities. The Bialowieza Primeval Forest as a case study. Springer Verlag. Berlin, Germany.

Jenkins, S.H. and P.E. Busher. 1979. *Castor canadensis*. Mammalian Species 120.

Jeo, R.M., Sanjayan, M.A. and D. Sizemore. 2000. A conservation area design for the central coast region of British Columbia, Canada. Special Report by Round River Conservation Studies, Salt Lake City, UT.

Johns, D. 2001. The Wildlands Project Outside North America, in A. Watson, Alan; Sproull, Janet, comps. 2001. Seventh World Wilderness Congress Symposium: Science and Stewardship to Protect and Sustain Wilderness Values; 2001 November 2-8; Port Elizabeth, South Africa.

Johnson, K.H., and Braun, C.E. 1999. Viability and Conservation of an Exploited Sage Grouse Population. Conservation Biology 13: 77-84.

Johnson, T.L. and D.M. Swift. 2000. A test of a habitat evaluation procedure for Rocky Mountain bighorn sheep. Restoration Ecology 8(4S): 47-56.

Johnstone, H.C. 2000. Temperature tolerances and habitat conditions for Bonneville cutthroat trout in the Thomas Fork of the Bear River, Wyoming. M.S. Thesis University of Wyoming. 107 pp.

Jones, J.R., N.V. DeByle, and D.M. Bowers. 1985. Insects and other invertebrates. Pages 107 - 114 in N.V. DeByle, and R.P. Winokur (eds.). Aspen: Ecology and Management in the Western United States. USDA Forest Service General Technical Report RM-119, Washington D.C.

Jones, K., and D. Cooper. 1993. Wetlands of Colorado. Colorado Department of Natural Resources, Denver, CO.

Jope, K.L. 1985. Implications of grizzly bear habituation to hikers. Wildlife Society Bulletin 13: 32-37.

Judd, S.L., R.R. Knight, and B.M. Blanchard. 1983. Denning of grizzly bears in the Yellowstone National Park area. International Conference on Bear Research and Management 6: 111-117.

Kane, D.M. and J.A. Litvaitis. 1992. Age and sex composition of live-captures and hunter-killed samples of black bears. Journal of Mammalogy 73(1): 215-217.

Kartesz, J.T. 1994. A synonymized checklist of the vascular flora of the United States, Canada, and Greenland. Second edition. Vol. 1., Checklist. Timber Press, Portland, Oregon.

Kay, C.E. 1990. Yellowstone's northern elk herd: a critical evaluation of the "natural regulation" paradigm. Ph.D. dissertation, Utah State University, Logan, UT.

Kay, C.E. and F.H. Wagner. 1994. Historic condition of woody vegetation on Yellowstone's northern range: A critical test of the "natural regulation" paradigm. Pp. 159-169 in D. Despain (ed.). Plants and their environments. Proceedings of the first biennial conference on the greater Yellowstone ecosystem. U.S. National Park Service Technical Report NPS/NRYELL/NRTR-93/xx.

Keesing, F. 2000. Cryptic consumers and the ecology of an African Savanna.

BioScience 50: 205-215.
Keith, L.B. 1983. Population dynamics of wolves. Pp. 66-77 in L. N. Carbyn (ed.). Wolves in Canada and Alaska. Canadian Wildlife Report Series No. 45. Ottawa, Ontario, Canada.
Kellert, S.R. 1984. Assessing wildlife and environmental values in cost-benefit analysis. Journal of Environmental Management 18: 355-363.
Kellert, S.R. 1996. The Value of Life. Washington DC: Island Press.
Kellert, S.R., M. Black, C.R. Rush, and A.J. Bath. 1996. Human culture and large carnivore conservation in North America. Conservation Biology 10: 977-990.
Kelly, W. E. 1980. Predator relationships. Pages 186-196 in G. Monson and L. Sumner, editors. The Desert Bighorn: Its Life History, Ecology and Management. Tucson, AZ: The University of Arizona Press.
Kerr, A. 1998a. The voluntary retirement option for federal public grazing permittees. Rangelands 20(5):26-29.
Kerr, A. 1998b. The voluntary retirement option for federal public land grazing permittees. Wild Earth 8(3):63-67.
Kerr, J.T. 1997. Species richness, endemism, and the choice of areas for conservation. Conservation Biology 11: 1094-1100.
Kershner, J.L. 1995. Bonneville cutthroat trout. In M.K. Young, tech. ed. Conservation assessment for inland cutthroat trout. General Technical Report RM-GTR-256. Pp. 28-35. Fort Collins, Colorado: United States Department of Agriculture, Forest Service, Rocky Mountain Experiment Station.
Kershner, J.L., C.M. Bischoff, and D.L. Horan. 1997. Population, habitat, and genetic characteristics of Colorado River cutthroat trout in wilderness and nonwilderness stream sections in the Uinta Mountains of Utah and Wyoming. North American Journal of Fisheries Management 17: 1134-1143.
Kindschy, R.R., C. Sundstrom, and J.D. Yoakum. 1982. Wildlife habitats in managed rangelands-the Great Basin of southern Oregon: Pronghorn. U.S.D.A. forest Service, Northwest Forest and Range Experiment Station, Portland, OR. General Technical Report PNW-145
King, J.A. 1955. Social behavior, social organization, and population dynamics in a black-tailed prairie dog town in the Black Hills of South Dakota. Contributions to the Laboratory of Vertebrate Biology 67: 1-128. Ann Arbor, MI.
King, M.M. and G. W. Workman. 1986. Response of desert bighorn sheep to human harassment: management implications. Transactions of 51st North American Wildlife and Natural Resources Conference: 74-84. Wildlife Management Institute, Washington, D.C.
Kipfmueller, K.F., and W.L. Baker. 1998. Fires and dwarf mistletoe in a Rocky Mountain lodgepole pine ecosystem. Forest Ecology and Management 108:77-84.
Kipfmueller, K.F., and W.L. Baker. 2000. A fire history of a subalpine forest in south-eastern WY. Journal of Biogeography 27: 71-85.
Kitchen, D.W. 1974. Social behavior and ecology of the pronghorn. Wildlife Monograph 38: 1-36.
Kitchen, D.W. and B.W. O'Gara. 1982. Pronghorn. Pp. 960-971 in J.A. Chapman and G.A. Feldhamer, (eds). Wild Mammals of North America. Baltimore, MD: John Hopkins Press.
Klein, R. G. and B. Edgar 2002. The Dawn Of Human Culture. New York, N.Y.: John Wiley & Sons.
Knick, S.T., and J.T. Rotenberry. 1995. Landscape characteristics of fragmented shrubsteppe habitats and breeding passerine birds. Conservation Biology 9: 1059-1071.
Knight, D.H. 1994. Mountains and plains: The ecology of Wyoming landscapes. Yale University, New Haven, CT.
Knight, D.H, and W.A. Reiners. 2000. Natural patterns in Southern Rocky Mountain landscapes and their relevance to forest management. Pp. 15-30 in Knight, R.L., F.W. Smith, S.W. Buskirk, W.H. Romme, and W.L. Baker, (eds.). Forest Fragmentation in the Southern Rocky Mountains. Boulder, CO: University Press of Colorado.
Knight, D.H, and L.L. Wallace. 1989. The Yellowstone fires: Issues in landscape ecology. BioScience 39: 700-706.
Knight, R.L. 2000. Forest fragmentation and outdoor recreation in the Southern Rocky Mountains. Pp. 135-153 in Knight, R.L., F.W. Smith, S.W. Buskirk, W.H. Romme, and W.L. Baker (eds.). Forest Fragmentation in the Southern Rocky Mountains. Boulder, CO: University Press of Colorado.
Knight, R.L. (ed.). 1995. Wildlife and Recreationists: Coexistence through Management and Research. Covelo, CA: Island Press.
Knight, R.R., B.M. Blanchard, and L.L. Eberhardt. 1988. Mortality patterns and population sinks for Yellowstone grizzly bears, 1973-1985. Wildlife Society Bulletin 16: 121-125.
Knight, R.R., B.M. Blanchard, and P. Schullery. 1999. Yellowstone bears. Pp. 51-75 in T.W. Clark, A.P. Curlee, S.C. Minta, and P.M. Kareiva (eds.). Carnivores in Ecosystems: The Yellowstone Experience. New Haven, CT: Yale University Press.
Knight, R.L, F.W. Smith, S.W. Buskirk, W.H. Romme, and W.L. Baker (eds.). 2000. Forest Fragmentation in the Southern Rocky Mountains. Boulder, CO: University Press of Colorado.
Kocher, S., E.H. Williams. 2000. The diversity and abundance of North American butterflies vary with habitat disturbance and geography. Journal of Biogeography 27: 785-794.
Koehler, G.M. 1990. Population and habitat characteristics of lynx and snowshoe hares in north central Washington. Canadian Journal of Zoology 68: 845-851.
Koehler, G.M. and K.B. Aubry. 1994. Lynx. Pages 74-98 in L. F. Ruggiero, K.B. Aubry, S.W. Buskirk, L.J. Lyon, and W.J. Zielinski, technical editors. The Scientific Basis for Conserving Forest Carnivores: American Marten, Fisher, Lynx, and Wolverine in the Western United States. USDA Forest Service, Rocky Mountain Forest and Range Experiment Station, Fort Collins, CO. General Technical Report RM-254.
Koehler, G.M. and M.G. Hornocker. 1977. Fire effects on marten habitat in Selway-Bitterroot wilderness. Journal of Wildlife Management 41: 500-505.
Koford, C.B. 1958. Prairie dogs, white faces, and blue gramma. Wildlife monograph 3: 1-78.
Korten, D.C. 1995. When Corporations Rule the World. West Hartford, CT: Kumarian Press.
Kotler, B.P., J.S. Brown, R.H. Slowtow, W.L. Goodfriend, and M. Strauss. 1993. The influence of snakes on the foraging behavior of gerbils. Oikos 67: 309-316.
Krausman, P.R. and B.D. Leopold. 1986. The importance of small populations of desert bighorn sheep. Transactions of 51st North American Wildlife and Natural Resources Conference: 52-61. Wildlife Management Institute, Washington, D.C.
Krausman, P.R. and D.M. Shackleton. 2000. Bighorn sheep. Pages 517-544 in S. Demarais and P. R. Krausman, editors. Ecology and management of large mammals in North America. Prentice Hall, Upper Saddle River, N.J.
Krausman, P.R., A.V. Sandoval, and R C. Etchberger. 1999. Natural history of desert bighorn sheep. Pages 139-191 in R. Valdez and P. R. Krausman, editors. Mountain Sheep of North America. Tucson, AZ: University of Arizona Press.
Krebs, C.J., S. Boutin, R. Boonstra, A.R.E. Sinclair, J.N.M. Smith, M.R.T. Dale, and R. Turkington. 1995. Impact of food and predation on the snowshoe hare cycle. Science 269: 1112-1114.
Krebs, C.J., R. Boonstra, S. Boutin, and A.R.E. Sinclair. 2001. What drives the 10-year cycle of snowshoe hares? BioScience 51: 25-35.
Kruse, C.G. 1998. Influences of non-native trout and geomorphology on distributions of indigenous trout in the Yellowstone River drainage of Wyoming. Ph.D dissertation, Department of Zoology and Physiology, University of Wyoming, December, 1998.
Kruse, C.G., W.A. Hubert, and F.J. Rahel. 2000. Status of Yellowstone cutthroat trout in Wyoming waters. North American Journal of Fisheries Management 20: 693-705.
Kunkel, K.E. 1997. Predation by wolves and other large carnivores in northwestern Montana and southeastern British Columbia. Ph.D. dissertation, University of Montana, Missoula, MT.
Kunkel, K.E. and D. H. Pletscher. 1999. Species-specific population dynamics of cervids in a multipredator ecosystem. Journal of Wildlife Management 63:1082-1093.
Kunkel, K.E. and D. H. Pletscher. 2001. Winter hunting patterns of wolves in and near Glacier National Park, Montana. Journal of Wildlife Management 65: 520-530.
Kunkel, K.E., T. K. Ruth, D. H. Pletscher, and M. G. Hornocker. 1999. Winter prey selection by wolves and cougars in and near Glacier National Park, Montana. Journal of Wildlife Management 63: 901-910.
Lacy, R.C. and T.W. Clark. 1993. Simulation modeling of American marten (Martes americana) populations: vulnerability to extinction. Great Basin Naturalist 53(3): 282-292.
Lacy, R.C. and T.W. Clark. 1989. Genetic variability in black-footed ferret populations: Past, present, and future. Pp. 83-103 in U.S. Seal, E.T. Thorne, M.A. Bogan, and S.H. Anderson. Conservation Biology of the Black-Footed Ferret. New Haven, CT: Yale University Press.
Lambeck, R.J. 1997. Focal species: A multi-species umbrella for nature conservation. Conservation Biology 11: 849-856.
Lande, R. 1988. Genetics and demography in biological conservation. Science 241: 1455-1460.
Lande, R. 1995. Mutation and conservation. Conservation Biology 9(4): 782-791.
Lande, R. and G. F. Barrowclough. 1987. Effective population size, genetic variation, and their use in population management. Pages 87-123 in M. E. Soulé, editor. Viable Populations for Conservation. Cambridge, MA: Cambridge University Press.
Lasswell, H.D. and A. Kaplan. 1950. Power and Society: A Framework for

Political Inquiry. New Haven, CT: Yale University Press.
Leakey, R. and R. Lewin 1995. The Sixth Extinction. New York, N.Y.: Doubleday.
LeCount, A.L. 1986 Black bear field guide. Special Report Number 16. Arizona Game and Fish Department, Phoenix.
LeCount, A.L. and J.C. Yarchin. 1990. Black bear habitat use in east-central Arizona. Arizona Game and Fish Department, Technical Report Number 4.
Leopold, A. 1966. A Sand County Almanac. New York, N.Y.: Ballantine Books.
Leopold, A. 1937. Conservationist in Mexico. American Forests, Vol. 43, March 1937. 118-120.
Leopold, A. 1972. Round River. New York, N.Y.:Oxford University Press.
Levins, R. 1969. Some demographic and genetic consequences of environmental heterogeneity for biological control. Bulletin of the Entomological Society of America (15): 237-240.
Licht, L. E. and K. P. Grant. 1997. The effects of ultraviolet radiation on the biology of amphibians. American Zoologist 37: 137-145.
Lindzey, F.G. and E.C. Meslow. 1977. Home range and habitat use by black bears in southwestern Washington. Journal of Wildlife Management 41: 413-425.
Linnell, J.D.C., R. Aanes, and R. Andersen. 1995. Who killed Bambi? The role of predation in the neonatal mortality of temperate ungulates. Wildlife Biology 1(4): 209-223.
Linnell, J.D.C., R. Aanes, J.E. Swenson, J. Odden, and M.E. Smith. 1997. Translocation of carnivores as a method for managing problem animals: A review. Biodiversity and Conservation 6: 1245-1257.
Logan, K.A. and L.L. Sweanor. 2001. Desert Puma: Evolutionary Ecology and Conservation of an Enduring Carnivore. Washington, D.C.: Island Press.
Logan, K.A., L.L. Sweanor, T.K. Ruth, and M.G. Hornocker. 1996. Cougars of the San Andres Mountains, New Mexico. Final Report to New Mexico Department of Game and Fish, Santa Fe, N.M.
Lovaas, A.L. and P.T. Bromley. 1972. Preliminary studies of pronghorn antelope/black-tail prairie dog relationships in Wind Cave National Park. Antelope States Workshop Proceedings 5: 115-152.
Luce, B., A. Cerovski, B. Oakleaf, J. Priday, and L.V. Fleet. 1999. Atlas of birds, mammals, reptiles, and amphibians in Wyoming. Wyoming Game and Fish Department, Lander. 190 pp.
Lynch, A.M. and T.W. Swetnam. 1992. Old-growth mixed-conifer and western spruce budworm in the Southern Rocky Mountains. Pp. 66-80 in M.R. Kaufmann, W.H. Moir, and R.L. Bassett (eds.). Old-Growth Forests in the Southwest and Rocky Mountain regions: Proceedings of a Workshop. USDA Forest Service General Technical Report RM-213, Rocky Mountain Forest and Range Experiment Station, Fort Collins, CO.
Lyon, L.J. 1983. Road density models describing habitat effectiveness for elk. Journal of Forestry 81:592-595.
Mac M.J., P.A. Opler, C.E. Puckett Haeker, and P.D. Doran. 1998. Status and trends of the nation's biological resources. U.S. Geological Survey, Washington D.C.
MacArthur, R.H. and E.R. Pianka. 1966. On optimal use of a patchy environment. American Naturalist 100: 603-609.
MacArthur, R.H. and E.O. Wilson. 1967. The theory of island biogeography. Princeton, NJ. Princeton University Press.
Mace, R.D. 1987. Food habits summary. In M.N. LeFranc, Jr. ed. Grizzly Bear Compendium. National Wildlife Federation, Washington, D.C.
Mace, R.D. and C.J. Jonkel. 1983. Local food habits of the grizzly bear in Montana. International Conference on Bear Research and Management 6: 105-110.
Mace, R.D. and J.S. Waller 1997. Spatial and temporal interaction of male and female grizzly bears in northwestern Montana. Journal of Wildlife Management 61: 39-52.
Mace, R.D., J.S. Waller, T.L. Manley, K. Ake, and W.T. Wittinger. 1999. Landscape evaluation of grizzly bear habitat in western Montana. Conservation Biology 13: 367-377.
Mace, R.D., J.S. Waller, T.L. Manley, J.L. Lyon, and H. Zuuring. 1996. Relationships among grizzly bears, roads, and habitat in the Swam Mountains, Montana. Journal of Applied Ecology 33: 1395-1404.
Mackie, R. J., K. L. Hamlin, and D. V. Pac. 1982. Mule Deer. Pages 862-877 in J. A. Chapman and G. A. Feldhammer (eds.). Wild Mammals of North America: Biology, Management, Economics. Baltimore, MD: Johns Hopkins University Press.
Madel, M.J. 1996. Rocky Mountain front grizzly bear management program: four-year progress report, 1991-1994. Montana Fish, Wildlife, and Parks. 80pp.
Madsen, T., R. Shine, M. Olssen, and H. Wittzell. 1999. Restoration of an inbred adder population. Nature 402: 34-35.
Manfredo, M.J., J.J. Vaske, G.E. Haas, and D. Fulton. 1993. The Colorado environmental poll. CEP #2. Human Dimensions in Natural Resources Unit, Colorado State University, Fort Collins, CO.
Martin, B., R. Edward, and A. Jones. 1999. Mapping A Future for Wolves in the Southern Rockies. Southern Rockies Wolf Tracks (The newsletter of Sinapu). Winter 1999, 7:1-12.
Martin, P.S. and R.G. Klein (eds.). 1984. Quaternary Extinctions: A Prehistoric Evolution. Tucson, AZ: The University of Arizona Press.
Martin, S.K. 1994. Feeding ecology of American martens. Pages 297-315 in S.W. Buskirk, A.S. Harestad, M.G. Raphael, and R.A. Powell, editors. Martens, Sables, and Fishers: Biology and Conservation. Ithaca, N.Y.: Comstock Publishing Associates.
Martinsen et al. 1998. Indirect interactions mediated by changing plant chemistry: beaver browsing benefits beetles. Ecology 79(1): 192-200.
Marvin, S. 1996. Possible changes in water yield and peakflows in response to forest management. Sierra Nevada Ecosystem Project: Final Report, V.III, Assessments, Comissioned Reports and Background Information, Univ. of Calif., Davis, CA.
Matthiassen, P. 1987. Wildlife in America. New York, NY. Viking Penguin Incorporated.
Mattson, D. J. 1997. Wilderness-dependent wildlife - the large and the carnivorous. International Journal of Wilderness 3: 34-38.
Mattson, D.J. 2000. Grizzly bear in R.P. Reading and B. Miller, editors, Endangered Animals. Westport, CT: Greenwood Press.
Mattson, D.J., B.M. Blanchard, and R.R. Knight. 1992a. Yellowstone grizzly bear mortality, human habitation, and whitebark pine seed crops. Journal of Wildlife Management 56: 432-442.
Mattson, D.J., R.R. Knight, B.M. Blanchard. 1992b. Cannibalism and predation on black bears by grizzly bears in the Yellowstone ecosystem 1975-1990. Journal of Mammalogy 73(2): 422-425.
Mattson, D.J., S. Herrero, R.G. Wright, and C.M. Pease. 1996. Science management of Rocky Mountain grizzly bears. Conservation Biology 10: 1013-1025.
Maxwell, J.R., C.J. Edwards, M.E. Jensen, S.J. Paustian, H. Parrott, and D. M. Hill. 1995. A hierarchical framework of aquatic ecological units in North America (nearctic zone). USDA Forest Service General Technical Report NC-176. USDA Forest Service North Central Forest Experimental Station, St. Paul, MN.
May, B.E., J.D. Leppink, and R.S. Wydowski. 1978. Distribution, systematics and biology of the Bonneville cutthroat trout. UDWR Publication 78-15, Ogden, UT.
Mayr, E. 2001. What Evolution Is. New York, N.Y.: Basic Books.
McClane, A.J. 1978. A Field Guide to Freshwater Fishes of North America. New York, N.Y.: Holt, Rinehart, and Winston.
McClellan, B., and D. Shackleton. 1988. Grizzly bears and resource-extraction industries: Effects of roads on behaviour, habitat use, and demography. Journal of Applied Ecology 25: 451-460.
McKelvy K.S., K.B. Aubry, and Y.K. Ortega. 1999a. History and distribution of Lynx in the contiguous United States. Pages 207-264 in L.R. Ruggiero, K.B. Aubry, S.W. Buskirk, G.M. Koehler, C.J. Krebs, K.S. McKelvey, and J.R. Squires. Ecology and Conservation of Lynx in the United States. Department of Agriculture, Forest Service, Rocky Mountain Research Station. General Technical Report RMRS-GTR-30WWW.
McKelvy K.S., S.W. Buskirk, and C.J. Krebs. 1999b. Theoretical insights into the population viability of lynx. Pages 21-38 in L.R. Ruggiero, K.B. Aubry, S.W. Buskirk, G.M. Koehler, C.J. Krebs, K.S. McKelvey, and J.R. Squires. Ecology and Conservation of Lynx in the United States. Department of Agriculture, Forest Service, Rocky Mountain Research Station. General Technical Report RMRS-GTR-30WWW.
McKelvy K.S., K.B. Aubry, J.K. Agee, S.W. Buskirk, L.F. Ruggiero, and G.M. Koehler. 1999c. Lynx conservation in an ecosystem management context. Pages 419-442 in L.R. Ruggiero, K.B. Aubry, S.W. Buskirk, G.M. Koehler, C.J. Krebs, K.S. McKelvey, and J.R. Squires. Ecology and Conservation of Lynx in the United States. Department of Agriculture, Forest Service, Rocky Mountain Research Station. General Technical Report RMRS-GTR-30WWW.
McLaren, B.E. and R.O. Peterson. 1994. Wolves, moose, and tree rings on Isle Royale. Science 266: 1555-1558.
McNab, W.H., and P.E. Avers. 1994. Ecological subregions of the Untied States: Section descriptions. U.S. Department of Agriculture, Forest Service. WO-WSA-5. Washington, D.C.
McNamee, T. 1997. The Return of the Wolf to Yellowstone. New York, N.Y.: Henry Holt & Company, Incorporated.
McShea, W. and J. Rappole. 1992. White-tailed deer as a keystone species within forested habitats of Virginia. Virginia Journal of Science 43: 177-186.
McShea, W., H.B. Underwood, and J.H. Rappole (eds.). 1997. The Science of Overabundance: Deer Ecology and Management. Washington D.C.: Smithsonian Institution Press.
Meadow, B. 2001. Wolf restoration issues, a report. Decision Research. Washington D.C.

Mech, L.D. 1977. Productivity, mortality, and population trends of wolves in northeastern Minnesota. Journal of Mammalogy 58: 559-574.
Mech, L.D. 1989. Wolf population survival in an area of high road density. American Midland Naturalist 121: 387-389.
Mech, L.D. 1990. Who's afraid of the big bad wolf? Audubon 92: 82-85.
Mech, L.D. 1995. The challenge and opportunity of recovering wolf populations. Conservation Biology 9: 270-278.
Mech, L.D. 1996a. A new era for carnivore conservation. Wildlife Society Bulletin 24: 397-401.
Mech, L.D. 1996b. Wolves and child lifting in India. International Wolf 6: 16.
Mech, L.D. 1998. Who's afraid of the big bad wolf? Revisited. International Wolf 8: 8-11.
Mech, L.D. 1999. Estimated costs of maintaining a recovered wolf population in agricultural regions of Minnesota. Wildlife Society Bulletin 26: 817-822.
Mech, L.D. 2001. Mech challenges study's pessimistic outlook. International Wolf 11:8.
Mech, L.D., S. H. Fritts, and M. E. Nelson. 1996. Wolf management in the 21st century: from public input to sterilization. Journal of Wildlife Research 1: 195-198.
Mech, L.D., E. K. Harper, T. J. Meier, and W. J. Paul. 2000. Assessing factors that may predispose Minnesota farms to wolf depredation on cattle. Wildlife Society Bulletin 28: 623-629.
Meehan, W.R. 1991. Introduction and Overview. In W.R. Meehan, ed. Influence of forest and rangeland management on salmonid fishes and their habitats. Pp. 1-15. Spec. Publ. 19. Bethesda, Maryland: American Fisheries Society.
Meehan, W.R. and T.C. Bjornn. 1991. Salmonid distributions and life histories. In W.R. Meehan, ed. Influence of Forest and Rangeland Management on Salmonid Fishes and Their Habitats. Pp. 47-82. Spec. Publ. 19. Bethesda, Maryland: American Fisheries Society.
Meffe, G. K. and C. R. Carroll. 1997. Principles of Conservation Biology. Second Edition. Sunderland, MA: Sinauer Associates, Inc.
Mehl, M.S. 1992. Old-growth descriptions for the major forest cover types in the Rocky Mountain region. Pp. 106-120 in M.R. Kaufmann, W.H. Moir, and R.L. Basset (eds.). Old-Growth Forest in the Southwest and Rocky Mountain Regions: Proceedings of a Workshop, March 9-13, 1992, Portal AZ. USDA Forest Service General technical report RM-213. USDA Forest Service Rocky Mountain Forest and Range Experiment Station, Fort Collins, CO.
Menge, B.A., Berlow, E.L., Blanchette, C.A., Navarrete, S.A., and S.B. Yamada. 1994. The keystone species concept: variation in interaction strength in a rocky intertidal habitat. Ecological Monographs 64: 249-286.
Merriam, C.H. 1890. Results of a biological survey of the San Francisco Mountain region and desert of the Little Colorado, Arizona. North American Fauna 3:1-136.
Merrill, E.H. and M.S. Boyce. 1991. Summer range and elk population dynamics in Yellowstone National Park. Pp. 263-274 in R.B. Keiter and M.S. Boyce (eds.). The Greater Yellowstone Ecosystem: Redefining America's Wilderness Heritage. New Haven, CT: Yale University Press.
Messier, F. 1985. Social organization, spatial distribution, and population density of wolves in relation to moose density. Canadian Journal of Zoology 63: 1068-1077.
Meyers, C. 2002. Hunters Can Reap an Abundance of Elk. Denver Post, 5 May 2002.
Miller, B., R. Reading, and S. Forrest. 1996. Prairie Night: Recovery of the Black-Footed Ferret and Other Endangered Species. Washington, D.C.: Smithsonian Institution Press.
Miller, B., R. Reading, J. Strittholt, C. Carroll, R. Noss, M. Soulé, O. Sanchez, J. Terborgh, D. Brightsmith, T. Cheeseman and D. Foreman. 1998. Using focal species in the design of nature reserve networks. Wild Earth 8: 81-92.
Miller, B., R. Reading, J. Hoogland, T. Clark, G. Ceballos, R. List, S. Forrest, L. Hanebury, P. Manzano, J. Pacheco, and D. Uresk. 2000. The role of prairie dogs as keystone species: Response to Stapp. Conservation Biology 14: 318-321.
Miller, B., B. Dugelby, D. Foreman, C. Marinez del Río, R. Noss, M. Phillips, R. Reading, M.E. Soulé, J. Terborgh, L. Willcox. 2001. The importance of large carnivores to healthy ecosystems. Endangered Species UPDATE 18: 202-210.
Miller, B., K. Kunkle, P. Paquet, R. Reading, and M.K. Phillips. In press a. The importance of large carnivores. In M.K. Phillips, B. Miller, and R. Reading (eds.). The Feasibility of Reintroducing Wolves into the Southern Rockies. Turner Endangered Species Fund and the Denver Zoological Foundation. Bozeman, MT and Denver, CO.
Miller, B., R.P. Reading, and H. Gosnell. In press b. Attitudes about wolves. In M.K. Phillips, B. Miller, and R. Reading (eds.). The Feasibility of Reintroducing Wolves Into the Southern Rockies. Turner Endangered Species Fund and the Denver Zoological Foundation. Bozeman, MT and Denver, CO.
Miller, R.F. and J.A. Rose. 1999. Fire history and western juniper encroachment in sagebrush steppe. Journal of Range Management 52: 550-559.
Mills, L.S., M.E. Soulé and D.F. Doak. 1993. The history and current status of the keystone species concept. Bioscience 43: 219-224.
Murie, A. 1948. Cattle on grizzly bear range. Journal of Wildlife Management 12: 57-72.
Murray, M.P. 1996. Natural processes: wilderness management unrealized. Natural Areas Journal 16: 55-61.
Mutel, C.F., and J.C. Emerick. 1992. From Grassland to Glacier: The Natural History of Colorado and the Surrounding Region. Boulder, CO: Johnson Books.
Naiman, R. J., C. A. Johnston, and J. C. Kelley. 1988. Alteration of North American streams by beaver. BioScience 38: 753-762.
Nash, R.F. 2001. Wilderness & The American Mind. Fourth Edition. New Haven, CT: Yale University Press.
Neely, B., P. Comer, C. Moritz, M. Lammert, R. Rondeau, C. Pague, G. Bell, H. Copeland, J. Humke, S. Spackman, T. Schulz, D. Theobald, and L. Valutis. 2001. Southern Rocky Mountains: An Ecoregional Assessment and Conservation Blueprint. Prepared by The Nature Conservancy with support from the U.S. Forest Service, Rocky Mountain Region, Colorado Division of Wildlife, and Bureau of Land Management.
Nelson, E.W. 1925. Status of the pronghorned antelope 1922-1924. U.S. Department of Agriculture Bulletin No. 1346: 1-64.
Nelson, M.E. and L.D. Mech. 1981. Deer social organization and wolf predation in northwestern Minnesota. Wildlife Monographs 77: 1-53.
New Mexico Department of Game and Fish. 2001a. Biota information system of New Mexico: version 7/2001 - American beaver (050115). Santa Fe, N.M.
New Mexico Department of Game and Fish. 2001b. Biota information system of New Mexico: version 7/2001 - lynx (050325). Santa Fe, N.M.
Newmark, W.D. 1987. Mammalian extinctions in western North American parks: A landbridge perspective. Nature 325: 430-432.
Newmark, W.D. 1995. Extinction of mammal populations in western North American national parks. Conservation Biology 9: 512-526.
Noel, T.J., P.F. Mahoney, and R.E. Stevens. 1994. Historical Atlas of Colorado. Norman, OK: University of Oklahoma Press.
Noss, R.F. 1983. A regional landscape approach to maintain diversity. BioScience. 33: 700-706.
Noss, R.F. 1987. Protecting Natural Areas in Fragmented Landscapes. Natural Areas Journal 7: 2-13.
Noss, R.F. 1992. The Wildlands Project Land Conservation Strategy. Wild Earth. Special Issue 1992.
Noss, R.F. 1999. A citizen's guide to ecosystem management. Biodiversity Legal Foundation, Louisville, CO.
Noss, R. F. and L. D. Harris. 1986. Nodes, networks, and MUMs: Preserving diversity at all scales. Environmental Management 10: 299-309.
Noss, R.F., and A.Y. Cooperrider. 1994. Saving Nature's Legacy: Protecting and Restoring Biodiversity. Washington DC: Island Press.
Noss, R.F, E.T. LaRoe, and J.M. Scott. 1995. Endangered Ecosystems of the United States: A Preliminary Assessment of Loss and Degradation. Biological Report 28. USDI National Biological Service, Washington, D.C.
Noss, R.F., H.B. Quigley, M.G. Hornocker, T. Merrill, and P.C. Paquet. 1996. Conservation biology and carnivore conservation in the Rocky Mountains. Conservation Biology 10(4): 949-963.
Noss, R.F., M.A. O'Connell, and D.D. Murphy. 1997. The Science of Conservation Planning: Habitat Conservation Under the Endangered Species Act. Washington, D.C.: Island Press.
Noss, R, E. Dinerstein, B. Gilbert, M. Gilpin, B. Miller, J. Terborgh, and S. Trombulak. 1999. Core areas in reserve design. Pp. 99-128 in. M.E. Soulé and J.Terborgh (eds.). Continental Conservation: Scientific Foundations of Regional Reserve Networks. Washington D.C.: Island Press.
Noss, R.F., C. Carroll, K. Vance-Borland, and G. Wuerthner. 2002. A multicriteria assessment of the irreplaceability and vulnerability of sites in the Greater Yellowstone Ecosystem. Conservation Biology 16: 895-908.
Nowak, R.M. 1978. Reclassification of the gray wolf in the United States and Mexico, with determination of critical habitat in Michigan and Minnesota. Federal Register 43: 9607-9615.
Nowak, R.M. 1991. Walker's Mammals of the World. 5th Edition. Baltimore, MD: John Hopkins Press.
Noyce, K.V. and D.L. Garshelis. 1997. Influence of natural food abundance on black bear harvests in Minnesota. Journal of Wildlife Management 61(4): 1067-1074.
Nudel, M.2001. Press release from the Land Trust Alliance (http://www.lta.org)
O'Gara, B.W. 1978. Antilocapra Americana. Mammalian Species 90: 1-7.
Odell, E.A., and R.L. Knight. 2001. Songbird and medium-sized mammal communities associated with exurban development in Pitkin County, Colorado. Conservation Biology 15:1143-1150.

Odum, E.P. 1993. Ecology and Our Endangered Life Support Systems. Second Edition. Sunderland, MA: Sinauer Associates, Inc.

Oksanen, L. and T. Oksanen. 2000. The logic and realism of the hypothesis of exploitation ecosystems. American Naturalist 155: 703-723.

Oksanen, T., L. Oksanen, M. Schneider, and M. Aunapuu. 2001. Regulation, cycles and stability in northern carnivore-herbivore systems: back to first principles. Oikos 94: 101-117.

Olson, J.A. and M.P. Zanna. 1993. Attitudes and attitude change. Annual Review of Psychology 44: 117-154.

Olson, R.A., and W.A. Gerhart. 1982. A physical and biological characterization of riparian habitats and its importance in Wyoming. Wyoming Game and Fish Department, Cheyenne, WY.

Osmundson, D.B., P. Nelson, K. Fenton, and D.W. Ryden. 1995. Relationship between flow and rare fish habitat in the "15-mile reach" of the Colorado River: Final report. U.S. Fish and Wildlife Service.

Paine, R.T. 1966. Food web complexity and species diversity. American Naturalist 100: 65-75.

Paine, R.T. 1980. Food webs: Linkage, interaction strength, and community infrastructure. Journal of Animal Ecology 49: 667-685.

Palomares, F. and M. Delibes. 1997. Predation upon European rabbits and their use of open and closed patches in Mediterranean habitats. Oikos 80: 407-410.

Palomares, F., and P. Gaona, P. Ferreras, and M. Delibes. 1995. Positive effects on game species of top predators by controlling smaller predator populations: An example with lynx, mongooses, and rabbits. Conservation Biology 9: 295-305.

Paquet, P.C. 1989. Behavioral ecology of wolves (Canis lupus) and coyotes (C. latrans) in Riding Mountain National Park, Manitoba. Ph.D. dissertation, University of Alberta, Edmonton, Alberta.

Paquet, P.C. 1991. Winter spatial relationships of wolves and coyotes in Riding Mountain National Park, Manitoba. Journal of Mammalogy 72: 397-401.

Paquet, P.C. 1992. Prey use strategies of sympatric wolves and coyotes in Riding Mountain National Park, Manitoba. Journal of Mammalogy 73: 337-343.

Paquet, P.C. and L.N. Carbyn. 1986. Wolves, Canis lupus, killing denning black bears, Ursus americanus, in the Riding Mountain National Park area. Canadian Field-Naturalist 100(3): 371-372.

Parendes, L.A. and J.A. Jones. 2000. Role of light availability and dispersal in exotic plant invasion along roads ad streams in the H.J. Andrews Experimental Forest, Oregon. Conservation Biology 14: 64-75.

Pasitschniak-Arts, M. 1993. Ursus arctos. Mammalian Species 439: 1-10.

Pastor, J., R.J. Naiman, and B. Dewey. 1988. Moose, microbes and boreal forests. BioScience 38: 770-777.

Pate, J., M.J. Manfredo, A.D. Bight, and G. Tischbein. 1996. Coloradans' attitudes toward reintroducing the gray wolf into Colorado. Wildlife Society Bulletin 24: 421-428.

Patnode, K.A. and M.N. LeFranc. 1987. Distribution. in M.N. LeFranc, Jr. ed. Grizzly bear compendium. National Wildlife Federation, Washington D.C.

Pearl, R.H. 1974. Geology of groundwater resources in Colorado. Colorado Geological Survey, Department of Natural Resources, Denver, CO.

Peek, J. M. and P. D. Dalke (eds.). 1982. Wildlife-livestock relationships. Symposium: Proceedings 10. University of Idaho, Forest Wildlife, and Range Experiment Station, Moscow, ID.

Peek, J.M., M.R. Pelton, H.D. Picton, J.W. Schoen, and P. Zager. 1987. Grizzly bear conservation and management: a review. Wildlife Society Bulletin 15: 160-169.

Pelton, M.R. 2000. Black Bear. Pages 389-408 in S. Demarais and P. R. Krausman editors. Ecology and Management of Large Mammals in North America. Upper Saddle River, N.J.: Prentice Hall.

Perry, D.A. 1994. Forest Ecosystems. Baltimore, MD: John Hopkins University Press.

Peterson, D. 1995. Ghost grizzlies. Boulder, CO. Johnson Press.

Peterson, R.O., J.D. Woolington, and T.N. Bailey. 1984. Wolves of the Kenai Peninsula, Alaska. Wildlife Monograph 88: 1-52.

Petty, R. E., D. T. Wegener, and L. R. Fabrigar. 1997. Attitudes and attitude change. Annual Review of Psychology 48: 609-674.

Phillips, M.K. and D.W. Smith. 1998. Gray wolves and private landowners in the Greater Yellowstone Area. Transactions of the North American Wildlife and Natural Resources Conference 63: 443-450.

Phillips, M.K. and B. Miller. In press. The extermination and recovery of the gray wolf. In M.K. Phillips, B. Miller, and R. Reading (eds.). The Feasibility of Reintroducing Wolves into the Southern Rockies. Turner Endangered Species Fund and the Denver Zoological Foundation. Bozeman, MT and Denver, CO.

Phillips, M.K., R. Edwards, B. Miller, and R.P. Reading (eds.), in press. The Feasibility of Reintroducing Wolves into the Southern Rockies. Turner Endangered Species Fund and the Denver Zoological Foundation. Bozeman, MT and Denver, CO.

Pickett, S.T.A., and P.S. White (eds.). 1985. The Ecology of Natural Disturbance and Patch Dynamics. Orlando, FL: Academic Press.

Pimm, S.L. 2001. The World According to Pimm. New York, N.Y.: McGraw-Hill.

Pletscher, D.H., R.R. Ream, D.K. Boyd, M W. Fairchild, and K.E. Kunkel. 1997. Population dynamics of a recolonizing wolf population. Journal of Wildlife Management 61: 459-465.

Pollock, M.M. and K. Sucking. 1998. Presettlement conditions of ponderosa pine forests in the American Southwest. Unpublished report.

Possingham, H. Ball, I. and Andelman, S. 2000. Mathematical methods for identifying representative reserve networks. Pages 291-305 in Quantitative Methods for Conservation Biology. Ferson, S. and Burgman, M. (eds). New York, N.Y.: Springer-Verlag.

Post, E., R.O. Peterson, N.C. Stenseth, and B.E. McLaren. 1999. Ecosystem consequences of wolf behavioural response to climate. Nature 401: 905-907.

Potvin, F., R. Courtois, and L. Belanger. 1999. Short-term response of wildlife to clear-cutting in Quebec boreal forest: multiscale effects and management implications. Canadian Journal of Forest Research 29(7): 1120-1127.

Powell, R.A. 1994. Structure and spacing of Martes populations. Pages 101-121 in S.W. Buskirk, A.S. Harestad, M.G. Raphael, and R.A. Powell, editors. Martens, Sables, and Fishers: Biology and Conservation. Ithaca, N.Y.: Comstock Publishing Associates.

Powell, R.A., J.W. Zimmerman, D.E. Seaman, and J.F. Gilliam. 1996. Demographic analysis of a hunted black bear population with access to a refuge. Conservation Biology 10(1): 224-234.

Powell, R.A., J.W. Zimmerman, and D.E. Seaman. 1997. Ecology and Behaviour of North American Black Bears: Home Ranges, Habitat and Social Organization. New York, N.Y.: Chapman and Hall.

Power, M.E., Tilman, D., Estes, J.A., Menge, B.A., Bond, W.J., Mills, L.S., Daily, G., Castilla, J.C., Lubchenco, J. and R.T. Paine. 1996. Challenges in the quest for keystones. Bioscience 46: 609-620.

Power, M.E, S.J. Kupferburg, G.W. Minshell, M.C. Molles, and M.S. Parker. 1997. Chapter III: Sustainability and western riparian ecosystems. Pp. 17-31 in W.L. Minckley (ed.). Aquatic Ecosystems Symposium: Report to the Western Water Policy Review Advisory Commission, August 1997. U.S. Geological Survey, Reston, VA.

Power, T.M. 1996. Lost Landscapes and Failed Economies: The Search for a Value of Place. Covelo, CA: Island Press.

Pressey, R.L., H. P. Possingham and C.R. Margules. 1996. Optimality in reserve selection algorithms: When does it matter and how much? Biological Conservation 76: 259-267.

Price, T., K. Walker, and J. Catlin. 1998. Utah Wilderness Inventory. Wild Earth 8 (4): 74-77.

Primack, R.B. 1998. Essentials of conservation biology, $2^{nd}$ ed. Sinauer Associates. Sunderland, Massachusetts, USA.

Raleigh, R.F. and D.A. Duff. 1981. Trout stream habitat improvement; ecology and management. Pages 67-77 in W. King, ed. Proceedings of a Wild Trout Symposium II. Yellowstone National Park, WY. Sept. 24-25, 1979.

Ramey, R.R., G. Luikart, and F. J. Singer. 2000. Genetic bottlenecks resulting from restoration efforts: the case of bighorn sheep in Badlands National Park. Restoration Ecology 8(4S): 85-90.

Raphael, M.G. 1994. Techniques for monitoring populations of fishers and American martens. Pages 224-240 in S.W. Buskirk, A.S. Harestad, M.G. Raphael, and R.A. Powell, editors. Martens, Sables, and Fishers: Biology and Conservation. Ithaca, N.Y.: Comstock Publishing Associates.

Rasker, R. 1994. Measuring Change in Rural Economies: A Workbook for Determining Demographic, Economic, and Fiscal Trends. Washington, D.C.: The Wilderness Society.

Reading, R.P. 1993. Toward an endangered species reintroduction paradigm: A case study of the black-footed ferret. Ph.D. dissertation, Yale University, New Haven, CT.

Reading, R.P. and B. Miller. 2000. Conclusions: Causes of endangerment and conflicts in recovery. Pp. 302-312 in R.P. Reading and B. Miller (eds.) Endangered Animals. Westport, CT: Greenwood Press.

Reading, R.P. and T.W. Clark. 1996. Carnivore reintroduction Pp. 296-336 in J.L. Gittleman (ed.). Carnivore behavior, ecology, and evolution. Ithaca, N.Y.: Comstock Publishing Association.

Reading, R.P, T.W. Clark, and S.R. Kellert. 1994. Attitudes and knowledge of people living in the Greater Yellowstone Ecosystem. Society and Natural Resources 7: 349-365.

Rebertus, A.J., T.T. Veblen, L.M. Roovers, and J.N. Mast. 1992. Structure and dynamics of old-growth Engelmann Spruce-subalpine fir in Colorado. Pp. 139-153 in M.R. Kaufmann, W.H. Moir, and R.L. Basset (eds.). Old-Growth Forest in the Southwest and Rocky Mountain Regions: Proceedings of a Workshop, March 9-13, 1992, Portal AZ. USDA Forest Service General technical report RM-213. USDA Forest Service Rocky Mountain Forest and

Range Experiment Station, Fort Collins, CO.
Reed, R.A., J. Johnson-Barnard, and W.L. Baker. 1996. Fragmentation of a forested Rocky Mountain landscape, 1950-1993. Biological Conservation 75:267-277.
Refsnider, R. 2000. Proposal to reclassify and remove the gray wolf from the list of endangered and threatened wildlife in portions of the conterminous United States. Federal Register 65: 43450-43496.
Reice, S.R. 1994. Nonequilibrium determinants of biological community structure. American Scientists 82: 424-435.
Reimchen, T.E. 1998. Diurnal and Nocturnal behavior of black bears, Ursus americanus, on bear trails. Canadian Field-Naturalist 112(4): 698-699.
Reinhart, D.P. and D.J. Mattson. 1990. Bear use of cutthroat trout spawning streams in Yellowstone National Park. International Conference on Bear Research and Management 8: 343-350.
Retzer, J. L., H. M. Swope, J. D. Remington, and W. H. Rutherford. 1956. Suitability of physical factors for beaver management in the Rocky Mountains of Colorado. Colorado Department of Game, Fish and Parks. Technical Bulletin 2.
Reynolds, R.T., R.T. Graham, M.H. Reiser, R.L. Bassett, P.L. Kennedy, D.A. Boyce Jr., G. Goodwin, R. Smith, and E.L. Fisher. 1992. Management recommendations for the northern goshawk in the southwestern United States. USDA Forest Service General Technical Report RM-253, Rocky Mountain Forest and Range Experiment Station, Fort Collins, CO.
Rhodes, J., M. Purser. 1998. Thinning for increase water yield in the Sierra Nevada: Free lunch or a pie in the sky? Pacific Rivers Council.
Ricketts, T.H., E. Dinerstein, D.M. Olson, and C. Loucks. 1999. Who's where in North America? Bioscience 49: 369-381.
Riebsame, W.E. 1997. Atlas of the New West. New York, N.Y.: W.W. Norton.
Ripple, W.J. and E.J. Larsen. 2000. Historic aspen recruitment, elk, and wolves in northern Yellowstone National Park. Biological Conservation 95: 361-370.
Robinson, W.B. 1953. Coyote control with Compound 1080 in national forests. Journal of Forestry 51: 880-885.
Robitaille, J.F. and K. Aubry. 2000. Occurrence and activity of American martens, *Martes Americana*, in relation to roads and other routes. Acta Theriologica 45(1): 137-143.
Rohlf, D.J. 1991. Six biological reasons why the Endangered Species Act doesn't work - and what to do about it. Conservation Biology 5: 273-282.
Rokeach, M. 1972. Beliefs, Attitudes, and Values: A theory of Organization and Change. San Francisco, CA: Josey-Bass, Incorporated.
Romme, W.H., and D.G. Despain. 1989. Historical perspective on the Yellowstone fires of 1988. Bioscience 39: 695-699.
Romme, W.H., M.G. Turner, L.L. Wallace, and J.S. Walker. 1995. Aspen, fire, and elk in the northern Yellowstone Park. Ecology 76:2097-2104.
Romme, W.H, M.L. Floyd, D. Hannah, and J.S. Redders. 2000. Using natural disturbance regimes as a basis for mitigating impacts of anthropogenic fragmentation. Pp. 377-400 in R.L. Knight, F.W. Smith, S.W. Buskirk, W.H. Romme, and W.L. Baker (eds.). Forest Fragmentation in the Southern Rocky Mountains. Boulder, CO: University Press of Colorado.
Romme, W.H., L. Floyd-Hanna, D.D. Hanna, and E. Bartlett. 2001. Aspen's Ecological Role in the West. Pp. 243-259 in W.D Shepperd, D. Binkley, D.L. Bartos, T.J. Stohlgren, and L.G. Eskew, compilers. Sustaining aspen in Western Landscapes: Symposium Proceedings; 13-15 June 2000; Grand Junction, CO. Proceedings RMRS-P-18. U.S. Department of Agriculture, Forest Service, Rocky Mountain Research Station. Fort Collins, CO.
Rosen, P.C., and C.R. Schwalbe. 1995. Bullfrogs: Introduced predators in Southwestern wetlands. Pp. 452-454 in E.T. LaRoe, G.S. Farris, C.E. Puckett, P.D. Doran, and M.J. Mac (eds.). Our living resources: A report to the nation on distribution, abundance, and health of U.S. plants, animals, and ecosystems. USDI National Biological Service, Washington, D.C.
Route, B. 1998. No easy answers: effects of wolf population expansion on deer hunting in northern Minnesota. International Wolf 8: 19-21.
Rudis, V.A. and J.B. Tansley. 1995. Regional assessment of remote forests and black bear habitat from forest resource surveys. Journal of Wildlife Management 59(1): 170-180.
Ruggiero, L.F., K.B. Aubry, S.W. Buskirk, G.M. Koehler, C.J. Krebs, K.S. McKelvey, and J.R. Squires. 1999. The scientific basis for lynx conservation: qualified insights. Pages 443-454 in L.R. Ruggiero, K.B. Aubry, S.W. Buskirk, G.M. Koehler, C.J. Krebs, K.S. McKelvey, and J.R. Squires. Ecology and Conservation of Lynx in the United States. Department of Agriculture, Forest Service, Rocky Mountain Research Station. General Technical Report RMRS-GTR-30WWW.
Russell, T.P. 1964. Antelope of New Mexico. Bulletin of New Mexico Department of Game and Fish 12: 1-103.
Rutherford, W. H. 1964. The beaver in Colorado. Colorado Department of Game, Fish and Parks. Technical Publication 17.
Rutledge, C.R., and T. McLendon. No Year. An assessment of exotic plant species of Rocky Mountain National Park. Department of Rangeland Ecosystem Science, Colorado State University. Northern Prairie Wildlife Research Center Home Page: http://www.npwrc.usgs.gov/resource/othrdata/ explant/explant.htm (Version 15DEC98).
Sackett, S.S., S. Haase, and M.G. Harrington. 1994. Restoration of southwestern ponderosa pine ecosystems with fire. Pp. 115-121 in W.W. Covington and L.F. DeBano (technical coordinators). Sustainable ecological systems: Implementing an ecological approach to land management. U.S. Forest Service General Technical Report RM-247.
Sampson, F.B., E.L. Adams, S. Hamilton, S.P. Nealey, R. Steele, and D. Van De Graaf. 1994. Assessing forest ecosystem health in the inland West. Forest Policy Center, Washington D.C.
Sampson, F.B., F.L. Knopf, and W.R. Ostlie. 1998. Pp. 437-472 in M.J. Mac, P.A. Opler, C.E. Puckettt Haeker, and P.D. Doran, (eds.). Status and Trends of the Nation's Biological Resources. U.S. Geological Survey, Washington D.C.
Schmid, J.M. and S.A. Mata. 1996. Natural variability of specific forest insect populations and their associated effects in Colorado. USDA Forest Service General Technical report RM-GTR-275. Rocky Mountain Forest and Range Experiment Station, Fort Collins, CO.
Schmitz, O.J. 1998. Direct and indirect effects of predation and predation risk in old-field interaction webs. American Naturalist 151: 327-340.
Schmitz, O.J, P.A. Hamback, and A.P. Beckerman. 2000. Trophic cascades in terrestrial systems: A review of the effects of carnivore removals on plants. American Naturalist 155: 141-153.
Schneider, R.R. and P. Yodzis. 1994. Extinction dynamics in the American marten (*Martes americana*). Conservation Biology 8(4): 1058-1068.
Schoener, T.W. and D.A. Spiller. 1999. Indirect effects in an experimentally staged invasion by a major predator. American Naturalist 153: 347-358.
Schoenwald-Cox, C., and M. Buechner. 1992. Park protection and public roads. Pp. 373-379 in P.L. Fielder and S.K. Jain (eds.). Conservation Biology: The Theory and Practice of Nature Conservation, Preservation, and Management. New York, N.Y.: Chapman and Hall.
Schultz, T.T. and W.C. Leininger. 1990. Differences in riparian vegetation structure between grazed areas and exclosures. Journal of Range Management 43: 295-299.
Schwartz, C.C. and A.W. Franzmann. 1991. Interrelationship of black bears to moose and forest succession in the northern coniferous forest. Wildlife Monographs 113.
Scott, D.W. 2001. A Wilderness-Forever Future. Pew Wilderness Center, Washington, D.C.
Seidel, J., B. Andree, S. Berlinger, K. Buell, G. Byrne, B. Gill, D. Kenvin, and D. Reed. 1998. Draft strategy for the conservation and reestablishment of lynx and wolverine in the Southern Rocky Mountains. Colorado Division of Wildlife, U.S. Forest Service, National Park Service, U.S. Fish and Wildlife Service, New Mexico Game and Fish Department, Wyoming Game and Fish Department. Technical Report, RM-GTR-256. Rocky Mountain Forest and Range Experimental Station, Fort Collins, CO.
Shackleton D. M., C. C. Shank, and B. M. Wikeem. 1999. Natural history of Rocky Mountain and California bighorn sheep. Pages 78-138 in R. Valdez and P. R. Krausman, editors. Mountain Sheep of North America. Tucson, AZ: University of Arizona Press.
Shaffer, M, and B. Stein. 2000. Safeguarding our precious heritage. Pp. 301-321 in B.A. Stein, L.S. Kutner, and J.S. Adams (eds.). Precious Heritage: The Status of Biodiversity in the United States. New York, N.Y.: Oxford University Press.
Shelford, V. E., editor, 1926. Naturalist's Guide to the Americas. Baltimore, MD: Williams and Wilkins.
Shelford, V. E. 1933. Ecological Society of America: A nature sanctuary plan unanimously adopted by the Society. December 28, 1932. Ecology 14: 240-245.
Shenk, T. 2003. Lynx Update, 26 February 2003 (e-mail from Colorado Division of Wildlife).
Shepard, B.B., B. Sanborn, L. Ulmer, and D.C. Lee. 1997. Status and risk of extinction for westslope cutthroat trout in the Upper Missouri River, Montana. North American Journal of Fisheries Management 17: 1158-1172.
Sherburne, S.S. and J.A. Bissonette. 1994. Marten subnivean access point use: response to subnivean prey levels. Journal of Wildlife Management 58(3): 400-405.
Shields, W.M. 1987. Dispersal and mating systems: investigating their causal connections. Pages 3-24 in B.D. Chepko-Sade and Z.T. Halpin, editors. Mammalian Dispersal Patterns: The Effects of Social Structure on Population Genetics. Chicago, IL: University of Chicago Press.
Shinneman, D.J., and W.L. Baker. 1997. Nonequilibrium dynamics between catastrophic disturbances and old-growth forests in ponderosa pine landscapes of the Black Hills. Conservation Biology 11: 1276-1288.
Shinneman, D.J and W.L. Baker. 2000. Impact of logging and roads on a Black Hills ponderosa pine forest landscape. Pp. 311-335 in Knight, R.L., F.W.

Smith, S.W. Buskirk, W.H. Romme, and W.L. Baker (eds.). Forest Fragmentation in the Southern Rocky Mountains. Boulder, CO: University Press of Colorado.

Shinneman, D. and B. Miller., in press. The natural landscape of the Southern Rockies ecosystem. In M.K. Phillips, B. Miller, and R. Reading (eds.). The feasibility of reintroducing wolves into the Southern Rockies. Turner Endangered Species Fund and the Denver Zoological Foundation. Bozeman, MT and Denver, CO.

Shinneman, D.J., R. McClelan, and R. Smith. 2000. The State of the Southern Rockies ecoregion. Southern Rockies Ecosystem Project, Nederland, CO.

Sigler, W.F. and J.W. Sigler. 1996. Fishes of Utah: A Natural History. Salt Lake City, UT: University of Utah Press.

Simberloff, D. and J. Cox. 1987. Consequences and costs of conservation corridors. Conservation Biology 1: 63-71.

Simberloff, D. J., D. Doak, M. Groom, S. Trombulak, A. Dobson, S. Gatewood, M. E. Soulé, M. Gilpin, C. Martinez del Rio, and L. Mills. 1999. Regional and Continental Restoration. Pp. 65-98 in M.E. Soulé and J. Terborgh (eds.). Continental Conservation. Washington, D.C.: Island Press.

Simon, N P.P., F.E. Schwab, M.I. LeCoure, and F.R. Phillips. 1999. Fall and winter diet of martens, *Martes americana*, in central Labrador related to small mammal densities. Canadian Field-Naturalist 113(4): 678-680.

Singer, F. and J.A. Mack. 1999. Predicting the effects of wildfire and carnivore predation on ungulates. Pp. 189-237 in T.W. Clark, A.P. Curlee, S.C. Minta, and P.M. Karieva (eds.). Carnivores in Ecosystems: The Yellowstone Experience. New Haven, CT: Yale University Press.

Singer, F.J. and L.C. Zeigenfuss (eds.). 2002. Ecological evaluation of abundance and effects of elk herbivory in Rocky Mountain National Park, Colorado, 1994-1999. Report 02-208. U.S. Department of Interior, U.S. Geological Survey, and Natural Resources Ecology Lab at Colorado State University. Fort Collins, CO.

Singer, F., A. Harting, K. K. Symonds, and M. B. Coughenour. 1997. Density dependence, compensation, and environmental effects on elk calf mortality in Yellowstone National Park. Journal of Wildlife Management 61: 12-25.

Singer, F. J., M. E. Moses, S. Bellew, and W. Sloan. 2000a. Correlates to colonizations of new patches by translocated populations of bighorn sheep. Restoration Ecology 8(4S): 66-74.

Singer, F. J., C. M. Papouchis, and K. K. Symonds. 2000b. Translocations as a tool for restoring populations of bighorn sheep. Restoration Ecology 8(4S): 6-13.

Singer, F. J., E. Williams, M. W. Miller, and L. C. Zeigenfuss. 2000c. Population growth, fecundity, and survivorship in recovering populations of bighorn sheep. Restoration Ecology 8(4S): 75-84.

Singer, F. J., V. C. Bleich, and M. A. Gudorf. 2000d. Restoration of bighorn sheep metapopulations in and near western national parks. Restoration Ecology 8(4S): 14-24.

Singleton, P.H., W. Gaines, and J.F. Lehmkuhl. 2001. Using weighted distance and least-cost path corridor analysis to evaluate regional-scale large carnivore habitat connectivity in Washington. Proceedings of the International Conference on Ecology and Transportation, Keystone, CO, September 24-27.

Slough, B.G. 1989. Movements and habitat use by transplanted marten in the Yukon Territory. Journal of Wildlife Management 53(4): 991-997.

Slough, B.G. 1994. Translocations of American martens: an evaluation of factors in success. Pages 165-178 in S.W. Buskirk, A.S. Harestad, M.G. Raphael, and R.A. Powell, editors. Martens, Sables, and Fishers: Biology and Conservation. Ithaca, N.Y.: Comstock Publishing Associates.

Slough, B. G. and R. M. F. S. Sadleir. 1977. A land capability classification system for beaver (*Castor canadensis*). Canadian Journal of Zoology 55(8): 1324-1335.

Smith, B.L. and S.H. Anderson. 1996. Patterns of neonatal mortality of elk in northwest Wyoming. Canadian Journal of Zoological 74:1229-1237.

Smith, D.W., L.D. Mech, M. Meagher, W.E. Clark, R. Jaffe, M.K. Phillips, and J.A. Mack. 2000. Wolf-bison interactions in Yellowstone National Park. Journal of Mammalogy 81: 1128-1135.

Smith, D.W., K.M. Murphy, and D.S. Guernsey. 2001. Yellowstone wolf project: Annual report, 1999. National Park Service, Yellowstone Center for Resources, Yellowstone National Park, WY.

Smith, J. 1998. Petition to list the Yellowstone cutthroat trout (*Oncorhynchus clarki bouvieri*) as threatened under the Endangered Species Act. Biodiversity Legal Foundation, Boulder, CO, Alliance for the Wild Rockies, Missoula, MT, Montana Ecosystems Defense Council, Bozeman, MT. 100 pp.

Smith, M.E. and E.H. Follman. 1993. Grizzly bear, Ursus arctos, predation of a denned adult black bear, U. americanus. Canadian Field-Naturalist 107(1): 97-99.

Smith, T. S., J.T. Flinders, and D. S. Winn. 1991. A habitat evaluation procedure for Rocky Mountain bighorn sheep in the Intermountain West. Great Basin Naturalist 51: 205-225.

Snyder, G. No Nature. 1992. New York and San Francisco: Pantheon Books.

Snyder, G.R. and H.A. Tanner. 1960. Cutthroat trout reproduction in inlets to Trappers Lake. Colorado Department of Game and fish, Technical Bulletin 7, Fort Collins, CO.

Somers, P., and L. Floyd-Hanna. 1996. Wetlands, riparian habitats, and rivers. Pp. 175-189 in R. Blair (ed.). The Western San Juan Mountains: Their Geology, Ecology and Human History. Niwot, CO: University Press of Colorado.

Soulé, M.E. 1983. What do we really know about extinction? in Schonewald-Cox, Christine M. et al., eds., Genetics and Conservation (Benjamin-Cummings, Menlo Park, CA, 1983), p. 116.

Soulé, M.E. 1995. An unflinching vision: networks of people defending networks of lands. Pp. 1-8 in D.A. Saunders, J.L. Craig, and E.M. Mattiske (eds.). Nature Conservation 4: The role of networks. Surrey Beatty and Sons Limited, Sydney, Australia.

Soulé, M.E. 1996. The end of evolution. World Conservation (Originally IUCN) Bulletin 1/96:24-25.

Soulé, M.E. 2001. Should wilderness be managed? Pp. 136-152 in T. Kerasote (ed.). Return of the wild. Washington DC. Island Press.

Soulé, M.E. and B.A. Wilcox 1980. Conservation Biology: Its Scope and Challenge. In: Soulé, M.E. and B.A. Wilcox, editors, Conservation Biology: An Evolutionary-Ecological Perspective. Sunderland, MA: Sinauer Associates.

Soulé, M.E., and D. Simberloff. 1986. What do genetics and ecology tell us about the design of nature reserves? Biological Conservation 35: 19-40.

Soulé, M.E. and R. Noss 1998. Rewilding and biodiversity: complementary goals for continental conservation. Wild Earth Vol. 8, No. 3: 18-28.

Soulé, M. and J. Terborgh (eds.). 1999. Continental conservation: scientific foundations of regional reserve networks. Covelo, CA: Island Press.

Soulé, M.E., D.T. Bolger, A.C. Alberts, R. Sauvajot, J. Wright, and S. Hill. 1988. Reconstructed dynamics of rapid extinctions of chaparral-rearing birds in urban habitat islands. Conservation Biology 2: 75-92.

Soulé, M.E., J.A. Estes, J. Berger, and C.M. del Rio. In press. Ecologically effective numbers of endangered keystone species: theory and practice. Conservation Biology.

Soulounias, N. 1988. Evidence from horn morphology on the phylogenetic relationships of the pronghorn (*Antilocapra americana*). Journal of Mammalogy, 69:140-143.

Soutiere, E.C. 1979. Effects of timber harvest on marten in Maine. Journal of Wildlife Management 43(4): 850-860.

Sovada, M.A., A.B. Sargeant, and J.W. Grier. 1995. Differential effects of coyotes and red foxes on duck nest success. Journal of Wildlife Management 59: 1-9.

Speas, C., M. Fowden, M. Gorges, T. Rinkes, G. Eaglin, and B. Wengert. 1994. Conservation plan for Colorado River cutthroat trout, Little Snake River drainage, Southeastern Wyoming. U.S. Forest Service Medicine Bow National Forest, Wyoming Game and Fish Department, and Bureau of Land Management Great Divide Resource Area.

Spencer, W.D. 1987. Seasonal rest-site preferences of pine martens in the northern Sierra Nevada. Journal of Wildlife Management 51(3): 616-621.

Squires, J.R. 2000. Food habits of northern goshawks nesting in south central Wyoming. Wilson Bulletin 112(4): 536-539.

Steventon, J.D. and J.T. Major. 1982. Marten use of habitat in a commercially clear-cut forest. Journal of Wildlife Management 46: 175-182.

Stohlgren, T.J. 1998. Rocky Mountains. Pp. 473-482 in M.J. Mac, P.A. Opler, C.E. Puckett Haeker, and P.D. Doran (eds.). Status and trends of the nation's biological resources. Volume 2. USDI, U.S. Geological Survey, Reston, VA.

Streeter, R.G. and C.E. Braun. 1968. Occurrence of pine marten, *Martes americana*, (*Carnivora: Mustelidae*) in Colorado alpine areas. The Southwestern Naturalist 13: 449-451.

Strickland, M. A. 1994. Harvest management of fishers and American martens. Pages 149-164 in S. W. Buskirk, A. S. Harestad, M. G. Raphael, and R. A. Powell, editors. Martens, Sables, and Fishers: Biology and Conservation. Ithaca, N.Y.: Comstock Publishing Associates.

Strickland, M.A. and C.W. Douglas. 1987. Marten. Pages 530-546 in Novak, M., J.A. Baker, and M.E. Obbard, editors. Wild Furbearer Management and Conservation in North America. North Bay, Ontario: Ontario Trappers Association.

Strickland, M.A., C.W. Douglas, M. Novak, and N.P. Hunziger. 1982. Marten. Pages 599-612 in Chapman, J.A. and G.A. Feldhamer, editors. Wild Mammals of North America. Baltimore, MA: The Johns Hopkins University Press.

Strittholt, J.R., and D.A. DellaSala. 2001. Importance of roadless areas in biodiversity conservation in forested ecosystems: a case study-Klamath-Siskiyou ecoregion, U.S.A. Conservation Biology 15: 1742-1754.

Stucky-Everson, V. 1997. Mushrooms of Colorado and the Southern Rocky Mountains. Denver Botanic Gardens, Denver, CO.

Stumpff, W.K. and J. Cooper. Rio Grande cutthroat trout. Pp. 74-86 in D.A. Duff, editor. Conservation assessment for inland cutthroat trout: distribu-

tion, status, and habitat management implications. U.S. Department of Agriculture, Forest Service, Intermountain Region, Ogden, UT.
Sun, L, D. Muller-Schwarze, and B. A. Schulte. 2000. Dispersal pattern and effective population size of the beaver. Canadian Journal of Zoology 78: 393-398.
Sundstrom, C. 1968. Water consumption by pronghorn antelope and distribution in Wyoming's Red Desert. Antelope Status Workshop Proceedings 3: 239-346.
Sutter, Paul S. 2002. Driven Wild. Seattle, WA: University of Washington Press.
Suzuki, N. and W. C. McComb. 1998. Habitat classification models for beaver (*Castor canadensis*) in the streams of the central Oregon Coast Range. Northwest Science 72(2): 102-110.
Sweitzer, H. A., S. H. Jenkins, and J. Berger. 1997. Near-extinction of porcupines by mountain lions and consequences of ecosystem change in the Great Basin desert. Conservation Biology 11(6): 1407-1417.
Swetnam, T. W. 1990. Fire history and climate in the southwestern United States. Pp. 6-17 in Effects of fire management on Southwestern natural resources. U.S. Forest Service GTR RM-191.
Swetnam, T.W. and J.L. Betancourt. 1998. Mesoscale disturbance and ecological response to decadal climatic variability in the American Southwest. Journal of Climate 11: 3128-3147.
Swetman, T.W. and A.M. Lynch. 1993. Multi-century, regional-scale patterns of western spruce budworm history. Ecological Monographs 63: 399-424.
Sydoriak, C., C. Allen, and B. Jacobs. 2000. Would Ecological Restoration Make the Bandelier Wilderness More or Less of a Wilderness? Wild Earth 10 (4): 83-90.
Talberth, J. and Karyn Moskowitz. 2000. The Economic Case Against National Forest Logging. Forest Conservation Council and Forest Guardians.
Taylor S.L. and S.W. Buskirk. 1994. Forest microenvironments and resting energetics of the American marten, *Martes americana*. Ecography 17(3): 249-256.
Tear, T.H., J.M. Scott, P.H. Hayward, and B. Griffith. 1993. Status and prospects for success of the endangered species act: A look at recovery plans. Science 262: 976-977.
Terborgh, J. 1988. The big things that run the world - a sequel to E. O. Wilson. Conservation Biology 2: 402-403.
Terborgh, J. 1999. Requiem for Nature. Covelo, CA: Island Press.
Terborgh, J. and B. Winter. 1980. Some causes of extinction. Pp. 119-133 in M.E. Soulé and B.A. Wilcox (eds.). Conservation Biology: An Evolutionary-Ecological Perspective. Sunderland, MA: Sinauer Associates.
Terborgh, J. and M. Soulé. 1999. Why we need large-scale networks and megareserves. Wild Earth 9: 66-72.
Terborgh, J., L. Lopez, J. Tello, D. Yu, and A.R. Bruni. 1997. Transitory states in relaxing land bridge islands. Pp. 256-274 in W.F. Laurance and R.O. Bierregaard Jr. (eds.). Tropical Rorest Remnants: Ecology, Management, and Conservation of Fragmented Communities. Chicago IL: University of Chicago Press.
Terborgh, J., J. Estes, P. Paquet, K. Ralls, D. Boyd-Heger, B. Miller, R. Noss. 1999. The role of top carnivores in regulating terrestrial ecosystems. Pages 39-64 in M. E. Soulé and J. Terborgh, editors. Continental Conservation: Scientific Foundations of Regional Reserve Design Networks. Covelo, CA: Island Press.
Terborgh, J., L. Lopez, P. Nuñez V., M. Rao, G. Shahabuddin, G. Orihuela, M. Riveros, R. Ascanio, G.H. Adler, T.D. Lambert, and L. Balbas. 2001. Ecological meltdown in predator-free forest fragments. Science 294: 1923-1925.
Terrell, B. 1971. Lynx. Colorado Outdoors 20: 19.
Tessler, A. and D.R. Shaffer. 1990. Attitudes and attitude change. Annual Review of Psychology 41: 479-523.
The Nature Conservancy. 1998. Ecoregion-based conservation in the Central Shortgrass Prairie. Central Shortgrass Prairie Ecoregional Planning Team, The Nature Conservancy, Boulder, CO.
The Wilderness Act. 1964. Public Law 88-577. 88th congress, Second Session.
The Wilderness Society. 1998. The Wilderness Act Handbook third edition (revised). The Wilderness Society, Washington, D.C.
Theobald, D.T. 2000. Fragmentation by inholdings and exurban development. Pp. 155-174 in R.L. Knight, F.W. Smith, S.W. Buskirk, W.H. Romme, and W.L. Baker (eds.). Forest fragmentation in the Southern Rocky mountains. Boulder, CO: University Press of Colorado.
Theobald, D.T. 2001. Technical description of mapping historical, current, and future housing densities in the US using Census block-groups. Natural Resource Ecology Laboratory, Colorado State University. Data and technical documentation available at:www.ndis.nrel.colostate.edu/davet/dev_patterns.htm.
Theobald, D.T., T. Hobbs, D. Scrupp, and L. O'Brien. 1998. An assessment of imperiled habitat in Colorado. Colorado Division of Wildlife, Unpublished Report.
Thiel, R.P. 1985. Relationship between road densities and wolf habitat suitability in Wisconsin. American Midland Naturalist 113: 404.
Thomas, J.W., R.G. Anderson, C. Maser, and E.L. Bull. 1979. Snags. Pp. 60-77 in J.W. Thomas (ed.). Wildlife habitats in managed forests: The Blue Mountains of Oregon and Washington. U.S. Department of Agriculture. Agriculture Handbook 553, Washington DC.
Thompson, I.D. and P.W. Colgan. 1987. Numerical responses of martens to a food shortage in north central Ontario (Canada). Journal of Wildlife Management 51(4): 824-835.
Thompson, I.D. and P.W. Colgan. 1994. Marten activity in uncut and logged boreal forests in Ontario. Journal of Wildlife Management 58(2): 280-288.
Thompson, I.D. and W.J. Curran. 1995. Habitat suitability for marten of second-growth balsam fir forest in Newfoundland. Canadian Journal of Zoology 73(11): 2059-2064.
Thompson, R.W. and J.C. Halfpenny. 1991. Canada lynx presence on Vail Ski Area and proposed expansion area, unpublished report. Western Ecosystems, Inc. Lafayette, CO.
Thompson, I.D. and A.S. Harestad. 1994. Effects of logging on American martens, and models for habitat management. Pages 355-367 in S.W. Buskirk, A.S. Harestad, M.G. Raphael, and R.A. Powell, editors. Martens, Sables, and Fishers: Biology and Conservation. Ithaca, N.Y.: Comstock Publishing Associates.
Thurow, R.F., C.E. Corsi, and V.K. Moore. 1988. Attributes of Yellowstone cutthroat trout redds in a tributary of the Snake River, Idaho. Transactions of the American Fisheries Society 123: 37-50.
Trombulak, S.C. and C.A. Frissell. 2000. Review of ecological effects of roads on terrestrial and aquatic communities. Conservation Biology 14: 18-35.
Turchin, P., L. Oksanen, P. Ekerholm, T. Oksanen, and H. Henttonen. 2000. Are lemmings prey or predators? Nature 405: 562-565.
Tysor, R.W., and C.A. Worley. 1992. Alien flora in grasslands adjacent to road and trail corridors in Glacier National Park, Montana (U.S.A.). Conservation Biology 6:253-262.
U.S. Department of Agriculture, National Agricultural Statistics Service. 1997. Census of Agriculture.
U.S. Deptartment of Agriculture, Natural Resources Conservation Service. 2002. The PLANTS Database, Version 3.5 http://plants.usda.gov. National Plant Data Center, Baton Rouge, LA 70874-4490 USA.
U.S. Fish and Wildlife Service. 1994. The reintroduction of gray wolves to Yellowstone National Park and central Idaho: final environmental impact statement. U.S. Fish and Wildlife Service, Denver, CO.
U.S. Fish and Wildlife Service. 1996. Reintroduction of the Mexican wolf within its historic range in the southwestern United States: final environmental impact statement. U.S. Fish and Wildlife Service, Albuquerque, N.M.
U.S. Fish and Wildlife Service. 2000. Memo from Region VI Director, 29 November, 1999. Denver, CO.
U.S. Fish and Wildlife Service. 2001a. State lists of Endangered species, December 31, 1999. Available: http://ecos.fws.gov/webpage/webpage_usa_lists.html
U.S. Fish and Wildlife Service. 2001b. Status Review for Bonneville cutthroat trout (*Oncorhynchus clarki utah*). United States Department of the Interior, U.S. Fish and Wildlife Service. Regions 1 and 6, Portland, Oregon and Denver, CO.
U.S. Fish and Wildlife Service. 2002. Rocky Mountain wolf recovery 2001 annual report. U.S. Fish and Wildlife Service, Ecological Services, Helena, MT.
U.S. Forest Service. 1997a. Forest Insect and Disease Conditions in the Rocky Mountain Region, 1996. USDA Forest Service Rocky Mountain Region Renewable Resources, Forest Health Management, Lakewood, CO.
U.S. Forest Service. 1997b. America's forests: 1997 health update. Available: http://www.fs.fed. us/foresthealth/fh_update/update97.
U.S. Forest Service. 1998. Final environmental impact statement for the Rio Grande National Forest Revised Land and Resource Management Plan.
U.S. Forest Service. 1999. Draft environmental impact statement for the White River National Forest Land and Resource Management Plan.
U.S. Forest Service. 2000. Environmental assessment for the upper South Platte watershed protection and restoration project. USDA Forest Service, Pike National Forest, South Platte Ranger District, Morrison, CO.
U.S. Forest Service and U.S. Bureau of Land Management. 1997. The assessment of ecosystem components in the interior Columbia Basin and portions of the Klamath and Great Basins, Vol. I-IV. PNW GTR 405. USFS. Walla Walla, WA.
U.S. General Accounting Office. 1988. More emphasis needed on declining and overstocked grazing allotments. GAO/RCED-88-80. Washington D.C. U.S. General Accounting Office.
U.S. Ninth Circuit Court of Appeals. 2001. United States Court of Appeals for the Ninth Circuit, Order Nos. 99-56362, 00-55496, D.C. No. CV-97-02330-TJW/LSP.

Unsworth, J.W., D. Pac, and G.C. White, and R.M. Bartmann. 1999 Mule. Deer survival in Colorado, Idaho, and Montana. Journal of Wildlife Management 63: 315-326.

Utah Department of Natural Resources. 1996. Conservation agreement and strategy for colorado river cutthroat trout in Utah - Draft. Utah Division of Wildlife Resources, Salt Lake City, Utah; US. Fish and Wildlife Service, Denver, Colorado; US Bureau of Land Management, Salt Lake City, UT; US Forest Service, Ogden, UT.

Utah Wilderness Coalition 1990. Wilderness at the Edge. Salt Lake City, UT.

Valdez, R. and P. R. Krausman. 1999. Description, distribution, and abundance of mountain sheep in North America. Pages 3-22 in R. Valdez and P. R. Krausman, editors. Mountain Sheep of North America. Tucson, AZ: University of Arizona Press.

Van Dyke, F.B., R.H. Brocke, H.G. Shaw, B.B. Ackerman, T.P. Hemker, and F.G. Lindsey. 1986. Reactions of mountain lions to logging and human activity. Journal of Wildlife Management 50: 95-102.

Van Eimeren, P. 1996. Westslope cutthroat trout, *Oncorhynchus clarki lewisi*. In: D.A. Duff, editor. Conservation assessment for inland cutthroat trout: distribution, status, and habitat management implications. U.S. Department of Agriculture, Forest Service, Intermountain Region, Ogden, UT.

VanderHeyden, M. and E.C. Meslow. 1999. Habitat selection by female black bears in the central Cascades of Oregon. Northwest Science 73(4): 283-294

Varley, J.D., and R.E. Gresswell. 1988. Ecology, status, and management of Yellowstone cutthroat trout. American Fisheries Society Symposium 4: 13-24.

Veblen, T.T. 2000. Disturbance patterns. Pp. 31-54 in R.L. Knight, F.W. Smith, S.W. Buskirk, W.H. Romme, and W.L. Baker (eds.). Forest Fragmentation in the Southern Rocky Mountains. Boulder, CO: University Press of Colorado.

Veblen, T.T. and D.C. Lorenz. 1991. The Colorado Front Range: A Century of Ecological Change. Salt Lake City, UT: University of Utah Press.

Veblen, T.T., K.S. Hadley, E.M. Nel, T. Kitzberger, M. Reid, and R. Villalba. 1994. Disturbance regime and disturbance interactions in a Rocky Mountain subalpine forest. Journal of Ecology 82: 125-135.

Veblen, T.T, T. Kitzberger, and J. Donnegan. 2000. Climatic and human influences on fire regimes in ponderosa pine forests in the Colorado Front Range. Ecological Applications 10: 1178-1195.

Verts, B.J. and L.N. Carraway. 1998. Land Mammals of Oregon. Berkeley, CA: University of California Press.

Vickery, P. D., M.L. Hunter Jr., and S.M. Melvin. 1994. Effects of habitat area on the distribution of grassland birds in Maine. Conservation Biology 8: 1087-1097.

Vint, M. M. Briggs, L. Carder, L. Probst. 1998. Conservation options for landowners. The Rincon Institute. Tucson, AZ.

Waldman, C. 1999. Encyclopedia of Native American Tribes. New York, N.Y.: Checkmark Books.

Walker, R., and L. Craighead. 1997. Analyzing wildlife movement corridors in Montana using GIS. Presented at the 1997 ESRI Users Conference. http://www.grizzlybear.org/col/lcpcor.html

Ward, P.D. 1997. The call of the distant mammoths: Why the Ice Age mammals disappeared. New York, NY. Copernicus.

Warren, M.L. and B.M. Burr. 1994. Status of freshwater fishes of the United States: overview of an imperiled fauna. Fisheries 19: 6-18.

Watson, J., D. Freudenberger, and D. Paul. 2001. An assessment of the focal species approach for conserving birds in variegated landscapes in southeastern Australia. Conservation Biology 15: 1364-1373.

Weber, W.A., and R.C. Wittman. 1992. Catalog of the Colorado flora: A Biodiversity Baseline. Niwot, CO: University Press of Colorado.

Weckwerth, R.P. and V.D. Hawley. 1962. Marten food habits and population fluctuations in Montana. Journal of Wildlife Management 26: 55-74.

Weddell, B.J. 1996. Geographic overview: Climate, phenology, and disturbance regimes in steppe and desert communities. Pages 3-12 in D.M. Finch (ed.). Ecosystem Disturbance and Wildlife Conservation in Western Grasslands: A symposium proceedings. USDA Forest Service General Technical Report RM-GTR-285, USDA Forest Service Rocky Mountain Forest and Range Research Station, Fort Collins, CO.

Wehausen, J. D. 1999. Rapid extinction of mountain sheep populations revisited. Conservation Biology 13(2): 378-384.

Weilgus, B. and F.L. Bunnell. 1994. Sexual segregation and female grizzly bear avoidance of males. Journal of Wildlife Management 58: 405-413.

Weilgus, B. and F.L. Bunnell. 2000. Possible negative effects of adult male mortality on female grizzly bear reproduction. Biological Conservation 93: 145-154.

Whicker, A.D. and J.K. Detling. 1988. Ecological consequences of prairie dog disturbances. BioScience 38: 778-785.

Whicker, A.D. and J.K. Detling. 1993. Control of grassland processes by prairie dogs. Pp. 18-27 in J.L. Oldemeyer, D.E. Biggins, B.J. Miller, and R. Crete. Management of Prairie Dog Complexes for the Reintroduction of the Black-Footed Ferret. U.S. Fish and Wildlife Service Biological Report 13. Washington D.C.

White, C.A., C.E. Olmsted, and C.E. Kay. 1998. Aspen, elk, and fire in the Rocky Mountain national parks of North America. Wildlife Society Bulletin 26: 449-462.

White, G.C., and R.M. Bartmann. 1998. Effect of density reduction on overwinter survival of free-ranging mule deer fawns. Journal of Wildlife Management 62: 214-225.

White, R. 1991. It's your misfortune and none of my own: a new history of the American West. Norman, OK: University of Oklahoma Press.

Wiens, J.A. 1997. The emerging role of patchiness in conservation biology. Pp. 93-107 in S.T.A. Pickett, R.S. Ostfeld, M. Shachak, and G.E. Likens (eds.). The ecological basis of conservation: heterogeneity, ecosystems, and biodiversity. New York, N.Y.: Chapman and Hall.

Wilcove, D.S., C.H. McLellan, and A.P. Dobson. 1986. Habitat fragmentation in the temperate zone. Pp. 237-256 in M.E. Soulé (ed.). Conservation Biology: The Science of Scarcity and Diversity. Sunderland, MA: Sinauer Associates.

Wilcove, D.S., D. Rothstein, D. Dubow, A. Philips, and E. Losos. 1998. Quantifying threats to imperiled species in the United States. BioScience 48: 607-615.

Wilcove, D.S., D.S. Voilcove, D. Rothstein, J. Dubow, A. Phillips, E. Losos. 2000. Leading threats to biodiversity-what's imperiling U.S. species. Pp. 237-254 in B.A. Stein, L.S. Kutner, and J.S. Adams (eds.). Precious Heritage-the Status of Biodiversity in the United States. The Nature Conservancy and the Association for Biodiversity information. New York, N.Y.: Oxford University Press.

Wilcox, B.A. and D.D. Murphy. 1985. Conservation strategy: The effects of fragmentation on extinction. American Naturalist 125: 879-887.

Wildlands Project. 2001. Great Dreams. 10th Anniversary Publication. Richmond, VT.

Wildlands Project/Wild Earth. 1992. Wild Earth Special Issue 1992.

Wilen, B.O. 1995. The nation's wetlands. Pp. 473-476 in E. T. LaRoe, G. S. Farris, C. E. Puckett, P. D. Doran, and M. J. Mac (eds.). Our living resources: a report to the nation on the distribution, abundance, and health of U.S. plants, animals, and ecosystems. U.S. Department of the Interior, National Biological Service, Washington, D.C.

Wiley, P. and R. Gottlieb. 1982. Empires in the Sun: The Rise of the New American West. Tucson, AZ: The University of Arizona Press.

Wilkinson, C.F. 1992. Crossing the Next Meridian: Land, Water and the Future of the West. Covelo, CA: Island Press.

Williams, R.M. 1965. Beaver habitat and management. Idaho Wildlife Review 17(4): 3-7.

Williams, R.M., Jr. 1979. Change and stability in values and value systems: A sociological perspective. Pp. 15-46, in M. Rokeach (ed.). Understanding Human Values: Individual and Societal. New York, N.Y.: The Free Press.

Wilson, E.O. 1992. The Diversity of Life. Cambridge, MA: Belknap Press of Harvard University Press.

Wilson, E.O. 2002. The Future of Life. New York, N.Y.: Alfred A. Knopf.

Witmer, G.W., S.K. Martin, and R.D. Sayler. 1998. Forest carnivore conservation and management in the interior Columbia Basin: issues and environmental correlates. USDA Forest Service, Pacific Northwest Research Station and USDI Bureau of Land Management. General Technical Report PNW-GTR-420.

Woodruffe, R., and J.R. Ginsberg. 1998. Edge effects and the extinction of populations inside protected areas. Science 280: 2126-2128.

Wright, H.A. and A.W. Bailey. 1982. Fire Ecology: United States and Southern Canada. New York, N.Y.: John Wiley and Sons.

Wuerthner, G. 2001. Rocky Mountain. Mechanicsburg, PA: Stackpole Books.

Wydeven, A.P. and R.B. Dahlgren. 1985. Ungulate habitat relationships in Wind Cave National Park. Journal of Wildlife Management 49: 805-813.

Yoakum, J.D. 1978. Pronghorn. Pp. 103-121 in J.L. Schmidt and D.L. Gilbert, editors. Big Game of North America. Harrisburg, PA: Stackpole Books.

Yoakum, J.D. 1986. Trends in pronghorn populations: 1925-1983. Pronghorn Antelope Workshop Proceedings 12: 77-85.

Yoakum, J.D. 1990. Food habits of the pronghorn. Pronghorn Antelope Workshop Proceedings 14: 102-111.

Yoakum, J.D. and B.W. O'Gara. 1990. Pronghorn/livestock relationships. North American Wildlife and Natural Resources Conference 55: 475-487.

Yoakum, J.D., B.W. O'Gara, and V.W. Howard. 1996. Pronghorn on the western rangelands. Pp. 211-226 in P.R. Kraussman, editor. Rangeland wildlife. Society for Range Management, Denver, CO.

Young, M.K. (ed.). 1995. Conservation assessment for inland cutthroat trout. USDA Forest Service General Technical Report, RM-GTR-256. Rocky Mountain Forest and Range Experiment Station, Fort Collins, CO.

Young, M.K. 1996. Summer movements and habitat use by Colorado River cutthroat trout (*Oncorhynchus clarki pleuriticus*) in small, montane streams.

Canadian Journal of Fisheries and Aquatic Sciences 53: 1403-1408.
Young, M.K., R.N. Schmal, T.W. Kohley, and V.G. Leonard. 1996. Colorado River cutthroat trout, Oncorhynchus clarki pleuriticus. In: D.A. Duff, editor. Conservation assessment for inland cutthroat trout: distribution, status, and habitat management implications. U.S. Department of Agriculture, Forest Service, Intermountain Region, Ogden, UT.
Young, M.K., A.L. Harig, B. Rosenland, and C. Kennedy. 2002. Recovery history of greenback cutthroat trout: Population characteristics, hatchery involvement, and bibliography. U.S. forest Service Rocky Mountain Research Station. RMRS-GTR-88WWW.
Zarn, M. 1981. Wild ungulate forage requirements-a review. U.S. Bureau of Land Management Special Report. Denver, CO.
Zeigenfuss, L.C., F.J. Singer, and M. A. Gudorf. 2000. Test of a modified habitat suitability model for bighorn sheep. Restoration Ecology 8(4S): 38-46.
Zielinski, W.J., and T.E. Kucera. 1995. American marten, fisher, lynx, and wolverine: survey methods for their detection. USDA, Forest Service, Pacific Southwest Research Station. General Technical Report PSW-GTR-157.
Zinn, H.C. and W.F. Andelt. 1999. Attitudes of Fort Collins, Colorado, residents toward prairie dogs. Wildlife Society Bulletin 27: 1098-1106.

# APPENDIX 1:

# FOCAL SPECIES OF THE SOUTHERN ROCKIES WILDLANDS NETWORK DESIGN

Dave Parsons, Brian Miller, Angie Young, Kyran Kunkle, Paul Paquet, Mike Phillips, and Mike Seidman

(Fielder 1997)

Black bear (*Ursus americanus*)

## American Marten *(Martes americana)*

### *1. Introduction*

The American marten (*Martes americana*) is a member of the weasel family (Mustelidae) and is slightly smaller than an average house cat. The marten is a habitat-specialist that lives in old-growth spruce-fir forests and preys primarily on small mammals. It requires habitat protection and forested linkages among patches of suitable habitat for its survival and persistence. Ecologically, the marten is an important predator of small mammals and may be an important distributor of seeds of fruit-bearing shrubs. It is an indicator of habitat and wilderness qualities within the Southern Rockies planning area.

### *2. Distribution*

*Historic*

American martens occupied temperate forests of the Rocky Mountains from northern New Mexico to Alaska; the Sierra Nevada, Cascade and Coastal ranges of California, Oregon, and Washington; most of Canada; and the Great Lakes and Northeast regions of the United States (Hall 1981). Their distribution is discontinuous in the Southern Rocky Mountains because climate change isolated suitable habitats on the tops of mountains following the Pleistocene epoch (Gibilisco 1994). The southern limit of marten distribution in the Southern Rocky Mountains roughly coincides with the southern limit of Engelmann spruce (*Picea engelmannii*, Buskirk and Ruggiero 1994). The resulting patchy distribution is common at the edge of a species' distribution (Gibilisco 1994). Findley et al. (1975) depict historic range of martens in New Mexico as far south as the

Mogollon Mountains, Black Range, Sacramento Mountains, and Guadalupe Mountains, but they provide no records of specimens south of Santa Fe. Bailey (1931) described marten distribution as uncommon in the Sangre de Cristo and San Juan Mountains, which he believed marked the southern limit of their range. He attributed their scarceness to trapping pressure for their valuable pelts.

Population declines resulted from exploitation of martens for their valuable fur and habitat alterations caused by logging (Strickland 1994). Martens were extirpated from many of the southern parts of their historic range (Strickland 1994), but have become reestablished (through recolonization or reintroduction) in many states and provinces (Gibilisco 1994).

*Current*

Marten populations at the edge of their distributional range are naturally unstable and are especially vulnerable to local extirpation as a result of overharvest, habitat alterations, and fragmentation of forested environments. This appears to be the situation in the Southern Rocky Mountains. The presence of American martens has been documented generally north of Santa Fe in the San Juan and Sangre de Cristo Mountains. Reports of martens in the Jemez Mountains have not been verified (B. Long, personal communication). While the presence of martens has been documented in the Southern Rockies, little is known about the status, trends, or potential long-term viability of extant populations.

*Potential*

Areas of approximately 40 km$^2$ (15.4 mi$^2$) or greater within high-elevation mesic coniferous forest types with the following characteristics generally represent potentially occupied range for American martens: 1) >20% of forest stand in old-growth age classes; 2) >30% canopy closure; 3) small openings with high understory plant diversity; 4) complex vertical and horizontal structure with an abundance of snags and large woody debris that breeches the snow surface in winter; 5) forested connectivity within and between patches of suitable habitat; and 6) an abundance of small mammalian prey.

## 3. Habitat

*General*

The following description of habitat requirements and preferences of American martens is specific to the intermountain West. Habitat relationships are somewhat different for midwestern and eastern populations. American martens are "habitat-specialists" that occupy structurally complex, late successional forests dominated by conifers (Koehler and Hornocker 1977, Buskirk and Powell 1994). Martens will use patches of preferred habitat that are interconnected by other forest types, but will avoid similar patches that are separated by large open areas (Buskirk and Powell 1994). This explains why marten populations no longer exist in many small mountain ranges in post-glacial time and why recolonization of potentially suitable, but isolated, habitats has not occurred (Buskirk and Powell 1994, Gibilisco 1994). Open areas with abundant cover (*e.g.* shrubs or fallen trees) are used by martens for foraging (especially in summer), but open grasslands and alpine tundra are avoided (Koehler and Hornocker 1977, Buskirk and Powell 1994).

*Preferred*

American martens prefer old-growth mesic coniferous forest types with complex physical structure, especially near the ground (Buskirk and Ruggiero 1994, Thompson and Curran 1995). Mesic sites support greater understory plant species diversity and higher vole populations (Koehler and Hornocker 1977). Small openings where large trees have fallen enhance understory plant and animal diversity (Thompson and Harestad 1944). Abundance of small mammals, especially voles (*Clethrionomys* spp., *Microtus* spp., and *Phenacomys* spp.) and pine squirrels (*Tamiasciurus* spp.), is directly associated with abundance of American martens (Strickland et al. 1982, Buskirk and Ruggiero 1994, Martin 1994). Optimum American marten habitat should include 20-25% of the forest at the landscape level in mature age classes at any given time; at least 50% conifers; large standing snags; and numerous fallen trees for subnivean (under the snow) foraging and rest sites (Soutiere 1979, Steventon and Major 1982, Buskirk et al. 1989, Berg and Kuehn 1994, Thompson and Harestad 1994). An abundance of large, near-ground woody structures provides martens with protection from predators, access to subnivean space where most prey are captured in winter, and protective thermal microenvironments (Buskirk and Ruggiero 1994).

In the Rocky Mountains, American martens prefer mesic high-elevation stands dominated by spruce (*Picea* spp.) and fir (*Abies* spp.) over stands dominated by dry-site species such as ponderosa pine (*Pinus ponderosa*) and lodgepole pine (*Pinus contorta*, Buskirk and Powell 1994). In more southern reaches of their range, martens may select riparian forests for resting during winter (Buskirk et al. 1989). Mature spruce-fir forests are preferred habitat for important marten prey species such as southern red-backed voles (*Clethrionomys gapperi*), which depend on large downed logs and stumps (coarse woody debris), dense forest canopy, and pine squirrels (*Tamiasciurus hudsonicus*, Martin 1994).

Martens avoid areas with less than 30% canopy closure and open areas with no overstory or shrub cover (Buskirk and

Powell 1994). Habitat that has become highly fragmented (*e.g.*, habitat patches <200 km$^2$ separated by >5 km zones with <50% overhead cover) is not likely to be used by martens (Buskirk and Powell 1994). Robitaille and Aubry (2000) observed significantly fewer marten tracks within 400 m of roads compared to transects 800-1,000 m from roads. This suggests a zone of reduced habitat suitability for martens near roads.

In areas where deep snow accumulates, American martens prefer cover types with sufficient vertical and horizontal structure (*e.g.*, closed, multi-layered coniferous forests) to prevent snow from packing hard and with near-ground structures that breech the snow surface and provide access to subnivean spaces (Buskirk and Powell 1994).

#### *Seasonal Habitat Preferences*

Generally, martens use a wider range of cover types within their established home ranges (including small open areas) in summer but strongly prefer old conifer-dominated stands in winter (Buskirk and Powell 1994, Buskirk and Ruggiero 1994). Habitat suitability for martens appears to be limited by the availability of preferred winter habitat within their home ranges (Buskirk and Ruggiero 1994).

#### *Resting Sites*

American martens use resting sites for thermoregulation (*i.e.*, to conserve energy by reducing heat loss in winter), resting in all seasons, and protection from predators (Buskirk and Ruggiero 1994). Generally, martens rest above ground in the warmer months and below the snow when temperatures fall below freezing (Buskirk et al. 1989). Martens remain in resting sites for several hours and often all day (Buskirk 1984). Energy conservation achieved through rest site selection reduces foraging needs by 1-3 voles per day and corresponding foraging effort for martens (Taylor and Buskirk 1994).

Favored resting sites include logs, stumps, snags, pine squirrel middens and tree nests, tree cavities, tree limbs, dense clumps of tree limbs ("witches' brooms") caused by tree parasites (*e.g.*, broom rust and dwarf mistletoe), ground burrows, rock fields, root wads, lumber piles, and buildings (Buskirk 1984, Spencer 1987, Buskirk et al. 1989, Buskirk and Powell 1994, Bull and Heater 2000). Resting sites were located under or adjacent to, as well as within cavities inside, logs, stumps, and the base of snags (Buskirk et al. 1989, Bull and Heater 2000). Most subnivean resting sites contained evidence of use by pine squirrels (Spencer 1987, Bull and Heater 2000). Large trees, logs, and snags are preferred over smaller ones (Spencer 1987).

Preferred resting sites are generally found within old-growth spruce-fir forests with relatively closed canopies and an abundance of standing snags in early stages of decay and coarse woody debris on the forest floor (Buskirk 1984, Buskirk et al. 1989, Buskirk and Powell 1994, Bull and Heater 2000). Coarse woody debris in addition to large trees with low limbs and snags provide important access points to subnivean resting and foraging sites (Buskirk and Ruggiero 1994). The distribution and abundance of preferred rest sites contribute to habitat quality and may limit the distribution and abundance of martens (Buskirk 1984, Spencer 1987, Taylor and Buskirk 1994).

#### *Den Sites*

American martens use two types of dens: natal (where young are born) and maternal (where young are reared). In order of observed importance, martens den in hollow logs, live tree cavities, rocks, snag cavities, stumps, slash piles, underground burrows, pine squirrel middens, man-made structures, and rootwads (Buskirk and Ruggiero 1994, Bull and Heater 2000). Den trees, snags, and logs were characterized by large diameters, emphasizing the importance of old-growth forests for martens (Buskirk and Ruggiero 1994, Bull and Heater 2000). The availability of structurally complex old-growth coniferous forests, which provide suitable natal dens, could have important implications for the conservation of American martens (Buskirk and Ruggiero 1994).

### *4. Food Habits / Hunting Behavior*

American martens are opportunistic feeders and adjust their dietary preferences seasonally to take advantage of prey availability (Martin 1994, Verts and Carraway 1998, Bull 2000). They eat small mammals, birds, insects, and fruits (Strickland et al. 1982, Martin 1994). Voles are prominent in their diet (Koehler and Hornocker 1977, Soutiere 1979, Strickland et al. 1982, Gordon 1986, Fitzgerald et al. 1994, Martin 1994, Thompson and Curran 1995, Simon et al. 1999, Bull 2000). In some areas, pine and ground (*Spermophilus* spp.) squirrels may be important prey in late winter and early spring (Buskirk and Ruggiero 1994). Birds, eggs, insects, and fruits are important foods in summer and fall when they are available and vulnerable (Soutiere 1979, Strickland et al. 1982). Vegetative, bird, and insect components of martens' overall diets are likely secondary in importance to mammalian prey (Martin 1994). Martens readily consume carrion, especially in winter (Strickland et al. 1982, Fitzgerald et al. 1994).

Martens pursue their prey on the ground and in trees (Fitzgerald et al. 1994). Activity periods vary seasonally, and martens are mostly crepuscular and nocturnal in winter with increased activity during daylight hours in the summer (Fitzgerald et al. 1994, Verts and Carraway 1998). In winter, much hunting occurs beneath the snow (Martin 1994).

Fallen logs, trees with low limbs, and tree stumps that break the snow surface provide access to this subnivean zone (Strickland et al. 1982) and are routinely investigated by hunting martens, presumably for olfactory cues of prey presence (Buskirk and Powell 1994). Sherburne and Bissonette (1994) found that American martens used subnivean areas with higher prey biomass and higher amounts of coarse woody debris. They speculated that martens are able to discriminate between subnivean access points with high prey levels and those with low prey levels. Food may be cached at resting sites (Fitzgerald et al. 1994). Typically, older forests with accumulated coarse woody debris provide the forest floor structure necessary to enable martens to forage effectively in the winter (Buskirk and Ruggiero 1994). Exceptions may include mid-aged stands where burns or windthrow provide coarse woody debris on the forest floor (S. Buskirk, personal communication).

## 5. Population Dynamics

### *Life History*

American martens are members of the weasel family (Mustelidae). They are slightly smaller than an average house cat with the following body dimensions: total length 460-750 mm (1.5-2.5 ft); length of tail 170-250 mm (0.6-0.8 ft); and weight 0.5-1.2 kg (1.1-2.6 lbs, Fitzgerald et al. 1994). Males are 20-40% larger than females (Buskirk and Ruggiero 1994). The following life history information is summarized from Verts and Carraway (1998) and Strickland et al. (1982). Martens are solitary, but males and females may pair briefly for mating; home ranges of males overlap little and home ranges of one or more females may occur within the home range of a male. Most female martens become sexually mature at about 15 months and produce their first litters at the age of 2 years. Males and females may mate with more than one partner and breeding occurs in the summer months. Total gestation ranges from 220-276 days; only one litter is produced per year; young are born in March and April; and average litters have about 3 young. Wild reproduction by 12 to 14-year-old females has been documented, although few martens (<15%) live beyond 4 years (Clark et al. 1989, Nowak 1991, Fitzgerald et al. 1994).

### *Population Density*

American martens occur at very low densities compared to other mammalian carnivores of similar size (Buskirk and Ruggiero 1994). While marten populations are constantly in flux due to changes in prey availability, average population density in suitable habitat is about 1.2 adults/km$^2$ (Strickland et al. 1982). In fall and winter, transient adults and juveniles may increase density to over 2 martens/km$^2$.

### *Home Range*

Mean home range size for males is 8.1 km$^2$ and for females is 2.3 km$^2$ (Powell 1994). Home range size is inversely related to prey availability and habitat quality and directly related to body size of the marten (Thompson and Colgan 1987, Buskirk and Ruggiero 1994).

### *Causes of Death*

About half of martens die in their first year of life; 15-20% die in their second year; 12-13% in their third year; and few martens (<15%) live beyond 4 years (Strickland et al. 1982, Bull and Heater 2001).

The average annual harvest of American martens in North America in the 1980s was 192,000 animals (Strickland 1994). Reduced popularity of furs in the 1990s has led to reduced trapping pressure. In harvested populations, trapping is the predominant cause of death for martens (Buskirk and Ruggiero 1994). Trapping mortality may be additive to natural mortality (Strickland 1994). Bull and Heater (2001) found predation to be the major cause of mortality of an untrapped marten population in Oregon.

Predation on martens by coyotes (*Canis latrans*), bobcats (*Lynx rufus*), red foxes (*Vulpes vulpes*), other martens, great-horned owls (*Bubo virginianus*), fishers (*Martes pennanti*), lynx (*Lynx canadensis*), mountain lions (*Puma concolor*), eagles (*Aquila chrysaetos*, *Haliaeetus leucocephalus*), and northern goshawks (*Accipiter gentilis*) has been documented (Strickland et al. 1982, Buskirk and Ruggiero 1994, Slough 1994, Squires 2000, Bull and Heater 1995, 2001). Bull and Heater (2001) observed that most predation occurred between May and August and that no predation occurred between December and February, when martens spend more time in dense cover and under the snow.

Diseases and parasites are not known to limit American marten populations (Strickland et al. 1982, Slough 1994), with the exception of canine distemper which was determined to be the cause of a dramatic die-off of martens on Newfoundland Island (Fredrickson 1990, cited in Buskirk and Ruggiero 1994).

In roaded areas, vehicle collisions cause some marten mortality (Buskirk and Ruggiero 1994). Bull and Heater (2001) documented hypothermia as a cause of mortality in martens.

### *Population Structure and Viability*

A metapopulation structure has not been described for marten populations; but, intuitively, this type of dispersed population structure seems plausible for the naturally and anthropogenically fragmented habitats that exist in the southern portions of its range (Buskirk and Ruggiero 1994).

Generally, marten population structure is characterized by non-overlapping territories of same-sex adults, larger male territories overlapping one or more female territories, pre-dispersal young occupying adult female territories, and dispersing juveniles often occupying marginal habitats (Buskirk and Ruggiero 1994). Martens are believed to be territorial; territory size and local population density are linked to food density; and intraspecific aggression has the potential to increase marten mortality rates (Fryxell et al. 1999, Bull and Heater 2001).

American marten populations fluctuate dramatically (up to an order of magnitude in less than a decade) in direct response to fluctuations in prey populations (Buskirk and Ruggiero 1994, Martin 1994, Powell 1994). American martens are easily trapped, have high pelt values, have relatively low recruitment rates, and are vulnerable to overharvesting (Strickland 1994). Because marten reproductive rates are low, they are slow to recover from population declines (Buskirk and Ruggiero 1994).

Slough (1994) proposed that populations in excess of 50 animals were necessary for short-term viability and that over 500 animals were required to assure long-term fitness and genetic adaptability of a marten population. Schneider and Yodzis (1994) predicted that populations with 75-125 females were likely to persist for 500 years. Assuming an even sex ratio for untrapped populations (Powell 1994), this would equate to an adult population of 150-250 martens, which accords with the estimate of a minimum of 237 martens for population viability by Thompson and Harestad (1994). Lacy and Clark (1993) found that all but one simulated populations of 50 and 100 martens had a 100% probability of extinction within 100 years when no immigration into the population was assumed. Simulated immigration of one pair of martens per year was sufficient to prevent extinction, even with specified levels of trapping and logging. Smaller populations benefit from periodic infusions of immigrants from larger, more genetically diverse populations.

Assuming a population density of 1.2 individuals/km$^2$, core populations in reserves of 200-400 km$^2$ would be necessary to support nearby subpopulations in patches of suitable habitat ranging from 40-200 km$^2$. Multiple populations of minimum viable size or larger linked by suitable migration corridors are necessary to further ensure long-term survival of American martens in a region (Thompson and Harestad 1994). Small populations are at increased risk of extinction from natural catastrophies (*e.g.*, severe storms and fires) and other random events that affect the survival and reproduction of individuals (*e.g.*, an unusually skewed sex ratio); environmental conditions or ecological relationships (*e.g.*, reduced food supply or increased population levels of competitors or predators); and genetic health of the population (*e.g.*, inbreeding and genetic drift, Meffe and Carroll 1997). Migrants from surviving populations can "rescue" declining populations or recolonize suitable habitats following local extirpations.

American marten populations decline following clear-cut logging at all but very small (0.5-3 ha) scales (Soutiere 1979, Buskirk and Ruggiero 1994, Thompson and Colgan 1994, Thompson and Harestad 1994, Forsey and Baggs 2001). Thompson and Harestad (1994) compared 10 studies of habitat selection by martens. Use/availability ratios were consistently less than 1 for shrub, sapling, and pole stages of forest regeneration. Only old-growth stands consistently had use/availability ratios that exceeded 1, indicating a preference or selection for these habitats by martens. Clear-cutting removes overhead cover, removes large-diameter coarse woody debris, and tends to convert mesic sites to xeric sites with associated changes in prey communities to less preferred species (Buskirk and Ruggiero 1994).

## 6. Movements

### *Dispersal*

Data on the dispersal and migrational movements of martens are scarce (Buskirk and Ruggiero 1994). American martens will travel through forested areas that are otherwise not preferred habitat (Buskirk and Powell 1994). Factors important to successful marten dispersal and population persistence of isolated populations include size of habitat islands, distance between habitat islands, distance to large core populations, and the nature of the zones that separate populations (Gibilisco 1994). Martens are known to traverse alpine tundra (1.1 km), forests with <25% canopy cover (>10 km), extensive burned forest (>20 km), and large rivers (>100 m, Slough 1989). Reported dispersal distances for juvenile martens range from 27-60+ km (Weckwerth and Hawley 1962, Strickland and Douglas 1987). No significant between-sex differences in dispersal distances have been reported.

### *Barriers to Movement*

American martens tend to avoid open areas, but may travel up to 3 km from forested cover provided some physical cover (*e.g.*, talus fields) is available (Streeter and Braun 1968, Buskirk and Powell 1994, Potvin et al. 1999). Distances of more than 5 km of unforested land below the conifer zone are believed to present a complete barrier to dispersal (Gibilisco 1994). Koehler and Hornocker (1977) observed that martens passed through, but did not hunt in, openings <100 m wide, and Soutiere (1979) documented martens crossing 200-m-wide openings. Martens avoided bait placed >23 m from the forest edge within openings in winter (Robinson 1953). Bull and Blumton (1999) observed that radio collared martens in northeastern Oregon avoided all harvested stands and stands with less than 50% canopy cover.

### *Use of Corridors*

Use of corridors by dispersing martens has not been documented. Gibilisco (1994) suggested that forested riparian zones may serve as dispersal corridors linking disjunct populations of martens. Martens' observed reluctance to cross areas lacking overhead cover suggests the importance of linkages with overhead cover to the colonization of patches of suitable habitat devoid of martens, the genetic and/or demographic augmentation of small subpopulations, and the persistence and viability of marten populations (Buskirk and Ruggiero 1994).

## 7. Ecology

### *Effects on Prey*

American marten populations tend to respond directly to changes in prey populations (Strickland et al. 1982, Thompson and Colgan 1987, Buskirk and Ruggiero 1994, Fryxell et al. 1999, Simon et al. 1999). Martens kill many small mammals but are only one of many predators of small mammals in the forest communities they occupy (Schneider and Yodzis 1994). There is no evidence to date that suggests that martens exert a regulatory effect on prey populations, except perhaps as a member of the larger suite of predators of small mammals.

### *Ecological Effects*

Because American martens forage heavily on fleshy fruits in the summer and fall, they may be important dispersers of the seeds of fruit-bearing shrubs, such as blueberries (*Vaccinium* spp.) and salmonberries, raspberries, blackberries, and thimbleberries (*Rubus* spp., Buskirk and Ruggiero 1994, Hickey et al. 1999). Other ecological effects of martens have not been elucidated.

## 8. Management Recommendations

### *Establish Refugia*

Many relict populations of American martens owe their existence to refuges where furbearers were protected or inaccessible. Refuges provide population reservoirs for dispersal to surrounding areas (Strickland 1994). However, much of the knowledge necessary to inform refuge design, such as the required sizes and distances separating refugia, is lacking (Buskirk and Ruggiero 1994). Based upon published population viability assessments, population density estimates, and dispersal distances, we recommend the establishment of refugia in all patches of suitable habitat of 40 $km^2$ or larger. Refugia in the size range of 200-400 $km^2$ or larger are preferred. Refugia smaller than 200 $km^2$ should be connected by forested cover with >50% canopy closure and no farther than 50 km from two or more other refugia and, preferably, one in excess of 200 $km^2$. In the absence of suitable linkage between refugia smaller than 200 $km^2$, supplementation of small populations with at least one pair of adults per year is recommended.

### *Protect Isolated Subpopulations*

Schneider and Yodzis (1994) suggest that a system of multiple marten subpopulations (supported through landscape-level habitat management) will be necessary to ensure the long-term persistence of martens in a given region. Management practices should consider the vulnerability of local, isolated marten subpopulations to extinction (Buskirk and Ruggiero 1994, Gibilisco 1994, Schneider and Yodzis 1994). Management considerations should include habitat protection and enhancement, prohibition of take, and subpopulation augmentation through translocation of martens from larger populations. Populations that have persisted since prehistoric times likely represent locally adapted forms and warrant greater protection than populations created through translocations (Buskirk and Ruggiero 1994).

### *Protect or Restore Corridors*

Because martens will not travel far from overhead forest cover, direct links of forest cover with greater than 50% canopy cover among suitable habitat blocks are essential and recommended (Witmer et al. 1998).

### *Reintroduce Martens to Suitable Habitats*

Reintroductions have successfully reestablished American marten populations in many areas (Slough 1994). All translocation attempts where 30 or more martens were released were successful. Quality of habitat is important to the success of reintroductions (Slough 1994). Reintroduction should be considered for areas with suitable habitat of sufficient size (>40 $km^2$) to support a subpopulation of martens (Buskirk and Ruggiero 1994).

### *Monitor Marten Populations*

Because fitness is difficult to assess in marten populations, population density is probably the most useful and attainable measure of population fitness and habitat quality (Buskirk and Ruggiero 1994). Martens are sensitive to habitat loss or degradation. Resource managers must implement effective monitoring strategies to detect effects of land management practices on habitat quality and numerical abundance and density of martens (Buskirk and Ruggiero,

Raphael 1994). Promising techniques for monitoring marten populations are snow tracking, sooted track plates, and baited remote camera stations (Bull et al. 1992, Buskirk and Ruggiero 1994, Raphael 1994, Zielinski and Kucera 1995).

#### *Forest Planning*

Forest plans should provide sufficient old-growth coniferous forest to ensure the long-term survival and viability of American martens. Thompson and Harestad (1994) recommend the following timber harvest strategy for mature conifer-dominated forests, which they predict will increase the forest's carrying capacity for American martens: dispersed timber removal in 1-3 ha cuts of less than 25% of total stems with no more than 20-30% total forest removal. We believe that these criteria should be applied over the maturation cycle for old-growth ("overmature") spruce-fir forests at the landscape scale. The proportion of old-growth spruce-fir forest should exceed 20% at the landscape scale at any given time. Forest management should promote coarse woody debris and snags larger than 80 cm in diameter in forest communities identified as winter resting habitat for martens (Buskirk et al. 1989, Witmer et al. 1998). The most important consideration for forest planners is the dynamic extent and configuration of the remaining forest following the application of a timber harvest prescription (Potvin et al. 1999).

#### *Fire*

Fire is an important agent in creating forest diversity. "A mosaic of forest communities supporting discontinuous fuel types can...be expected to result in smaller and generally cooler fires, which would result in less marten habitat being replaced though time and space"Koehler and Hornocker (1977:504). High-intensity fires that consume coarse woody debris and large snags are not beneficial to martens (Strickland et al. 1982). Martens tend to avoid large openings created by fire.

#### *Educate Managers and the Public*

Wildlife managers and policy makers need a thorough understanding of marten ecology in order to establish appropriate policies and make sound management decisions. In addition, the public needs accurate information and knowledge about martens to inform their opinions and values and their understanding of appropriate management measures. Knowledge is the key to informed conservation actions and advocacy by both agencies and the public.

### *9. Justification*

The American marten was selected as a habitat and wilderness quality indicator species.

**Habitat Quality Indicator:** Decline of American marten populations because of habitat loss is indicative of more insidious forest management problems—specifically, the inability to regenerate forest ecosystems to prior levels of complexity and the general lack of long-term, broad-scale vision in forest planning (Thompson and Harestad 1994).

**Wilderness Quality Indicator:** The American marten is dependent on the presence of old-growth, unlogged forests for habitat. The best protection for such forests is formal designation as Wilderness Areas.

However, this species was not selected for inclusion in the initial SITES modeling because its home range and dispersal needs require a fine resolution, the details of which would be lost at an ecoregional scale.

## Beaver (*Castor canadensis*)

### *1. Introduction*

The beaver (*Castor canadensis*) is the largest native rodent (order Rodentia) and only representative of its family (Castoridae) and genus (*Castor*) in North America. It is a riparian-obligate species that is widely distributed in low-gradient streams and rivers and lakes throughout the Southern Rockies. Ecologically, beavers are herbivores and potential prey for large and medium sized predators; but, most importantly, the beaver is a keystone species that significantly increases biological diversity and ecological productivity and stability through its activities. They are common throughout the Southern Rockies, but are most abundant in the subalpine zone. The beaver's popularity with the general public and nature enthusiasts fosters considerable public interest and support for nature conservation.

### *2. Distribution*

#### *Historic*

Beavers historically occurred in streams, ponds, and lake shores throughout North America, except in the arctic tundra, Florida peninsula, and southwestern deserts (Jenkins and Busher 1979). Populations were extirpated or reduced to near extinction throughout most of the beaver's historic range by the early 1900s, but reintroduction programs have successfully restored beavers to many parts of their former range (Jenkins and Busher 1979, Naiman et al. 1988). However, present populations represent a small fraction of historical numbers of beavers (Naiman et al. 1988). The situation was similar in the Southern Rockies where beavers were historically abundant in most watersheds, but were eliminated by trapping from many streams and rivers and reduced to extremely low numbers elsewhere by the end of the 19th century (Bailey 1931).

#### *Current*

Beavers are distributed throughout much of their former range in the Southern Rockies, and they are mostly found in the subalpine zone.

#### *Potential*

Although there is potential to restore and expand beaver populations, many landowners in the Southern Rockies consider them a pest.

### 3. Habitat

#### *General*

The following attributes contribute to habitat suitability for beavers: (1) stable aquatic systems with adequate, permanent water; (2) channel gradients less than 15%; (3) wide valley floors, and (4) adequate supplies of quality food resources (Williams 1965, cited in Allen 1983, Suzuki and McComb 1998). Beavers are stimulated to build dams by the sound of running water (Jenkins and Busher 1979). Suitability of beaver habitat may be reduced by nearby railways, roads, and land clearing activities (Slough and Sadleir 1977, cited in Allen 1983). Streams and lakes exhibiting extreme annual fluctuation in flow volumes or water levels have little value as beaver habitat (Allen 1983).

#### *Preferred*

The following information is from Allen (1983), unless otherwise cited. Beavers require a permanent supply of water and stable water levels, which beavers can control on small streams, ponds, and lakes by constructing dams. Larger rivers and lakes where water level control is beyond the capability of beavers are usually unsuitable, as are swift streams that lack suitable dwelling sites during periods of high and low water.

In lotic systems, stream gradient is the most significant factor affecting habitat suitability. In Colorado, most beaver colonies occupied stream valleys with less than 6% gradient; 90% of all colonies observed were in reaches with less than 12% gradient; and no colonies were found in streams with gradients that equaled or exceeded 15% (Retzer et al. 1956, cited in Allen 1983). Only streams with valleys that are wider than the stream channel provide suitable habitat for beavers. Valley widths that equal or exceed 46 m are considered most suitable. Flat flood plains allow for the construction of extensive canal systems which aid beavers in accessing and transporting food. Marshes, ponds, and lakes associated with adequate food resources also provide suitable habitat for beavers.

Food availability is an important determinant of habitat suitability, especially palatable woody plants that can be cached for winter consumption (Allen 1983, Jenkins and Busher 1979. Food preferences are discussed in more detail below.

#### *Special Features*

Beavers construct lodges or burrows which provide secure cover for escape from predators, resting, thermal regulation, and reproduction (Jenkins and Busher 1979). Lodges may be constructed against a shoreline (sometimes over a bank burrow) or in open water, and entrances are below the water line (Allen 1983). Beavers often modify their habitats by constructing dams and extensive canal systems. Generally, dam building is limited to 1st to 4th order streams that are 15 m or less wide and have gradients of 4% or less (Gurnell 1998). Hydrogeomorphological effects of dam building by beavers include: stream flow stabilization, attenuation of peak flood flows; raised water tables; increased hydrologic complexity in riparian areas; increased stream channel complexity; increased sediment storage and decreased sediment yields by streams; sorting of bed sediments creating greater benthic substrate diversity; increased decomposition of organic matter and release of nutrients within the stream ecosystem; and, generally, increased lotic and riparian habitat diversity and stability (Gurnell 1998).

### 4. Food Habits / Foraging Behavior

Beavers appear to prefer herbaceous vegetation (such as forbs, grasses, and aquatic vegetation, including their roots and tubers) over woody plants when it is available (Allen 1983, Howard and Larson 1985). They eat the bark, buds, leaves, and twigs of a wide variety of woody plants and may show strong local preferences for particular plant species (Fitzgerald et al. 1994, Breck et al. 2003). Commonly preferred woody species include quaking aspen (*Populus tremuloides*), willows (*Salix* spp.), cottonwoods (*Populus* spp.), and alders (*Alnus* spp., Allen 1983, Breck et al. 2003). These species sprout vigorously after fires (New Mexico Department of Game and Fish 2001a). Both terrestrial and aquatic herbaceous plants are eaten during twilight and darkness in spring and summer (Fitzgerald et al. 1994). Beavers are coprophagous, re-ingesting their feces to achieve more complete digestion of foods consumed (Fitzgerald et al. 1994). Most feeding by beavers occurs within 30 m of water, but foraging can extend out to about 100 m (Allen 1983, Fitzgerald et al. 1994). Harvested food is typically cached and consumed under the ice near lodges or burrows in winter (Fitzgerald et al. 1994).

Beavers cut trees year round, but tree cutting peaks in late fall, when herbaceous vegetation is less available and food is being stored under water for winter, and again in early spring (Allen 1983).

## 5. Population Dynamics

### *Life History*

The beaver is the largest rodent in North America, weighing 15-30 kg, with males slightly larger than females (Jenkins and Busher 1979, Fitzgerald et al. 1994). These semiaquatic mammals build dams, lodges, and canal systems to create a secure, aquatic environment which provides shelter and a system for floating and storing food and moving construction materials (Fitzgerald et al. 1994). Beavers are social animals and live in colonies of 4-8 animals comprised of an adult pair, yearling offspring, and young of the year (Jenkins and Busher 1979, Fitzgerald et al. 1994). Beavers are monogamous, become sexually mature at 1.5 years of age but first reproduce at about 3 years of age, mate in late winter, and bear average litters of 3-4 kits (Jenkins and Busher 1979). Beavers may live up to 21 years, but most do not live beyond 10 years in the wild (Jenkins and Busher 1979).

### *Population Density*

Typical population densities for beaver are 0.4-0.8/km$^2$ (Jenkins and Busher 1979). Breck et al. (2001) found densities of bank-dwelling beavers at 2.5 beavers per km of shore on the Green River (or 0.5 colonies per km) and 1.75 beavers per km of the Yampa River (or 0.35 colonies per km).

### *Home Range*

In suitable habitats, beaver colonies tend to be spaced about 1 km apart along streams, and colonies defend a territory of about 2-3 ha (Fitzgerald et al. 1994).

### *Causes of Death*

Predators of beavers include wolf (*Canis lupus*), coyote (*Canis latrans*), bears (*Ursus* spp.), river otter (*Lontra canadensis*), mink (*Mustela vison*), lynx (*Lynx canadensis*), bobcat (*Lynx rufus*), and mountain lion (*Puma concolor*, Jenkins and Busher 1979, Fitzgerald et al. 1994). Other causes of death include hunting and trapping by humans, starvation, drowning during floods, and epizootics of tularemia (Fitzgerald et al. 1994). Average annual mortality is about 30% (Jenkins and Busher 1979, Fitzgerald et al. 1994).

### *Population Structure and Viability*

Beaver populations consist of nonoverlapping colonies, each defending their territory (Allen 1983). Typically, a colonized area consists of a series of ponds of varying ages, sizes, and depths (Rutherford 1964, cited by Allen 1983). Colony density increases with watershed size (Howard and Larson 1985). Fryxell (2001) demonstrated that a regional population comprised of several individual beaver colonies is similar in many respects to a metapopulation, where individual colonies are analogous to subpopulations within a metapopulation structure. Colonies occupying the most suitable habitats are more reproductively successful and provide a source of dispersing beavers. Dispersers periodically recolonize vacant habitats where former resident colonies have gone extinct due to negative net production over time or various stochastic events. Thus, while many individual colonies are randomly becoming extirpated and subsequently recolonized, other colonies tend to exhibit long-term persistence, and relative stability is maintained throughout the larger population (Fryxell 2001). Estimates of minimum viable population sizes for beavers are not available.

## 6. Movements

### *Dispersal*

Typically, all beavers (most at two years of age) disperse from their natal colonies in April and May, but it may take several months for dispersers to ultimately settle in suitable habitats (Fitzgerald et al. 1994, Sun et al. 2000). In New York, the predominant direction of dispersal was downstream and the mean dispersal distance for females (10 km) was about three times that of males (Sun et al. 2000). Sun et al. (2000) found that most dispersers moved only short distances and 73% colonized the closest neighboring suitable site. Fitzgerald et al. (1994) noted an average dispersal distance is about 7 km and that most dispersal movements are less than 16 km.

### *Migration*

Beavers are nonmigratory. However, entire beaver colonies may move to new stream sections when local food supplies become scarce (Fitzgerald et al. 1994).

### *Barriers to Movement*

We found no discussion in the literature about barriers to the movement of beavers.

### *Use of Corridors*

Most authors suggest that beavers usually follow watercourses when dispersing.

## 7. Ecology

### *Interspecies Interactions*

Beaver ponds may be beneficial or harmful to various species of fish, and they provide or enhance habitats for waterfowl and amphibians (Jenkins and Busher 1979, Fitzgerald et al. 1994). Beavers create habitat mosaics that increase arthropod diversity, which may benefit insectivorous birds and mammals (Martinsen et al. 1998). Beavers influence plant and animal community composition and diversity both in the vicinity of their activities and downstream (Naiman et al. 1988). Water management by beavers aids recovery of overgrazed, eroded stream banks, which improves habitats for many species (Jenkins and Busher 1979, Albert and Trimble 2000). Beavers are prey for various large and medium sized predators (see above).

Beavers compete for food resources with elk (*Cervus elaphus*) and cattle, but the effects of these interactions have not been extensively researched and reported in the literature. Obviously, riparian areas where livestock or elk browsing has significantly eliminated woody riparian vegetation (and prohibits its recovery) have little habitat value for beavers. In Yellowstone National Park, beaver live where elk densities are lower; elk numbers may be part of multiple causes behind lower beaver numbers, including less woody riparian browse, fewer large fires, a more arid climate, and fewer wolves (Romme et al. 1995, Singer and Mack 1999). Several of these factors are interrelated.

### *Ecological Effects*

Beavers are classic "keystone" species in that their activities significantly affect ecosystem function to a degree that is disproportionate to their numerical abundance (Miller et al. 1998). These effects stem from the beaver's ability to physically alter aquatic systems and from their concentrated herbivory near water bodies (Jenkins and Busher 1979, Naiman et al. 1988, Gurnell 1998). Naiman et al. (1988) and Gurnell (1998) identified the following ecological effects: stabilization of stream flows; increased wetted surface area (*i.e.* benthic habitat); elevation of water tables causing changes in floodplain plant communities; creation of forest openings; creation of conditions favoring wildlife that depend upon ponds, pond edges, dead trees, or other new habitats created by beavers; enhancement or degradation of conditions for various species of fish; replacement of lotic invertebrate taxa (*e.g.*, shredders and scrapers) by lentic forms (*e.g.*, collectors and predators); increased invertebrate biomass; increased plankton productivity; reduced stream turbidity; increased nutrient availability; increased carbon turnover time; increased nitrogen fixation by microbes; increased aerobic respiration; increased methane production; reduced spring and summer oxygen levels in beaver ponds; and increased ecosystem resistance to perturbations. Beaver ponds undergo predictable succession over long time periods from open water ponds to marshes to seasonally flooded meadows (Naiman et al. 1988). Many mountain meadows are the result of past beaver activities (New Mexico Department of Game and Fish 2001a). Removal of beavers from an area can cause their dams to deteriorate, increasing runoff and gully formation, lowering water tables, and reducing biological diversity (New Mexico Department of Game and Fish 2001a). Beavers are being used as ecological restoration agents in degraded riparian ecosystems (Albert and Trimble (2000).

Martinsen et al. (1998) discovered a most unusual relationship between beavers and beetles. Resprout growth following the cutting of cottonwood trees by beavers contained twice the level of defensive (against mammalian herbivory) chemicals as normal juvenile cottonwoods. Leaf beetles fed preferentially on cottonwood resprout growth, thus bioaccumulating the protective chemicals (phenolic glycosides) that they used for their own defense.

## 8. Management Recommendations

The recommendations of Bailey (1931: 219) remain valid and appropriate today:

> *On almost all the mountain streams they should be protected and encouraged. A series of beaver ponds and dams along the headwaters of a mountain stream would hold back large quantities of mountain water during the dangerous flood season and equalize the flow of the streams so that during the driest seasons the water supply would be greatly increased in the valleys. Beaver ponds not only hold water but distribute it through the surrounding soil for long distances, acting as enormous sponges as well as reservoirs. A series of ponds also increases the fishing capacity and furnishes a safe retreat for the smaller trout and protection from their enemies. In addition a protected beaver colony is one of the most interesting features of mountain or forest, as with protection the animals become less wary and more diurnal in their habits so that they can be readily observed and studied by those traveling and camping in wild regions.*

### *Restore Beaver Colonies*

Beavers should be restored to all drainages where they historically occurred, especially in headwater and low-order streams. Reintroductions should proceed immediately in stream reaches with suitable habitat.

*Improve Potential Beaver Habitat*

Trees and shrubs preferred by beavers for food and construction materials tend to sprout vigorously after fires. Restoration of natural fire regimes and prescribed natural fires should be incorporated into management recommendations for beavers. Domestic ungulates should be excluded from riparian areas or managed in ways that promote the recovery of woody riparian plants, especially aspen, willow, cottonwood, and alder.

*Use Beavers as Agents for Restoring Streams and Watersheds*

Beavers have been shown to be effective agents for restoring degraded streams and watersheds and should be used in this capacity wherever feasible. In critical reaches, managers should recognize the beaver's ability to improve its own habitat and consider supplemental feeding during the establishment period. If trying to restore both beavers and cottonwood along a river, one should consider the effect of beaver herbivory on cottonwood saplings (Breck et al. 2003).

*Close Areas to Harvesting of Beavers*

Legal beaver harvests tend to be small and of little economic consequence compared to the potential ecological benefits and services provided by beavers. Beaver harvesting should be prohibited in all core and compatible use areas identified in this plan, except where significant economic impact may result from their activities.

*Educate Managers and the Public*

Wildlife managers and policy makers need a thorough understanding of beaver ecology in order to establish appropriate policies and make sound management decisions. In addition, the public needs accurate information and knowledge about beavers to inform their opinions and values and their understanding of appropriate management measures. Knowledge is the key to informed conservation actions and advocacy by both agencies and the public.

### 9. Justification

The beaver was selected as a habitat quality indicator, keystone species, and prey species.

**Habitat Quality Indicator:** Beavers prefer aspens, willows, cottonwoods and alders, which occur in high-quality riparian ecosystems (Allen 1983). Degraded riparian ecosystems, usually the result of overgrazing by domestic ungulates, are generally unsuitable for beavers. Beavers enhance the quality of habitats they occupy for themselves and many other species (Naiman et al. 1988).

**Keystone:** The beaver is a keystone riparian species because its activities substantially alter landscapes and create new ecosystems (Collen and Gibson 2001). Ecosystem productivity and biological diversity are also enhanced by the activities of beavers (Naiman et al. 1988).

**Prey:** Beaver are prey for gray wolf, mountain lion, black bear, grizzly bear, coyote, lynx, river otter, and mink (Jenkins and Busher 1979, Fitzgerald et al. 1994).

However, this species was not selected for inclusion in the initial SITES modeling because their preferred habitat consists of narrow woody riparian corridors that are currently poorly represented in GIS data at the ecoregional scale. A new mapping effort is currently underway within Colorado which would greatly improve the quality of available riparian GIS data for the state. However, it will take several more years to complete, and equivalent efforts have not been attempted for Wyoming and New Mexico.

## Bighorn Sheep (*Ovis canadensis*)

### 1. Introduction

The bighorn sheep (*Ovis canadensis*) is one of only two native members of the mammalian family Bovidae in the Southern Rockies (the other being the bison [*Bison bison*]). Ecologically, bighorns are herbivores that occupy a unique grazing niche on steep slopes and are occasional prey for large predators, especially mountain lions (*Puma concolor*). Bighorns require interconnected clusters of suitable habitat for population viability and migrate between seasonal and special-use ranges. The Rocky Mountain bighorn (*O. c. canadensis*) is classified in the Southern Rockies as a game species. Although there are about 6,000 bighorn sheep in Colorado alone, not all bighorn populations in the Southern Rockies are doing well. The popularity of bighorn sheep with both trophy hunters and nonconsumptive nature enthusiasts fosters considerable public interest and support for nature conservation. It is the state mammal for Colorado.

### 2. Distribution

*Historic*

Bighorn sheep probably evolved from Asian ancestors that migrated to North America across the Bering Land Bridge during the Pleistocene epoch (Krausman and Shackleton 2000). Historically, bighorn sheep were distributed throughout the mountainous regions of western North America and adjacent river valleys and prairies from southwestern Canada to northwestern Mexico, including most mountain ranges within the Southern Rockies planning area (Findley et al. 1975, Hall 1981, Shackleton et al. 1999). Once relatively abundant, bighorn sheep are now one of the rarest ungulates in North America (Valdez and Krausman 1999).

Northern New Mexico and southern Wyoming hold the Rocky Mountain bighorn, but there are two subspecies of bighorn in Colorado: the native Rocky Mountain bighorn and the desert bighorn (*O. c. nelsoni*), introduced near Colorado National Monument in 1979. Herds are widely scattered throughout the mountains and foothills (Findley et al. 1975, Hall 1981). During the late 1800s, disease, competition from domestic livestock, and indiscriminate hunting caused a drastic reduction of bighorn numbers. Indeed, unregulated harvesting caused the extirpation of Rocky Mountain bighorn sheep from northern New Mexico by 1903 (Frey and Yates 1996). Some restoration efforts have reestablished populations of Rocky Mountain bighorns within their former range, although at present they are still greatly reduced from historic numbers.

#### *Current*

As of 2002, there are small populations of Rocky Mountain bighorn sheep in northern New Mexico and southern Wyoming. About 6,000 bighorn sheep live in Colorado.

#### *Potential*

If reasons for past extirpations can be fully understood and addressed, potential exists to restore and expand populations of native bighorn sheep (Singer et al. 2000a, d).

### 3. Habitat

#### *General*

Typical habitat for bighorn sheep occurs in river canyons and benches, foothills, and mountains on or near rugged terrain with steep slopes. Bighorn sheep seldom venture far from steep, rugged escape terrain, which provides protection from terrestrial predators. In addition to rugged physical terrain, bighorns require high-visibility habitats dominated by grasses and low shrubs (Fitzgerald et al. 1994, Krausman et al. 1999, Shackleton et al. 1999). Fire suppression has resulted in degradation of some bighorn sheep habitat by the encroachment of dense, tall shrubs and conifers (Singer et al. 2000b).

#### *Preferred*

Essential habitat components for bighorn sheep are food, water, open space, and escape terrain (Krausman et al. 1999). Food preferences are discussed in the following section. Smith et al. (1991) developed habitat suitability criteria for Rocky Mountain bighorn sheep that were further tested and refined by Johnson and Swift (2000) and Zeigenfuss et al. (2000). Unless otherwise attributed, the following habitat preferences are from these sources.

Bighorn sheep require escape terrain to avoid predation. Optimal escape terrain is defined as having slopes between 27-85° with occasional rock outcrops. Bighorns use zones within 300 m of escape terrain for foraging. Areas 1,000 m wide are considered available for foraging when escape terrain is present on two or more sides. Ewes prefer steeper, more rugged slopes of 2 ha or more in total area for lambing and rearing of lambs during their first week of life (Smith et al. 1991). Zeigenfuss et al. (2000) found that the proportion of habitat that is suitable for lambing was one of the best predictors of population size and success. Suitable lambing habitat for Rocky Mountain bighorn sheep may require water within 1,000 m, and southerly aspects are preferred (Smith et al. 1991, Zeigenfuss et al. 2000).

Bighorn sheep can meet a considerable amount of their water requirement from metabolic processes (oxidation) or from direct consumption of succulent vegetation, snow, or ice (Krausman et al. 1999, Krausman and Shackleton 2000). However, during hot, dry seasons when water loss due to thermoregulation is high and vegetative moisture content is low, nearby (< 3.2 km) water is considered essential for bighorns (Smith et al. 1991, Krausman et al. 1999, Krausman and Shackleton 2000).

Bighorn sheep tend to be intolerant of humans and their activities and have abandoned home ranges following increases in human activity (Krausman et al. 1999). Areas within 150 m of high-use areas are considered unsuitable habitat (Smith et al. 1991). King and Workman (1986) found that the flight response and time spent in behaviors associated with wariness of desert bighorn sheep in southeastern Utah increased with increasing levels of human disturbance. They speculated that increased energy expenditures and reduced foraging efficiency caused by human disturbance may predispose bighorns to reduced fitness and reproductive success and increased mortality, especially during critical time periods, such as lambing. However, bighorn sheep apparently can and do habituate to many predictable, nonthreatening human activities, such as rock quarries, busy highways, and residential areas (Fitzgerald et al. 1994, Johnson and Swift 2000).

Areas with high densities of cattle, feral burros, or elk (*Cervus elaphus*) are considered unsuitable for bighorn sheep because of food competition, disease transmission, and possibly behavioral avoidance of cattle by bighorns (Smith et al. 1991, Krausman et al. 1999, Krausman and Shackleton 2000). Domestic sheep and goats, and exotic relatives of bighorn sheep, such as mouflon sheep (*Ovis ammon musimom*), barbary sheep (*Ammotragus lervia*), and ibex (*Capra ibex*) carry diseases that are often lethal to bighorns, compete for limited food resources, and may cause genetic pollution through interbreeding with bighorns (Smith et al. 1991). Areas less than 20 km from populations of domestic or exotic sheep or goats are unsuitable because of transmissible diseases for which native bighorns lack immunity (Smith et al. 1991, Singer et al. 2000a).

Open space or habitat openness is important because bighorn sheep primarily rely upon sight to detect predators (Krausman et al. 1999). Horizontal visibility should exceed 62% of a square meter target when viewed at 28 m with the observer's eyes 90 cm above the ground. Invasion of tall shrubs or trees diminishes the value of bighorn sheep habitat, and fire is often prescribed to improve bighorn habitat (Krausman and Shackleton 2000).

#### *Seasonal and Special Ranges*

Seasonal and special "ranges" to which adult bighorns show a high degree of fidelity may include summer, winter, spring (lambing), fall (rutting), and salt lick ranges (Shackleton et al. 1999). Bighorns are unable to forage efficiently where snow depths exceed 25 cm. Under these conditions, they seek southerly slopes or wind-swept ridges near suitable escape terrain where forage remains available, or they migrate to lower elevations (Krausman and Shackleton 2000). Habitat suitability parameters for seasonal and special ranges are similar to those described above, and, often, a lack of a suitable seasonal range (especially winter range) or special range (especially lambing range) limits overall suitability of an area for bighorn sheep (Smith et al. 1991). A variety of environmental or behavioral stimuli may dictate seasonal habitat shifts. For example, desert bighorn sheep may respond to seasonal water availability (*e.g.* ephemeral water catchments) and use areas that are unsuitable during dry seasons (Krausman and Shackleton 2000). Ewes may make seasonal movements to more suitable lambing areas as defined above. Seasonal and special habitat requirements must be considered when assessing overall habitat suitability for bighorn sheep.

### 4. Food Habits

Bighorn sheep occupy a unique grazing niche on steep slopes (Singer et al. 2000c). They tend to be more opportunistic than preferential in their choice of foods, eating whatever palatable foods are available (Krausman and Shackleton 2000). In terms of plant types, Rocky Mountain bighorns tend to show a preference for forbs, grasses, and browse in that order. The order of preference is reversed for desert bighorn sheep, with no clear preference between grasses and forbs (Krausman and Shackleton 2000). Variations from these general preferences are common, especially when examined seasonally.

### 5. Population Dynamics

#### *Life History*

Bighorn sheep are a wild, native member of the bovine family (Bovidae). Adult male Rocky mountain bighorns weigh about 79 kg on average and females weigh about 59 kg; large Rocky Mountain bighorn rams may weigh up to 145 kg (Valdez and Krausman 1999). Both male and female bighorns reach sexual maturity at about 18 months, but in the wild females first mate at 2.5 years or later, and males do not fully participate in the rut until 7-8 years of age (Krausman and Shackleton 2000). Bighorn sheep in expanding populations may reproduce at younger ages. Breeding in Rocky Mountain bighorn sheep usually peaks between mid-November and mid-December (Krausman and Shackleton 2000). Males fight by clashing horns during the rut to compete for the opportunity to breed females. Larger-horned males generally do most of the breeding and are preferred by females (Krausman et al. 1999). Gestation lasts about 175 days and lambs weigh 3-5.5 kg at birth (Krausman and Shackleton 2000). Ewes usually bear single lambs; twinning is extremely rare. Most bighorn sheep die before age 10, a few may survive to age 15 (Krausman et al. 1999).

Like most ungulates, bighorn sheep occur in groups of 2 to 100 individuals. Except for juvenile males remaining with their mothers, groups are sexually and spatially segregated for all seasons except the rut (Krausman and Shackleton 2000). Bighorn sheep are active during daylight hours (diurnal). Feeding behavior is most active in early morning and late afternoon and bouts of resting and feeding alternate during the day (Krausman et al. 1999, Shackleton et al. 1999).

#### *Population Density*

Reported densities of bighorn sheep populations range from 0.33 to 7.7 sheep per km$^2$ (Zeigenfuss et al. 2000). Zeigenfuss et al. (2000) found that population densities of successful or marginally successful translocated populations of bighorn sheep were between 0.57 and 1.53 sheep per km$^2$, and that populations exceeding 3.0 sheep per km$^2$ were released into patches of less than 10 km$^2$ of suitable habitat. None of these higher-density populations exceeded 40 individuals. This suggests that higher reported bighorn sheep population densities may be artifacts of selected restoration sites being too small. Zeigenfuss et al. (2000) recommend the use of density estimates of 1.47 Rocky Mountain bighorns/km$^2$ and 0.33 desert bighorns/km$^2$ when planning restoration projects.

### *Home Range*

Information on home range size of Rocky Mountain bighorn sheep is limited (Shackleton et al. 1999), and seasonal movements complicate home range measurement. However, it is well known that mature Rocky Mountain bighorns exhibit a high degree of fidelity to established seasonal ranges and return to them yearly (Shackleton et al. 1999). Being herd animals, many sheep would share home ranges.

### *Causes of Death*

As with all ungulates, mortality rates are often highest (40-90%) during the first year of life. Predation is a common cause of lamb mortality. Other causes of lamb mortality may include pneumonia, severe weather, inbreeding depression, poor maternal nutrition, inattentive mothers, and human disturbance (Krausman and Shackleton 2000). Potential predators of bighorn sheep of various ages include coyote (*Canis latrans*), mountain lion, wolf (*Canis lupus*), bobcat (*Lynx rufus*), gray fox (*Urocyon cinereoargenteus*), golden eagle (*Aquila chrysaetos*), and humans (Kelly 1980, Fitzgerald et al. 1994). Bighorn sheep are well adapted for avoidance of predation, provided that escape terrain is nearby; and predation usually has little affect on population survival (Kelly 1980, Krausman et al. 1999).

Disease-related "die-offs" can kill 35-75% of a sheep population in a single year and can suppress recruitment for an additional 3-7 years (Gross et al. 2000). While bighorn sheep suffer from a variety of diseases, they are particularly susceptible to pneumonic pasteurellosis (*Pasteurella* spp.) that they readily contract from domestic sheep, goats, and exotic relatives (Smith et al. 1991, Gross et al. 2000). Usually all bighorn sheep die when placed in pens with domestic sheep, and evidence strongly suggests the transmission of fatal diseases from domestic to wild sheep in the wild (Singer et al. 2000a). About 85% of the San Andres Mountain (NM) herd of desert bighorns died from a virulent outbreak of contagious ecthyma caused by scabies mites (*Psoroptes ovis*). Various sources of stress and/or the presence of lungworm (*Protostrongylus* spp.) likely predispose bighorn sheep to disease die-offs (Krausman and Shackleton 2000).

Other causes of bighorn sheep mortality include vehicle collisions, natural accidents, drowning, and fence entanglement (Krausman et al. 1999).

### *Population Structure and Viability*

Suitable habitat for bighorn sheep tends to be comprised of climax vegetation types that change slowly and occur in "islands" within and among mountain ranges (Douglas and Leslie 1999, Singer et al. 2000c). Natural habitat patchiness tends to cause individual herds to be small and disjunct in their distribution. Low reproductive potential, the tendency of ewes to have strong fidelity to their natal home ranges, the distance between patches of suitable habitat, and barriers to dispersal between patches result in infrequent colonizations of vacant habitats by bighorns (Bleich et al. 1990, Douglas and Leslie 1999, Singer et al. 2000c). However, Bleich et al. (1990) suggest that natural local extinctions and recolonizations may have been more common than previously thought.

Thus, bighorn sheep appear to be predisposed to a metapopulation structure, where limited interchange (mostly by rams) occurs among smaller, geographically separated subpopulations (Lande and Barrowclough 1987, Bleich et al. 1990, Singer et al. 2000c). Singer et al. (2000c) speculated that bighorns historically existed mostly in metapopulations and that human disturbances have accelerated extinction rates, causing a current state of population disequilibrium and the existence of unoccupied habitat.

Ramey et al. (2000) advise that the establishment of single isolated populations of bighorn sheep with no potential for genetic exchange with other populations is of little long-term conservation value. They advocate the establishment of metapopulations. Indeed, genetic problems may be contributing to declines in some of the small, translocated herds (Berger 1990, Fitzsimmons et al. 1997).

A present-day example of a potentially viable bighorn sheep metapopulation exists in southeastern California near the community of Twentynine Palms. There, about 1,000 bighorn sheep occupy 15 of 31 mountain ranges, 10 currently vacant ranges previously supported bighorns, only 8 of the 15 subpopulations currently exceed 50 sheep, and bighorn movements of 6-20 km have been documented between 11 pairs of mountain ranges (Bleich et al. 1990). Reestablishment of bighorn populations in key mountain ranges is critical to the restoration of connectivity among currently isolated subpopulations in this region. This metapopulation is bounded by significant barriers to dispersal and migration (*i.e.*, interstate highways and the Colorado River).

Because ideal situations rarely exist in nature, the "genetically effective population size" ($N_e$) is probably always smaller than the actual census size ($N_c$) of the population (Meffe and Carroll 1997:172). A genetically effective population is generally defined as an ideal, stable population with randomly mating individuals, even sex ratio, equal birth rates among females, and nonoverlapping generations (Meffe and Carroll 1997). Douglas and Leslie (1999) believe that habitat fragmentation and local extinctions of subpopulations has led to effective population sizes that may be orders of magnitude smaller than the census count, and that small populations cannot persist without reproductive interactions with nearby populations. The need to preserve and restore functioning metapopulations is underscored by the fact that about two-thirds of all populations of Rocky Mountain and California bighorn sheep (*O. c. californiana*) populations have less than 100 individuals (Gross et al. 2000). Krausman and Leopold (1986) reported that only 7 of 59 populations of desert bighorn sheep in Arizona exceeded 100 animals in size.

While admitting that the concept of minimum viable population size in bighorn sheep is controversial, Singer et al. (2000b) considered metapopulations of 300-500 animals to be viable for at least 200 years, populations of 100-299 animals to be secure for shorter time periods, populations of 75-99 animals to be moderately secure, and populations of less than 50 animals to be vulnerable to extinction. They recommended restoring single populations of 125 or more animals in clustered restoration sites 12-25 km apart, sufficient to support a combined metapopulation of more than 400 animals.

Singer et al. (2000c) concluded that ultimate population size of translocated bighorns was correlated with $N_e$ of the founding population, number of source populations represented in the founding group, and early contact with a second population. In the populations they studied, successful colonizations of suitable habitat patches occurred every 13.5 years on average for increasing populations and every 22 years for all populations (increasing and decreasing). This indicates that time frames for natural recolonization of vacant habitat patches are quite long.

Bleich et al. (1990) defined the general requirements for a bighorn sheep preserve size as (1) sufficient suitable unoccupied habitat for future establishment of subpopulations and (2) large enough to support populations with long-term genetic health and the potential for long-term evolutionary processes to occur. In a study of reintroduced bighorn sheep populations in the Intermountain West, Singer et al. (2000c) found that the mean patch size for populations that increased to more than 100 individuals was 490 km$^2$; whereas, the mean patch size for populations that declined to less than 30 bighorns or went extinct was 60 km$^2$.

Zeigenfuss et al. (2000) recommend a minimum patch size of 200 km$^2$ and Singer et al. (2000c) recommend more than 400 km$^2$ for bighorn restoration projects. These recommendations stand in sharp contrast to findings by Johnson and Swift (2000) that populations of 125 or more Rocky Mountain bighorns have persisted for 20-52 years in areas with 6-9 km$^2$ of suitable habitat; and to the recommendation of Smith et al. (1991) of 17 km$^2$ of suitable habitat for successful ($\geq$ 125 bighorns) restoration projects. The findings and recommendations of Johnson and Swift (2000) and Smith et al. (1991) suggest population densities of 7-21 bighorns/ km$^2$, which are much higher than population densities reported by other authors (Zeigenfuss et al. 2000).

Bleich et al. (1990) recommended that small, isolated patches of suitable habitat should be recognized as potentially important habitat and included within preserve designs as potential seasonal habitats and "stepping stones" for migration or dispersal movements. Krausman and Leopold (1986) cautioned that the importance of small populations should not be discounted in bighorn sheep conservation programs. And Smith et al. (1991) advised that subpopulations of less than 125 bighorns can contribute to the formation of viable metapopulations, provided movement among populations can occur.

We base our design recommendations on mean population densities of 1.0 bighorn/km$^2$ of suitable habitat, preferred minimum subpopulation size of 125 bighorns, minimum metapopulation size of 400 bighorns, and minimum inter-patch distances within a metapopulation of 20 km. Thus, optimum subpopulation habitat patch sizes should be at least 125 km$^2$ with at least 360 ha of suitable lambing habitat (Smith et al. 1991). However, habitat patches of 10 km$^2$ or more near similar sized or larger patches or within potential migration or dispersal corridors should be protected for their potential to provide seasonal, special, or temporary habitats; habitat for small interconnected subpopulations; and "stepping stones" for migration or dispersal movements.

### *6. Movements*

#### *Dispersal*

Dispersal is important to bighorn sheep for recolonization of unoccupied suitable habitat patches, gene flow among subpopulations, maintenance of the evolutionary potential of metapopulations, and for discovery of newly-created suitable habitat caused by fires or removal of livestock (Bleich et al. 1990, Singer et al. 2000a). Generally, bighorn sheep are considered to be poor dispersers and colonizers of unoccupied habitats (Gross et al. 2000); but nevertheless, moderate rates of dispersal are critical to the survival and long-term persistence of self-perpetuating metapopulations of bighorn sheep. Singer et al. (2000c) found that most colonizations of new habitat (n=24) by dispersing bighorns from 31 translocated populations were from patches 10-15 km distant; and Gross et al. (2000) computed that the maximum probability of dispersal was to a patch 12.3 km away. Both ewes and rams disperse but rams disperse more frequently than ewes and generally disperse longer distances (Bleich et al. 1990, Douglas and Leslie 1999, Wehausen 1999).

#### *Migration*

Under natural conditions most bighorn sheep populations (rams and ewes) migrate seasonally among 2-7 different seasonal ranges separated by distances of 8-18 km with elevation changes that may exceed 1,000 m (Shackleton et al. 1999, Krausman and Shackleton 2000, Singer et al. 2000b). Factors influencing migration or movements in bighorn sheep may include home range knowledge, water and forage availability, location of mineral licks, snow accumulation, lambing and mating activities, season, presence of biting insects, topography, and age and sex class status (Krausman et al. 1999, Krausman and Shackleton 2000). In a review of 100 bighorn sheep translocation projects, Singer et al. (2000a) found that 100% of fully migratory populations were successful while only 65% of nonmigratory populations were successful. Rocky Mountain bighorns may migrate as far as 64 km annually (Singer et al. 2000a). The absence of migration in some present-day herds may be the result of human-caused habitat alterations, fragmentation, and barriers within traditional migration routes (Singer et al. 2000a).

### *Barriers to Movement*

Barriers that impede movement of bighorn sheep include sheer cliffs, wide valley floors, swift or wide rivers, lakes or reservoirs, dense vegetation with low horizontal visibility (< 30%), fences (if not designed for wildlife passage), motorized recreational activities, domestic livestock (especially sheep), concrete lined canals, highways (state, federal, and interstate) with 600 or more vehicles per day, and high-density centers of human activity (Bleich et al. 1990, Smith et al. 1991, Fitzgerald et al.1994, Johnson and Swift 2000, Zeigenfuss et al. 2000).

### *Use of Corridors*

Bighorn sheep habitat usually occurs as small, isolated patches within a larger matrix of unsuitable habitat. By necessity, dispersing or migrating bighorns usually must move through areas of unsuitable habitat (Bleich et al. 1990, Krausman and Shackleton 2000). Bleich et al. (1990) summarize the documentation of such movements by numerous researchers. Singer et al. (2000c) compared corridors used by bighorns to potential corridors not used by bighorns. They found bighorns recolonized vacant habitat patches by using corridors with fewer water barriers, more open vegetation, and more rugged, broken terrain. They observed much higher rates of colonizations by bighorn sheep than observed by previous researchers, which supports increased attention by managers to the protection and restoration of linkages among potentially suitable habitat patches (Bleich et al. 1990, Krausman and Shackleton 2000, Singer et al. 2000c).

## 7. Ecology

### *Interspecies Interactions*

Like other large ungulates, bighorns consume grasses, forbs, and shrubs (Krausman and Shackleton 2000). Considerable diet overlap exists between bighorns and elk, mule deer (*Odocoileus hemionus*), cattle, domestic sheep, goats, horses, and burros (Krausman et al. 1999, Shackleton et al. 1999). Competition for food resources may affect various population parameters of bighorn sheep, such as lamb survival and recruitment, but such interactions have not been confirmed (Shackleton et al. 1999). Krausman et al. (1999) attribute the precipitous historic decline of desert bighorn sheep populations in the western U.S. to overgrazing by domestic livestock, but admit that such a cause-and-effect relationship is difficult to document. One study in Nevada documented that bighorn sheep density was three times higher in ungrazed versus grazed (by domestic cattle) habitats (Krausman et al. 1999). Krausman et al. (1999) predicted that competition for food resources by exotic species such as barbary sheep and Persian wild goats (*Capra aegagrus*) will eventually extirpate bighorn sheep populations using the same range. Population declines of bighorns also have been attributed to disease transmission from domestic sheep and goats and exotic wildlife (Krausman et al. 1999, Shackleton et al. 1999).

Some large predators prey on bighorn sheep, but the bighorns' adaptations for predator avoidance—keen eyesight and sure-footedness in steep escape terrain—appear quite effective. Predation is unlikely to limit bighorn populations in areas with adequate escape terrain, but may explain their absence in areas deficient in escape terrain (Krausman et al. 1999, Shackleton et al. 1999). However, predation by mountain lions has been documented or suspected to be a major limiting factor in some bighorn populations, especially recently translocated populations (Krausman et al. 1999). In most (perhaps all) cases in which mountain lions endanger small populations of bighorn sheep, the impact of predation has been exacerbated by disease (*e.g.*, scabies and pneumonia – Logan and Sweanor 2001), woody plant invasion due to overgrazing or fire suppression (Sweitzer et al. 1997), or artificially high lion populations subsidized by year-round livestock operations (E. Rominger, New Mexico Department of Game and Fish, personal communication). Short-term control of mountain lions may be necessary as an emergency measure to prevent extinction of some populations of bighorns, but long-term conservation programs should follow a holistic approach that asks deeper questions as to why the population is small and vulnerable.

### *Ecological Effects*

Bighorn sheep occupy a unique grazing niche on steep

slopes (Singer et al. 2000b). Grazing by bighorns affects the availability of vegetation and successional processes within plant communities (Singer et al. 2000b). In addition to carrying out the ecological function of herbivory, bighorns are prey for large carnivores such as wolves, mountain lions, and bears (*Ursus* spp., Singer et al. 2000b).

## 8. Management Recommendations

Krausman and Shackleton (2000:533) summarize management needs for bighorn sheep as follows: *Habitat for bighorn sheep still exists in the west, but managers (and the public) have to ensure that sufficient habitat is protected, that movement corridors remain open, that human disturbance is reduced or kept to a minimum, and that transmission of diseases from livestock is eliminated. Only if these are accomplished will efforts to enhance viable populations of bighorn have a chance to be successful* (Underlining was added for emphasis).

### *Reestablish Populations*

Transplantation has been proven to be an effective means of establishing bighorn sheep in vacant suitable habitats (Douglas and Leslie 1999). Over half of extant populations of bighorns in the western U.S. result from translocations (Singer et al. 2000a). Singer et al. (2000a) developed the following seven point restoration plan: (1) survey existing populations, (2) conduct GIS-based habitat assessments, (3) review habitat assessments with a scientific advisory panel, (4) convene interagency panel to discuss metapopulation management and plan restoration projects, (5) draft interagency restoration and management plans, (6) conduct restoration activities, such as translocations, and (7) monitor populations and evaluate success of restoration efforts. Singer et al. (2000a) noted that translocation success was about twice as high when founders came from indigenous populations as compared to previously translocated herds. They further recommended that the founder herd size be 41 or more animals, that translocation areas be at least 20 km from domestic sheep, and that migration routes among seasonal ranges be free of barriers to bighorn movements. Singer et al. (2000c) recommended that founders be selected from more than one source population, and (Singer et al. 2000d) cautioned that removals of reintroduction stock should be only from increasing source populations and should not exceed 5% of the source population annually.

### *Protect Habitats*

Bighorn sheep require specialized habitats that are in short supply and bighorn populations are few and generally small. No further degradation of suitable bighorn sheep habitat should occur.

### *Improve Habitats*

Many habitats may be suitable for bighorns with certain improvements. Where forests and dense shrubs have invaded otherwise suitable bighorn habitat, prescribed fires that mimic natural historical fire regimes may reestablish necessary forage quantity and quality and visibility for bighorns. Development of reliable water sources may improve habitats for bighorns (Krausman and Shackleton 2000); but Krausman and Leopold (1986) caution that providing water sources may increase competition for food resources by attracting other ungulates, such as mule deer. Additionally, opportunities for predation may increase at or near water sources (Z. Parsons, personal communication).

### *Minimize Human Disturbances*

Managers should monitor levels and effects of human activities in crucial bighorn habitats and implement measures to reduce or eliminate human disturbances that may jeopardize the long-term health and persistence of bighorn populations (Krausman and Shackleton 2000).

### *Protect Linkages Between Habitat Patches*

Singer et al. (2000a) found that migrating sheep populations are more successful than those that don't or can't migrate. Most bighorn sheep populations exist in the form of metapopulations, which require periodic movement of individuals among subpopulations comprising the metapopulation for long-term viability and persistence (Bleich et al. 1990, Singer et al. 2000b). Managers should recognize key habitat linkages and focus on preventing or eliminating barriers to critical bighorn movements between patches of suitable habitats.

### *Remove Domestic Sheep*

The presence of domestic sheep within 20 km of existing bighorn herds or habitats suitable for the restoration of bighorns poses serious problems for bighorn sheep protection and restoration efforts (Singer et al. 2000a). With domestic sheep nearby, the potential for transmission of lethal diseases to bighorn herds is very high, because rams wander widely during rut and have been known to breed with domestic sheep and carry diseases back to their herd (Gross et al. 2000, Ramey et al. 2000). Ramey et al. (2000) suggest that the most cost-effective solution may be to buy out and retire domestic sheep allotments on public lands. Similar incentives may induce private livestock operators to stop grazing domestic sheep near bighorn herds or habitats suitable for bighorn restoration.

### *Eliminate Cattle Grazing in Bighorn Habitat*

A high degree of dietary overlap exists between bighorn sheep and cattle (Krausman et al. 1999, Shackleton et al.

1999). Singer et al. (2000a) found that translocated populations of bighorns increased faster on ranges where cattle were absent and that translocation success rate declined by 27% when cattle grazed restoration areas. A study in Nevada documented that bighorn sheep density was three times higher in ungrazed versus grazed (by domestic cattle) habitats (Krausman et al. 1999). Given the fragile nature of bighorn populations, every advantage should be sought to enhance populations and the success of reestablishment efforts.

*Interagency Coordination*

Douglas and Leslie (1999) advised that, where bighorn sheep populations cross jurisdictional boundaries (which is nearly always the case), agencies and private landowners develop joint management plans and strategies to ensure long-term persistence of bighorn sheep populations.

### 9. Justification

The bighorn sheep was selected as a flagship, habitat quality indicator, and wilderness quality indicator species.

**Flagship:** The bighorn sheep is a majestic, charismatic animal with considerable public appeal among both hunters and nature enthusiasts. The bighorn sheep is a focus of a major foundation—the Foundation for North American Wild Sheep—that is allowed to auction special hunting permits to benefit conservation programs for the species. The most recent permit for the opportunity to hunt one bighorn ram in New Mexico sold for $75,000. The restoration of bighorns from their near extirpation in the early 1900s is an ongoing conservation priority. Conservation programs that benefit bighorn sheep should readily capture the public's attention and interest.

**Habitat Quality Indicator:** Bighorn sheep are habitat specialists that graze in steep, rugged terrain. While naturally fragmented, habitats required by bighorns have become increasingly fragmented by human activities. In addition, important dispersal and migration routes have been rendered ineffective by human-caused barriers to bighorn movements. Bighorn sheep are intolerant of human disturbances and activities, except in situations where the predictability of nonthreatening human activity results in habituation by bighorns. The presence and persistence of bighorn populations is a clear indicator of the quality of the interconnected steep-sloped habitats they require.

**Wilderness Quality Indicator:** *Mountain sheep epitomize wilderness. They occupy some of the most inaccessible, rugged, and spectacular habitats in North America. Their ability to negotiate precipitous terrain is legendary....Only the most adventurous and hardy outdoor enthusiasts dare to tread in such hostile habitats of temperature extremes and rugged terrain* (Valdez and Krausman 1999:3).

This species was not selected for inclusion in the initial SITES modeling because suitable areas for this species are already covered by modeling done for wolves and bears.

## Black Bear (*Ursus americanus*)

### 1. Introduction

The black bear (*Ursus americanus*) is the largest extant member of the order Carnivora in the Southern Rockies planning area (given that grizzly bears, *Ursus arctos*, have not been positively identified in the Southern Rockies since 1979). Contrary to its taxonomic classification, however, it is mostly herbivorous in its dietary habits. Black bears are common and widely distributed in large tracts of forests and woodlands and immediately adjacent shrublands. They require large areas of suitable habitat and safe, densely forested linkages among habitat patches for population viability. If necessary, black bears migrate seasonally in search of high-quality foods. The black bear is classified as a game species, and most populations within the Southern Rockies planning area are hunted. Populations in northern New Mexico and Colorado are relatively secure. Populations in southern Wyoming have lower numbers and bears of smaller size. The black bear is adversely affected by human encroachment and habitat fragmentation, and its popularity with both hunters and nonconsumptive users fosters considerable public interest and support for nature conservation.

### 2. Distribution

*Historic*

Historically, black bears were widely distributed throughout all major forested regions (deciduous and coniferous) in North America (Hall 1981). Fossil evidence suggests that they were not common in open habitats, such as grasslands, shrublands, and desert areas. Nowak (1991) speculated that black bears avoided open habitats because of a lack of trees, which provided a means of escaping predation by grizzly bears. In support of this theory, Nowak (1991) noted range extensions of black bears following the extirpation of grizzly bears. In the early 1900s, black bear distribution in the Southern Rockies was greatly reduced by unregulated hunting, trapping, and use of poisons by both private individuals and government predator control agents.

*Current*

Black bears are currently distributed throughout the wooded foothills and coniferous forests of nearly all major mountain ranges in the Southern Rockies planning area. Exurban housing development, however, causes conflict between humans and bears, a conflict that bears nearly always lose.

*Potential*

Presently, little potential exists for significant range expansion of black bears in the Southern Rockies planning area. The Colorado Division of Wildlife estimates there are roughly 8,000 to 12,000 throughout Colorado, and the New Mexico Department of Game and Fish estimates about 5,000 bears throughout that state.

## 3. Habitat

*General*

Black bears show a strong tendency to select closed forest and woodland habitat types, and more than 80% of all bear locations are in these types. Aspen (*Populus tremuloides*) and Gambel oak (*Quercus gambelii*) are preferred. Shrublands received limited use and open grasslands and woodlands were generally avoided by black bears, except within 500 m of closed-canopy habitat edges. These observations were consistent with other studies of black bear habitat use (Lindzey and Meslow 1977, LeCount and Yarchin 1990, both as cited in Costello et al. 2001). Habitat use may vary seasonally depending upon the availability and location of preferred foods. Harding (2000) observed substantial use of sage (*Artemisia* spp.) steppe habitats in Utah during spring and early summer.

As with other large wildlife, black bears exhibit a behavioral avoidance of roads, especially in areas open to hunting (Brody and Pelton 1989, Powell et al. 1996). Bears may react to increases in road density by shifting home ranges to areas of lower road densities (Brody and Pelton 1989, VanderHeyden and Meslow 1999). Roads increase mortality through the facilitation of legal and illegal killing and vehicle collisions (Brody and Pelton 1989, Powell et al. 1996, Powell et al. 1997, Pelton 2000). In North Carolina, 50% and 75% of legally harvested black bears are killed within 0.8 km and 1.6 km of roads passable by four-wheel-drive vehicles (Powell et al. 1997). The New Mexico Department of Game and Fish (2000) recommends that road densities not exceed 0.8 km/km$^2$ in black bear habitat. Roadless areas smaller than average bear home ranges likely have insufficient escape value for bears (Powell et al. 1997). Powell et al. (1997) state that large roadless areas are an essential component of suitable bear habitat in the Southern Appalachian Region. We suspect that roadless areas are equally important to bear survival and habitat suitability in the Southern Rockies. On the other hand, lightly used or seasonally closed roads may serve as travel corridors and seasonal feeding sites for bears (Pelton 2000).

*Preferred*

Preferred habitats of black bears in the Southern Rockies include subalpine coniferous (spruce-fir) forests, subalpine aspen forests, upper montane coniferous forests, lower montane coniferous forests, and Rocky Mountain closed coniferous forest. Shrublands and grasslands adjacent to these closed-canopy forest types provide additional habitat, and may be especially important in spring and early summer when bears forage primarily on grass and ants (Harding 2000). Forest openings produce important foods for bears (Boileau et al. 1994, Verts and Carraway 1998, Pelton 2000). However, these openings should be small and occur within a matrix of large contiguous forested tracts with minimum human disturbances (Pelton 2000).

Water is an important component of black bear habitat (Verts and Caraway 1998, Vanderheyden and Meslow 1999). Five additional components of preferred habitat are: escape cover; sources of hard or soft mast foods in fall; spring and summer feeding areas; movement corridors; and winter denning habitat (Pelton 2000). Chaparral scrub provides important escape cover in Southwestern habitats (Pelton 2000). The absence or failure of mast crops significantly reduces subsequent reproductive performance (often causing population-wide reproductive failure) of female black bears (Pelton 2000, Costello et al. 2001). Spring and summer foods are important to the recovery of black bears emerging from winter dens, especially females with cubs. Grasses, berries, and insects, especially ants, bees, and wasps, provide important nutrients at this time of year (Verts and Carraway 1998, Harding 2000, Pelton 2000). Seasonally available foods are often widely distributed throughout bear home ranges. Movement corridors with thick cover along ridge tops, saddles, side drainages, streams, and rivers provide safe passage to important food resources (Pelton 2000).

*Den Sites*

Over half of 390 dens visited during the Costello et al. (2001) study were either excavations under rock or rock cavities. About one-third of the dens were excavations under trees or in natural tree cavities. The most common habitat types at den sites were mixed conifer forests (45%), piñon-juniper woodlands (21%), and spruce-fir forests (13%). Most tree cavity dens (83%) were in mixed conifer habitats. All tree cavity dens were used by females (Costello et al. 2001). This finding is consistent with Pelton's (2000:399) statement that "female bears prefer cavities in large, standing, dead or live trees." Of dens in piñon-juniper habitats, 88% were associated with rocks. The mean slope at den sites was 28 degrees and most dens were located at the upper (42%) or mid (37%) areas of slopes. Only 8% of dens were on flat sites—ridge tops or valley bottoms. Ground dens are associated with thick understory cover (Pelton 2000). Dens are sometimes re-used, but most black bears selected new den sites each year (Costello et al. 2001).

Bears are highly likely to abandon and change dens following human disturbance (Goodrich and Berger 1994). Periods of activity during winter (when food is generally unavailable) may increase body attrition resulting in reproductive failure, starvation, or poor nutritional/physical condition upon emergence from dens. Poor condition may predispose black bears to other mortality factors. Secure den sites can reduce the effect of human disturbance upon denning black bears (Pelton 2000).

## 4. Food Habits / Foraging Behavior

Black bears in the Southern Rockies consumed mostly plant matter—sometimes as much as 90% of the diet according to the Colorado Division of Wildlife. Grasses, sedges, and forbs dominated the diet from den emergence to mid summer. Ants (Formicidae) and soft mast (*e.g.*, fruits and berries) increased in importance from mid summer to the availability of hard mast (acorns and piñon nuts), and both hard and soft mast were consumed from mid September to den entrance. Except for ants, animal foods seemed less important to the diet of black bears in the Southern Rockies.

Carrion is eaten by bears and may be an important food item in the spring (Fitzgerald et al. 1994, Verts and Carraway 1998, Pelton 2000). Newborn ungulates (*e.g.* deer, *Odocoileus* spp., and elk, *Cervus elaphus*) are sometimes preyed upon by black bears but are able to escape such predation attempts by the age of about 2 weeks (Fitzgerald et al. 1994, Linnell et al. 1995, Verts and Carraway 1998, Pelton 2000). Domestic livestock are sometimes killed by black bears (Fitzgerald et al. 1994). While black bears are classified as carnivores, they function ecologically as omnivores; but mostly consume vegetable matter, especially herbaceous material, fruits, seeds, and nuts (Pelton 2000).

Black bears can become a nuisance in apiaries, fruit orchards, grain crops, garbage cans or dumps, and at bird feeders; and conflicts are common where bear habitats meet human settlements (Pelton 2000). Bears often peel the bark from trees to feed on the underlying cambium layer, causing damage and often death of the tree (LeCount 1986, Nowak 1991, Verts and Carraway 1998). This feeding behavior may be a response to the lack of more preferred foods.

Black bears in the Southern Rockies appear to be more wary of humans than the northern black bears, perhaps as a result of heavy persecution.

## 5. Population Dynamics

### *Life History*

Black bears are the largest extant carnivore in the Southern Rockies planning area. Adult males and females weigh an average of 104 kg and 64 kg, respectively. Adult males in Colorado may weigh as much as 225 kg. Black bears exhibit black and brown color phases, and brown-phase bears can appear dark brown, light brown, cinnamon, and blonde. In New Mexico and Colorado the brown phase dominates (Peterson 1995).

Except for females with cubs or yearlings, black bears tend to be solitary in their social organization (Fitzgerald et al. 1994) and polygynous in their reproductive habitats (Pelton 2000). Adult males and females associate solely for the purpose of mating (Pelton 2000). Home ranges of males overlap considerably and include the home ranges of one or more females, which may also overlap (Pelton 2000, F. Lindzey, personal communication). Black bears tend to congregate and exhibit intraspecific tolerance in areas where food is abundant (*e.g.*, berry patches and garbage dumps, Verts and Carraway 1998).

Black bears hibernate. In the Southern Rockies, most black bears enter hibernation dens between mid October and mid November and emerge in April and May. They spend as much as 200 days without eating, drinking, urinating, or defecating; their bodies recycle protein and their energy comes from fat reserves (Colorado Wildlife Commission 1994 Bear Facts, Colorado Division of Wildlife web site). Bears do not eat much for the first two weeks after they emerge from hibernation.

Most female black bears reach reproductive maturity at 3 years of age (range 1-5). The observed mean age at first reproduction for female bears in New Mexico was 5.7 years. Most breeding occurs in June and July; implantation of the embryo is delayed; and cubs are born in hibernation dens in late January or early February. Litter sizes range from 1 to 3 cubs (mean = 1.8), and the mean natality rate for females ≥ 4 years old was 0.77 cubs/female/year. Cubs remain with their mothers for about 16 to 18 months and den with their mothers the first winter following their birth. Prolonged parental care of cubs sets the birth interval for females at a minimum of 2 years. Reproductive success of black bears is strongly influenced by the previous season's production of acorns and berries. Failed mast production for two consecutive years usually results in complete reproductive failure.

Black bears may live more than 20 years, but few live longer than 10-12 years in the wild (Fitzgerald et al. 1994, Pelton 2000). In the New Mexico study, the oldest female documented was 27 years old and the oldest male was 23 (Costello et al. 2001). Two females produced litters at the age of 22 years. Ages of black bears in the New Mexico study were estimated from cementum annuli (growth marks in teeth) and, thus, were not known ages. There is still some debate as to the accuracy of cementum annuli techniques for aging black bears.

### *Population Density*

Costello et al. (2001) estimated black bear densities (excluding cubs) to be 17/100 $km^2$ in the Sangre de Cristo Mountains (a non-hunted area). Comparable black bear densities were reported by Fitzgerald et al. (1994) for Colorado. However, black bear population densities are expected to vary with habitat quality (L. Harding, personal communication).

### *Home Range*

Mean primary home ranges (excludes long-range movements) of male black bears were 3-5 times larger than female home ranges. In the Sangre de Cristo Mountains, female primary home ranges averaged 24 $km^2$ and male home ranges averaged about 131 $km^2$ (Costello et al. 2001). Increased movements occurred in the fall when bears were in search of highly-preferred hard mast (*e.g.*, acorns) foods prior to hibernation. When long-range movements were included in home range calculations some home ranges exceeded 3,000 $km^2$. Arid conditions, coupled with livestock grazing, may limit the availability of grasses to bears, and compel individuals to search more widely for other foods (Costello et al. 2001). In the New Mexico study, adult bears of both sexes and subadult females exhibited a "high degree of home range fidelity" (Costello et al. 2001:86).

### *Causes of Death*

Most mortality of adult black bears is human-caused by legal hunter kills, illegal kills, depredation kills, and automobile collisions (Fitzgerald et al. 1994). Legal kills of bears by hunters tend to be higher in years with shortages of natural foods, presumably because bears travel farther (and for more of the time) and often leave secure habitats in search of food. Natural causes of mortality include predation (by other large carnivores), diseases and parasites (not considered a major factor), and starvation; but natural mortality rates for adult bears are low compared to human causes (Paquet and Carbyn 1986, Schwartz and Franzmann 1991, Mattson et al. 1992a, Smith and Follman 1993, Pelton 2000,Costello et al. 2001). Mortality among yearling bears was mostly from natural causes, but human causes are also a factor. Mortality factors for cubs include predation, automobile collisions (Costello et al. 2001), and infanticide by adult males of unrelated cubs (Pelton 2000).

Cub mortality is high and was 44% in an area where roads were closed in Colorado (Fitzgerald et al. 1994). The next highest mortality (>35%) occurs among yearling and subadult bears from the time they leave their mothers until they have successfully established home ranges of their own (Pelton 2000). As bears age, mortality rates decline; however, the mortality rate of males (26%) is about double that of females (17%, Pelton 2000).

### *Population Structure and Viability*

Male:female ratios for all age classes were approximately 1:2; adult females comprised 35% of the population and adult males about 20% (Costello et al. 2001). The remaining 45% of the population was comprised of subadult (2-4 years old) and yearling bears. Cubs were not considered as recruited into the population unless they survived their first year.

For identifying small tracts of suitable black bear habitat in New Mexico, Costello et al. (2001:95) designated a "minimum sustainable population" as 50 bears, but their use of this terminology was not intended to denote population viability (C. Costello, personal communication). A more appropriate interpretation of this term would be a subpopulation inhabiting a relatively isolated tract of suitable habitat with potential connectivity with larger habitat blocks. According to the authors, such a population could be supported within about 300 $km^2$ of contiguous suitable habitat in northern portions of New Mexico. They also consider patches of suitable habitat sufficient in size to support 1-2 bears (>20 $km^2$) as potentially important habitat, provided these patches lie within 15 km of patches greater than 300 $km^2$ in size.

Powell et al. (1996) studied black bears residing both inside and outside the approximately 260-$km^2$ Pisgah Bear Sanctuary in North Carolina. They determined that the sanctuary provided increased protection for bears residing within it and provided dispersing bears for hunters outside it, but that its population may have been declining. Poaching occurred along roads within the sanctuary. No transmitter-collared male bears had home ranges that were totally confined within the boundaries of the sanctuary. Powell et al. (1996) concluded that sanctuaries are appropriate for managing black bears, but that sanctuary sizes need to be larger than the one they studied and that roads should be eliminated within sanctuaries. The beneficial effect of road elimination and increased size of sanctuaries for bears derives from decreased human access to bears and their habitat (Powell et al. 1996). Similarly, Rudis and Tansley (1995) recommended the retention or restoration of an extensive complex of interconnected blocks of remote (*i.e.*, inaccessible) forests for the protection and conservation of black bears in Florida.

Because ideal situations rarely exist in nature, the "genetically effective population size" ($N_e$) is probably always smaller than the actual census size ($N_c$) of the population (Meffe and Carroll 1997:172). A genetically effective population is generally defined as an ideal, stable population with randomly mating individuals, even sex ratio, equal birth rates among females, and nonoverlapping generations (Meffe and Carroll 1997). Using population age-sex data presented by Costello et al. (2001) and a formula for calculating $N_e$ presented by Meffe and Carroll (1997:173), the genetically effective population size of black bear populations in the southern part of the Southern Rockies is about one/half the actual number of bears present, excluding cubs.

A major goal of the Southern Rockies Wildlands Network Vision is to preserve or restore evolutionary processes in natural systems, which requires long-term population viability for resident species. Genetic variation within and among individuals comprising populations of animals is the currency of evolution. In short, genes that confer advantages to individuals are selected over time (because their carriers are more fit and survive longer) and non-advantageous genes are not selected as often. As a general rule, the preservation of "evolutionarily important amounts of quantitative genetic variation" requires effective population sizes of "at least several hundreds of individuals" (Lande and Barrowclough 1987:119). Franklin (1980) recommended an effective population size of 500 for long-term conservation. Rudis and Tansley (1995) recommend a metapopulation of five interconnected subpopulations of black bears totaling 1,000-1,500 animals as the minimum conservation goal for Florida. Following the "precautionary principle" (Meffe and Carroll 1997:546), we recommend core refugia that are large enough to support an effective population of at least 500 black bears. This translates to an actual population of about 1,000 bears, excluding cubs. Such populations would have a high probability of long-term viability and persistence.

Using density estimates of Costello et al. (2001), core areas of suitable black bear habitat should equal or exceed 6,000 $km^2$ in the southern parts of the Southern Rockies planning area. Alternatively, subpopulations comprising a metapopulation of 1,000-1,500 bears (Rudis and Tansley 1995) should each contain at least 50 bears ($N_e \geq 25$) and be interconnected by functional linkages. Per Costello et al. (2001), smaller subpopulations should not be discounted as lacking conservation value, especially if linked to larger subpopulations.

## 6. Movements

Black bears move for the purposes of finding food, dispersing as subadults, finding mates, and finding suitable dens (Pelton 2000). They often use long-established trails consisting of individual foot impressions in the substrate (LeCount 1986, Reimchen 1998). These trails may represent "corridors of least resistance" (Reimchen 1998:698) through natural habitats. Reimchen (1998) observed that during daylight, bears consistently moved off trails into surrounding forested habitats when encountering humans or other bears, but exhibited much higher fidelity to trails and much greater tolerance of human presence at night. Trails may serve an ecological role of facilitating bi-directional movements by black bears through habitats during darkness (Reimchen 1998).

### *Dispersal*

In the New Mexico study, there was no dispersal among transmitter-collared females (N = 21) with known natal ranges. The tendency of females to establish home ranges in or near their natal home range (Pelton 2000) significantly affects the rate of recolonization of extirpated subpopulations. If distances between core populations and isolated habitats are large, translocations (especially of females) may be necessary to achieve population re-establishment. In contrast, all male bears that were monitored until at least age 4 dispersed. Five male dispersal movements were documented. The age of dispersing males ranged from 1.5-3 years, and dispersal distances ranged from 25-60 km (15.5-37.3 mi). In their study of black bears in Alaska, Schwartz and Franzmann (1991) also observed that all surviving subadult males dispersed.

### *Migration*

Bears exhibit little migratory behavior except in search of seasonally abundant foods (Schwartz and Franzmann 1991, Fitzgerald et al. 1994). In Colorado, bears moved from 13 to 36 km between summer and fall feeding areas (Beck 1991, as cited in Fitzgerald et al. 1994). Comparable migration distances (17-31 $km^2$) were reported by Schwartz and Franzmann (1991) in Alaska.

### *Barriers to Movement*

Brody and Pelton (1989) found that bears avoided roads in areas open to hunting, but that in many protected areas bears were attracted to roads by the presence of human food. Thus, they postulate that the response of black bears to roads is primarily a learned behavior. In North Carolina, the frequency of road crossings by black bears was inversely related to traffic volume, with the greatest road avoidance occurring at an interstate (multi-lane) highway. Use of highway underpasses by black bears has been documented in Florida (Foster and Humphrey 1995).

Rudis and Tansley (1995) state that black bears avoid human contact. This suggests that black bears are not likely to move through areas occupied by humans unless the density of humans is low.

*Use of Corridors*

Rudis and Tansley (1995) describe short distance movement corridors as being characterized by dense vegetation and widths of 10-60 m. They suggest that long-distance travel or dispersal may require large blocks of contiguous forested habitat. However, we found no documentation of dispersing bears using corridors to move between blocks of suitable habitat.

## 7. Ecology

*Interspecies Interactions*

Interactions between black bears and other species are not well documented in the literature. There is little evidence to suggest that black bears limit populations of other species through competition or predation. Black bears were the leading cause of mortality to elk calves in the Jackson elk herd (Smith and Anderson 1996). In Alaska, where high levels of predation on moose (*Alces alces*) calves by black bears were documented, habitat quality was determined to be the ultimate factor controlling moose densities and population trends (Schwartz and Franzmann 1991).

*Ecological Effects*

Black bears function ecologically as omnivores, but mostly consume herbaceous material, fruits, seeds, and nuts (Pelton 2000). Energy obtained by bears from consumed plant material and some animals (alive and dead) is passed on to predators and scavengers completing the ecological cycling of energy and nutrients (Odum 1993). Seeds consumed by bears are dispersed when bears defecate, often considerable distances from where they were eaten. In their occasional role as carnivores, black bears may belong to a suite of top predators (potentially including mountain lions [*Puma concolor*], gray wolves [*Canis lupus*], and grizzly bears) that collectively regulate other animal and plant populations and ecological processes (Gassaway et al. 1992, Terborgh et al. 1999).

## 8. Management Recommendations

*Establish Refugia*

Black bears suffer from high levels of human-caused mortality in areas with roads or access by off-road vehicles. Legal protection does not eliminate poaching. Persistent, viable populations are essential to the long-term survival and conservation of this species and would serve as important sources of dispersing bears (primarily males) for rescuing declining populations in surrounding areas. Protected areas in the range of 6,000 km$^2$, or larger, characterized by preferred black bear habitat should be considered for refuge designation. Designated refuges should be roadless, or have low densities of roads with an access management plan that protects bears during critical periods. Hunting should be prohibited or at least severely restricted in refuge areas. Natural disturbance regimes should be allowed to operate within refugia, and prescribed fires may be a useful tool for improving black bear habitat. Management practices should consider the vulnerability of local, isolated black bear populations to extinction.

*Protect and Establish Travel Corridors*

Thick, continuous cover should be retained and encouraged along ridge tops, gaps, ravines, and riparian corridors and around water sources. This will provide safe access for bears to seasonal food and water resources and for dispersing bears to patches of suitable unoccupied habitat. Corridors should include underpasses integrated with roadside fencing where high-speed roads cross corridors.

*Road Closures*

Roads are the most significant contributing factor to black bear mortality, because they provide human access to occupied habitats, and humans cause most bear mortality. Nonessential roads (*e.g.*, off-road vehicle trails and dirt roads) should be closed in critical core areas and linkages between them where existing road densities exceed suitability thresholds (0.8 km/km$^2$) or illegal killing is documented.

*Retain Large Trees, Snags, and Fallen Logs*

Large trees, snags, and logs provide important denning sites, especially for female black bears. Thus, retention and perpetuation of old-growth forests through forest planning will likely enhance survival and reproductive success. In addition, some slash from logged areas should be piled in or near the edge of standing timber and left to provide additional den sites for bears.

*Conduct Annual Mast Surveys*

Production of acorns and berries is correlated to black bear reproductive success. Thus, reproductive declines or failures can be reasonably predicted. Hunt quotas should be adjusted to avoid over-hunting following reproductive failures. The lag effect of missing age cohorts on future reproductive potential of black bear populations should be considered in hunt quota adjustments.

*Monitoring*

Black bear populations are particularly sensitive to overharvesting, especially of females. Monitoring is important for the establishment of sustainable harvest limits. The accuracy of various population estimating methods and indices varies considerably, potentially causing erroneous interpretation and ill-advised management decisions (Kane and Litvaitis 1992, Noyce and Garshelis 1997). Managers should include multiple estimators and indices in ongoing black bear monitoring programs (Kane and Litvaitis 1992, Pelton 2000).

*Regional or Multi-jurisdictional Planning*

Conservation planning for large carnivores must be conducted over vast spatial scales and must consider connectivity among local subpopulations (Noss et al. 1996). Land areas large enough to support a viable metapopulation of black bears will likely encompass a multitude of jurisdictions. Establishment of regional planning authorities through appropriate means (*e.g.*, legislative or administrative) should be encouraged and pursued.

*Educate Managers and the Public*

Wildlife managers and policy makers need a thorough understanding of black bear ecology in order to establish appropriate policies and make sound management decisions. In addition, the public needs accurate information and knowledge about black bears to inform their behavior in bear country, their opinions, their values, and their understanding of appropriate management measures. Knowledge is the key to informed conservation actions and advocacy by both agencies and the public. They would do well to read Ghost Grizzlies by Peterson (1995). In particular, chapter 11 outlines the history of black bear management in Colorado.

### *9. Justification*

The black bear was selected as a habitat quality indicator and umbrella species.

**Habitat Quality Indicator:** Black bears occur in low densities, require large contiguous or connected areas for population persistence, avoid roads and open areas, have a relatively low reproductive potential, and subadult females rarely disperse from their natal areas. These qualities make black bears particularly sensitive to habitat fragmentation, which is a threat to habitat quality throughout the planning area. Black bears rely upon a variety of seasonally-available foods, which contribute to the quality of habitats for a variety of other species in the ecosystem.

**Umbrella:** Black bears have large area requirements for population persistence. They require a variety of forest, woodland, and shrubland habitats for feeding, hiding, and denning. Protection of habitat for viable populations of black bears will, by inclusion, protect the habitats of many other species.

Black bears are included as an input into the initial SITES analysis as a surrogate for grizzly bears because: 1) wide-ranging carnivores are a primary focus of this network design, 2) much more data exists for black bears than grizzlies in the Southern Rockies, and 3) grizzly habitat needs within the Southern Rockies are considered to be comparable to those of black bears, as discussed in Chapter 7.

## Grizzly Bear (*Ursus arctos*)

### *1. Introduction*

Grizzly bears (*Ursus arctos*) once ranged throughout central and western North America from northern Mexico to the Arctic Ocean. During the late 1800s and early 1900s, much of the habitat throughout grizzly bear range was eliminated, and grizzly bear numbers declined due to indiscriminate killing. Grizzly bears south of the Canadian border were listed as threatened under the Endangered Species Act in 1975. Grizzly bear habitat includes avalanche chutes, rocklands, shrublands, grasslands, and riparian areas. The home ranges of grizzly bears are not static and vary among age/sex cohorts. The diet contains fruits, herbaceous plants, and meat. Movements are mainly a result of their dietary needs.

Protection and management of adult females should be the primary management goal for recovering populations of grizzly bears. Reducing human-grizzly bear conflict is also a serious concern. Grizzly bears need conservation and protection, and road creation in grizzly habitat needs to be controlled and reduced.

### *2. Distribution*

*Historic*

Grizzly bears once ranged throughout most of central and western North America (Patnode and LeFranc 1987). In Wyoming and Colorado, grizzly bears inhabited prairie grasslands to alpine tundra and were commonly sighted and hunted by the first settlers in the West (Fitzgerald et al. 1994, Luce et al. 1999). Habitat conversion and persecution because of fear or conflict with livestock produced a drastic decline of numbers, particularly in the contiguous U.S. Throughout North America, the range of the species has been retreating steadily northward (Craighead and Mitchell 1982).

*Current*

The grizzly bear has been extirpated from most of its

original range including the states of Texas (1890), North Dakota (1897), California (1922), Utah (1923), Oregon (1931), New Mexico (1933), Arizona (1935), and Colorado (1979, Patnode and LeFranc 1987). The last known specimen in Colorado was collected on the northeastern edge of the San Juan Mountains (Fitzgerald et al. 1994, Peterson 1995). The isolated habitat fragments that continue to support grizzly bears make up less than 2% of their original habitat in the contiguous 48 states. All remaining grizzly bears in the lower 48 states are managed in the six ecosystems of Yellowstone, Northern Continental Divide, Cabinet/Yaak, Selkirk Mountains, North Cascades, and Selway/Bitteroot (Patnode and LeFranc 1987).

### *Potential*

Grizzly bears can be re-established in areas with low road density and low human population numbers, although protection of grizzly bears requires extensive public education efforts. Large core areas would be necessary, and they should be linked to facilitate dispersal. The San Juan Mountains present perhaps the best opportunity to reintroduce grizzly bears in the Southern Rockies, and the Colorado Grizzly Project has presented such a proposal (Andromidas 2001). This proposal outlines legal considerations, recovery goals, habitat suitability and linkages, habitat restoration, research needs, education, and translocation options (Andromidas 2001). For a copy of the proposal, contact Sinapu (grizzly@sinapu.org). There may, however, be a small number of grizzlies that persist in the San Juans (Peterson 1995). That must be investigated and considered before reintroduction plans are enacted.

## 3. Habitat

### *General*

Grizzly bears can potentially occupy nearly all types of temperate, boreal, and arctic regions where humans are scarce (Mattson 2000). The exceptions are hot deserts and grasslands without ungulates (Mattson 2000). While grizzly bears can adapt to the presence of humans, they cannot adapt to intensive use and modification of habitat by humans or persecution by humans (Craighead and Mitchell 1982). Already, 89% of grizzly bear mortality in the Yellowstone Ecosystem is human-caused (Woodroffe and Ginsberg 1998).

In northwestern Montana, grizzly bears often use avalanche chutes, grasslands, rocklands, shrublands, and riparian zones with high canopy cover of deciduous shrubs (Mace and Waller 1997, Mace et al. 1999). These grizzly bears select low temperate and temperate elevation zones over subalpine during all seasons (Mace et al. 1996). Avalanche chutes are important habitat during spring, but in summer, grizzly bears select for more cover types with shrublands and cutting units especially important (Mace et al. 1996). Avalanche chutes, shrublands, and nonvegetated cover types are selected in the autumn (Mace et al. 1996). Males are more likely to use resource rich areas at low elevations while females use sites at higher elevations in the summer (McLellan and Shackleton 1988). In Wyoming, habitat includes coniferous forests, mountain-foothills, shrublands, riparian shrub, and mountain-foothill grasslands (Luce et al. 1999).

There is little base of knowledge on grizzly bears in the Southern Rockies, but years of field work have led Tom Beck to suggest that grizzlies in the Southern Rockies behaved very differently from grizzly bears of the Northern Rockies (Peterson 1995). He stated that grizzly bears of the San Juans probably behaved more like black bears (*Ursus americanus*) of the San Juans than they did like northern grizzlies (Peterson 1995). In that case, any remaining grizzly bears in the San Juan Mountains may be selecting closed forest and woodland habitat types, such as aspen (*Populus tremuloides*) and Gambel oak (*Quercus gambelii*).

### *Denning*

The adaptive behavior of hibernating in winter dens aids survival during harsh conditions. The time of denning varies with region and area, but Yellowstone grizzly bears often dig and enter dens from October through mid-November (Craighead and Mitchell 1982, Judd et al. 1983). Bears usually enter dens during a snowstorm that provides insulation and protection (Craighead and Craighead 1972). Young are born in mid-winter, and begin life in the den. In Yellowstone, males emerge from hibernation in mid-February to late March; next to leave the den are single females, then females with yearlings and two-year olds, and finally females with new cubs emerge in early to mid-April (Judd et al. 1983).

Preferred den sites have moderate tree cover, 26-75% canopy cover, and 30°-60° slopes (Judd et al. 1983, Craighead and Craighead 1972). Fallen trees are used as support for some den roofs (Judd et al. 1983) and bedchambers are lined with bedding, usually boughs (Craighead and Craighead 1972). Dens in Yellowstone are generally located in altitudes ranging from 2,300 to 2,800 m (Craighead and Craighead 1972). Dens occur on all aspects but are most common on northern exposures that accumulate snow from prevailing southwest winds, thus insulating dens from extreme winter temperatures (Judd et al. 1983, Craighead and Craighead 1972). Although Yellowstone grizzly bears may hibernate in natural cavities and previously used dens, they are more likely to dig new dens each winter (Judd et al. 1983).

## 4. Food Habits

Grizzly bears eat both plant and animal foods. In Montana, the fruits of blueberry (*Vaccinium* spp.) and buffaloberry (*Shepherdia* spp.) are important energy sources and are used for fat deposition (Mace 1987, Mace and Jonkel 1983). Grasses, sedges, and forbs such as horsetails (*Equisetum* spp.), clover (*Trifolium* spp.), and dandelions (*Taraxacum* spp.) are important during all seasons. Whitebark pine (*Pinus albicaulis*) nuts and biscuit root (*Lomatium* spp.) roots are also consumed (Mace and Jonkel 1983, Mace 1987). According to Craighead and Mitchell (1982), the subalpine zone rated highest as an energy source for grizzly bears with the temperate zone second and the alpine zone third.

In the Greater Yellowstone Ecosystem, whitebark pine nuts, cutthroat trout (*Oncorhynchus clarki*) in shallow streams, army cutworm moths (*Euxoa auxiliaris*), and winter-killed elk (*Cervus elaphus*) and bison (*Bison bison*) carcasses are all critical food sources (Reinhart and Mattson 1990, Mattson et al. 1992b, Knight et al. 1999, D. Mattson pers. comm.). Three of the four food sources, however, are threatened. Whitebark pine is vulnerable to white pine blister rust (*Cronartium ribicola*), an exotic fungus; army cutworm moths are declining because of pesticide use at lower elevations; and cutthroat trout are being replaced through competition and predation by introduced lake trout (*Salvelinus namaycush*). Lake trout inhabit deeper waters and are unavailable to grizzly bears. Winter-killed elk and bison carcasses remain an abundant source of food for grizzly bears.

A grizzly will also steal a carcass from other predators, including wolves, if given the opportunity (Hornbeck and Horejsi 1986). Whether carcasses from wolf (*Canis lupus*) kills will help grizzly and black bears is still unknown. Grizzly bears prey upon animals weakened by severe weather, malnutrition, disease, injuries, and old age; but they rarely kill healthy ungulates (Pasitschniak-Arts 1993). Grizzly bears cache a carcass to possibly slow decomposition and to hide it from other scavengers (Pasitschniak-Arts 1993).

In Yellowstone National Park, grizzly bears migrate to winter feeding grounds of elk and other ungulates between March and May (Pasitschniak-Arts 1983). Females will use low elevation habitats in the spring when snow first melts and succulent vegetation first appears (Mace et al. 1999). In late summer, adult females with cubs will move to mid or high elevation habitats to avoid males (Weilgus and Bunnell 1994, Mace et al. 1999). Females may also avoid forested habitat in the spring and fall because of the increased chance of contact with males preying on and consuming elk calves and carrion (Weilgus and Bunnell 1994). Grizzly bears in Yellowstone migrate to higher elevations during July and August when the tourist season peaks and the most people are using hiking trails. That is also a period when army cutworm moths are abundant on talus slopes.

If any grizzly bears remain in the San Juan Mountains, and they are behaving like the black bears of the region, then undoubtedly Gambel oak acorns are an important food source. The San Juan Mountains have ample bear food (Andromidas 2001).

## 5. Population Dynamics

### *Life History*

The home ranges of grizzly bears are not static and may actually increase throughout a bear's life (Knight et al. 1988). The home ranges of females are smaller than males (Craighead and Mitchell 1982, Mace and Waller 1997), which may decrease the chance of a female encountering an aggressive male and threatening her own safety and that of her young (Pasitschniak-Arts 1993). In Montana, subadult ranges were also larger than those of adult females (Mace and Waller 1997). Home ranges in Montana ranged from 35 km$^2$ for a subadult female to 1,114 km$^2$ for an adult male (Mace and Waller 1997). The life range of one male in Yellowstone National Park exceeded 2,600 km$^2$ (Craighead and Mitchell 1982). The size of a grizzly bear home range is mostly determined by the availability of food (Pasitschniak-Arts 1993). The concept of home range implies that an animal spends the majority of its time within a definable area. Because the grizzly bear is a long-lived and large omnivore, one should probably exercise caution when interpreting home ranges for life history information (Craighead and Mitchell 1982).

Adult grizzlies are solitary except during mating season and during cub rearing, for the females (Pasitschniak-Arts 1993). Mating behavior begins as early as mid-May. Females typically do not become sexually mature until around 4.5 years of age (although it can be earlier), and some do not become mature until they are 7.5-8.5 years of age. Females breed every two years and usually produce two cubs (Craighead et al. 1969).

### *Population Viability*

Grizzly bears are very difficult to census, and thus density estimates over large geographic areas should be treated with some caution (Craighead and Mitchell 1982). Techniques, however, have been improving and are becoming more reliable.

Maintaining a viable population of grizzly bears requires large areas with minimal human disturbance, human tolerance, protection of all adult females, and low mortality of males. Human impact has often left only refuges of isolated habitat for grizzly bears, where they exist in small numbers. Such populations are subject to genetic drift (a process whereby small populations can lose genetic diversity to chance events), and that loss may reduce adaptability to environmental change, increase susceptibility to disease, and lower reproductive potential (Frankel and Soulé 1981).

Frankel and Soulé (1981) suggested that an effective population of 500 individuals may be necessary for adaptive evolution over the long term. In the absence of evidence, many geneticists endorse this general guideline. It should be remembered that an effective population is nearly always smaller than the actual population in the wild. Fluctuating populations, fluctuating sex ratios, reproductive variability, and behaviors like dominance and dispersal hinder the assumption of random mating (Lande 1988). For grizzly bears, a ratio of 4 actual individuals in the wild equals only 1 individual for an effective population (Allendorf et al. 1991). Thus an effective population of 500 may require 2,000 (or more) individuals in the wild. Because of the space necessary for even a small population of grizzly bears, this emphasizes the necessity of linkages among protected areas throughout the entire Rocky Mountain range.

Protecting female grizzlies should be the first priority and will be especially critical to the survival of small populations (Pasitschniak-Arts 1993, Peek et al. 1987). Indeed, long-term conservation of the grizzly bear is driven by habitat protection and limiting human caused mortality in balance with reproduction and natural mortality. However, trophy hunting of adult males may have caused or contributed to a population decline in southwestern Alberta, so male mortality should also be closely monitored (Weilgus and Bunnell 2000).

#### *Mortality*

Mortality of grizzly bears in the modern U.S. is rarely natural (Knight et al. 1988, Woodroffe and Ginsberg 1998) even though at present, there is no legal hunting in the lower 48 states. There are mortality quotas for all populations. The quotas designate allowable human caused mortality to meet recovery criteria. In the Greater Yellowstone Ecosystem, the major sources of mortality are legally sanctioned management killings due to conflict with human foods, conflict with human development, frequent contact with campgrounds, and repeated property damage (Peek et al. 1987, Knight et al. 1988). Deaths, legal and illegal, related to killing livestock also have an effect on grizzly bear numbers (Peek et al. 1987). In some cases, grizzly bears are killed because of conflict with elk hunters over a carcass.

## *6. Human Land Use and Grizzly Bears*

Managers should monitor levels and effects of human activities in crucial grizzly bear areas and implement measures to reduce or eliminate human disturbances that may jeopardize the long-term health and persistence of grizzly bear populations.

Intensive forest development and other resource extraction industries usually result in increased road access. Roads increase the amount of grizzly mortality by fostering human and grizzly encounters and illegal harvest (Peek et al. 1987, McLellan and Shackleton 1988, Andromidas 2001). Roads also reduce usable grizzly bear habitat. Areas within 100 m of a primary, secondary, or tertiary road will likely have reduced use by a grizzly bear because of the disturbance of passing vehicles (McLellan and Shackleton 1988). Thus, some roads would have to be closed. Habitat along unused roads or low use roads had neutral to positive selection by bears in Montana while habitat along roads traveled by more than ten vehicles/day experienced negative selection by grizzlies (Mace et al. 1996). However, habitat near a road and the road itself may be used by bears at night if there is little to no traffic (McLellan and Shackleton 1988).

As recreational hiking and other pastimes lead increasing numbers of people into wilderness areas, encounters with grizzly bears are a concern. Grizzly encounters with humans could be successfully decreased with a greater understanding of how to coexist, including properly storing food, garbage, and pet food, wearing bear bells, carrying bear spray, and not venturing into grizzly bear habitat (Peek et al. 1987). In Glacier National Park, Montana, only hikers not wearing a bear bell were charged and full charges occurred mainly on trails with little use. This suggests that habituation of grizzly bears to hikers decreases the number of fear-induced charges and any resulting injuries (Jope 1985).

Livestock grazing allotments on public land create problems for grizzly bears due to the possibility of grizzly bears killing livestock and the resulting negative sentiments toward the bear. Primarily livestock calves and yearlings are killed by grizzly bears (Murie 1948). Older males are more likely to become habitual predators of cattle (Anderson et al. 1997). Livestock grazing may become a conflict with grizzly recovery as bears expand into the recovery zone in the Southern Rockies (Andromidas 2001).

In core grizzly bear zones, there should be no livestock grazing. Buying and retiring allotments on public lands may be a cost-effective step. The Interagency Grizzly Bear Management Guidelines have established standards to reduce bear/livestock conflicts in areas not deemed critical to grizzly bears (Andromidas 2001). In such areas, ranchers need to remove cattle carcasses from the grazing allotments. Removing carcasses has proven to decrease the number of grizzly bears in an area and decreases the risk of depredation (Anderson et al. 1997). Translocating problem bears may provide temporary relief during the grazing season, although translocated bears usually return to their original home range (Anderson et al. 1997). Translocations of large carnivores should be a last resort at resolving conflict, particularly with a slow-reproducing species that lives at low density. Changing grazing practices through incentives and education is preferable.

## 7. Ecology in the Southern Rockies

While a few grizzly bears may persist in the San Juan Mountains of the Southern Rockies, we learned little about their ecology before their numbers were decimated. Information from the Northern Rocky Mountains of the U.S. is helpful, but may not always be directly applicable to the situation in the Southern Rockies (Peterson 1995).

## 8. Management Recommendations

### *Protect habitat*

Specific habitat needs must be addressed for survival of the grizzly bear (Andromidas 2001). Avalanche chutes along with montane grasslands and shrublands should be identified and protected as critical grizzly habitat (Mace and Waller 1997). Meadows and grasslands adjacent to forested areas need protection for the herbaceous component of the diet (Mace 1987). However, identification and protection of only habitats with excellent food abundance and availability could leave habitat used by females unidentified and unprotected (Weilgus and Bunnell 1994). Because grizzlies often utilize large fallen trees as roof supports in their dens, logging which eliminates old growth trees could destroy denning sites (Judd et al. 1983). Forest management practices in areas with grizzlies need to maintain diversity in stand structure to ensure all foraging needs are met (Peek et al. 1987).

Grizzly bears require habitats that are in decreasing supply. They need large areas with a minimum of human disturbance. Degradation of suitable grizzly bear habitat should not be allowed to continue. Indeed, grizzly bear habitats should be expanded and improved (Andromidas 2001). Because of grizzly bears' large size, wide-ranging movements, and low densities, it is imperative to protect linkages between core grizzly bear areas (Andromidas 2001). Managers should recognize key habitat linkages and focus on preventing or eliminating barriers to grizzly bear movements between patches of suitable habitats.

### *Interagency Coordination*

Grizzly bear populations nearly always move across jurisdictional boundaries, and agencies should develop joint management plans and strategies to ensure long-term persistence of grizzly bears.

### *Human Education*

Educating the public on the ecology of the grizzly bear may help people avoid potential seasonal habitats and conflict (Madel 1996). Restricting human traffic to spring and winter ranges with carrion should reduce negative encounters (Peek et al. 1987). In Kananaskis Provincial Park, a transect method rated grizzly bear habitat use and potential use, and that method outlined an area for human recreation. This method kept people away from grizzly bear habitat, reduced bear harassment, and enhanced human safety (Herrero et al. 1983).

Public tolerance for the grizzly bear would also increase their survival (Knight et al. 1988). Wildlife managers need to be attentive to both the needs of the grizzly bear and the needs of the public, to maintain positive public views toward this species. For example, during the development of a grizzly management plan in Montana, one of the primary responses of the public was the need for agency personnel to respond more quickly to nuisance reports and to quickly resolve conflicts (Madel 1996).

Mattson et al. (1996) outlined seven management strategies for improving grizzly conservation, which summarize the above concerns: (1) The conditioning of grizzly bears to human foods should be minimized by proper human sanitation facilities in all areas where humans and grizzlies come into contact. (2) Human facilities should be located in or relocated to areas that receive little to no grizzly use such as travel, bedding, foraging, and protection from other bears to minimize habituation of grizzly bears to humans and conflict. (3) Limit human activity in grizzly bear habitat to minimize habituation and conflict. (4) Limit human access to grizzly bear habitat through road and trail closure. (5) Reduce the number of armed people in grizzly bear habitat, especially when food or odors that may attract a grizzly bear are present. (6) Balance management to favor survival of females and limited and controlled mortality of males. (7) All back-country users and local residents need to be educated on grizzly bears to minimize conflicts (Mattson et al. 1996). Andromidas (2001) and Horejsi et al. (1998) also outline management recommendations for conservation.

## 9. Justification

Protecting a viable population of grizzly bears in the Southern Rockies would restore an essential link toward ecosystem health of the area. It would also counter the steady northward retreat of the grizzly bear range. The grizzly bear was selected as an umbrella, keystone, flagship, and wilderness quality indicator.

**Umbrella:** Grizzly bears have large area requirements for population persistence. They require a variety of habitats and foods for persistence, as well as large areas with minimal human disturbance. Protection of sufficient area and habitat for viable populations of grizzly bears will, by inclusion, protect the habitats of many other species.

**Flagship:** The grizzly bear is a majestic, charismatic animal with considerable public appeal among nature enthusiasts. Indeed, grizzlies, more than any other mammal, represent the wildness of western North America. Conservation programs that benefit grizzly bears should readily capture the public's attention and interest.

**Keystone:** Grizzly bears, particularly in conjunction with wolves, play a significant role in top-down ecosystem regulation. Their presence both represents and maintains a healthy ecosystem.

**Wilderness Quality Indicator:** Grizzly bears define wildness. They occupy some of the most rugged and spectacular habitats in North America. While all carnivores can adapt to human presence without disturbance, at least to some degree, the grizzly bear has shown repeated intolerance when humans alter and degrade habitats. Grizzly bears certainly benefit from areas with low densities of roads and humans. Thus, large core areas with such wilderness qualities will be required for preservation of grizzly bears for the foreseeable future.

Grizzly bears are included indirectly in the initial SITES and Least Cost Path analyses via modeling for black bears, as discussed in Chapter 7.

## Canada Lynx (*Lynx canadensis*)

### *1. Introduction*

The Canada lynx (*Lynx canadensis*) is one of three members of the cat family (Felidae) that are native to the southern Rocky Mountains. It requires large expanses of high-elevation boreal forest habitat and is highly dependent upon snowshoe hares (*Lepus americanus*) as its prey base. Ideal habitat for lynx is old-growth forest subjected to natural disturbance regimes resulting in a dynamic heterogeneous mosaic of various forest successional stages. The lynx requires habitat protection and forested linkages among patches of suitable habitat for its survival and persistence. Ecologically, the lynx is an important predator of snowshoe hares and pine squirrels (*Tamiasciurus hudsonicus*). It is a good indicator of wilderness habitat quality within the Southern Rocky Mountain ecoregion. Canada lynx are federally listed as a threatened species and are protected throughout the region.

### *2. Distribution*

#### *Historic*

Lynx are primarily associated with forested habitats in arctic and boreal regions of North America, but also occurred in the northern contiguous United States (Koehler and Aubry 1994, McKelvey et al. 1999a). See McKelvey et al. (1999a:244) for a map of documented lynx occurrences from 1842 to 1998 in the contiguous United States. The southern-most extension of their historic range included the Southern Rocky Mountains, with documented occurrences as far south as the Colorado-New Mexico border (McKelvey et al. 1999a). There are no specimens from or verified records of lynx occurrence in New Mexico (McKelvey et al. 1999a, New Mexico Department of Game and Fish 2001b). McKelvey et al. (1999a:230-231) provides an account of documented lynx records in Colorado.

Unlike other western montane regions, the boreal forest habitat in Colorado is isolated from similar habitat in Utah and Wyoming by more than 150 km of lower elevation habitats in the Green River Valley and Wyoming Basin (Findley and Anderson 1956). There are four specimens from the late 1800s: one without a specific collecting locality, one from Cumbres County near the New Mexico border; one from Breckenridge; Summit County; and one from Colorado Springs, El Paso County. Edwin Carter's taxidermy notes in the Denver Museum of Natural History included a lynx trapped in Soda Gulch, Clear Creek County in 1878. Museum specimens were also found from Grand Lake, Grand County in 1904-1905; Jefferson, Park County in 1912; and southwestern Gunnison County in 1925. Terrell (1971) reported one lynx trapped at Red Cliff, Eagle County in 1929 and one at Marble, Gunnison County in 1931. Through interviews with trappers, James Halfpenny concluded that reports of three lynx being trapped in Eagle County in 1930 and 1936 were reliable.

In 1969 a specimen was trapped near Leadville, Lake County (Terrell 1971). In 1972, a lynx specimen was trapped on Guanella Pass, Clear Creek County and, in 1974, two were trapped (one is preserved as a specimen) on the north side of Vail Mountain, Eagle County. Since that time, only tracks have been found, including three sets on the Frying Pan River, Eagle and Pitkin Counties, and five sets near Mt. Evans, Clear Creek County.

Seidel et al. (1998) concluded that no viable populations of lynx remained in Colorado and that they were most likely extirpated from the state. Koehler and Aubry (1994) theorized that viable lynx populations may never have occurred in the fragmented boreal habitats of the Southern Rocky Mountains. They based this hypothesis on the paucity of historical records and the long-distance dispersal capabilities of lynx, suggesting that lynx periodically emigrated to Colorado from more northern lynx populations when these populations were at high levels.

*Current*

Since 1999, 129 lynx have been released on the San Juan and Rio Grande National Forests in southwestern Colorado. There are 45 known mortalities with an additional 20 being unaccounted for. Most lynx are living in or near the release area in the San Juan Mountains. As of 24 May 2003, there are no known lynx living north of I-70. Survival rates have improved considerably in the last two years and 16 lynx kittens were found in the spring of 2003, the first evidence of reproduction and a major milestone for the program. Up to 130 more lynx are planned to be released in the state over the next four years (Colorado Division of Wildlife 2003).

*Potential*

The Colorado Division of Wildlife plans to identify and map potentially suitable lynx habitat remaining in Colorado (Seidel et al. 1998).

## 3. Habitat

*General*

The following information regarding lynx habitat is taken from Koehler and Aubry (1994), unless otherwise cited. Throughout their distribution, lynx occur in boreal forest vegetative communities. Suitable habitats in the Southern Rocky Mountains are at high elevations and are characterized by Engelmann spruce (*Picea engelmannii*), subalpine fir (*Abies lasiocarpa*), aspen (*Populus tremuloides*), and lodgepole pine (*Pinus contorta*). Typical lynx habitat is characterized by continuous forest stands of varying ages with low topographic relief. Late successional or old-growth forests provide denning sites and hiding cover for kittens. Early successional forests support high prey densities and provide important foraging area for lynx. Snowshoe hare presence and abundance is a major determinant of suitable lynx habitat. Intermediate successional forests provide cover for travel by lynx and provide important connectivity among favorable patches of denning and foraging habitat. Frequent, small-scale disturbances tend to improve habitat for lynx.

Habitat conditions near roads may favor snowshoe hares and, thus, attract lynx, increasing their vulnerability to illegal killing and vehicle collisions. Roads may also increase access of competing carnivores (*e.g.*, coyote [*Canis latrans*] and bobcat [*Lynx rufus*]) to winter habitats used by lynx (Aubry et al. 1999a). Limited field research suggests that high-use roads may affect the spatial distribution of lynx by truncating home ranges (Aubry et al. 1999a). Apparently, lynx can tolerate moderate levels of human disturbance (Aubry et al. 1999a).

*Preferred*

Preferred foraging habitat for lynx is early successional forests where snowshoe hares are abundant. These habitats may result from natural or human-set fires, logging, windthrow, or tree diseases. Such stands provide preferred habitat for snowshoe hares from about 10 to 30 years following their establishment. Dense conifer stands are also an important habitat component for snowshoe hares and, therefore, lynx.

Preferred denning habitats are mature or old-growth coniferous forests stands greater than 1 ha in size with abundant large woody debris on the forest floor. Fallen trees and upturned stumps enhance the vertical and horizontal complexity of the habitat and provide important thermal and hiding cover for kittens. Other important attributes of denning habitat are multiple suitable denning sites, minimal human disturbance, proximity to preferred foraging habitat, and connectivity among preferred use areas. Suitable denning areas also provide safe havens for adult lynx. Den sites include caves and cavities under ledges, trees, and logs (Fitzgerald et al. 1994).

Shrubs and small-diameter trees provide important habitat for snowshoe hares. A shrub stage is generally lacking in the regeneration cycle of southern boreal forests, but occurs in canopy gaps within old-growth forest and in riparian areas (Buskirk et al. 1999). Pine squirrels, the second most important prey item of lynx, require mature, cone-producing conifers that are also abundant in old-growth forests (Buskirk et al. 1999). Since the two most important prey species of lynx are likely to be most abundant in old-growth forests in the Southern Rocky Mountains, it follows that lynx will be most abundant in old-growth forests as well.

Lynx tend to not cross openings greater than 100 m in width and do not hunt in open areas. Travel corridors are important for lynx movements within their home ranges and for dispersal movements. Suitable travel corridors are characterized by woody vegetation greater than 2 m in height with a closed canopy and close proximity to foraging habitat.

## 4. Food Habits

The following information is taken from Aubry et al. (1999a) unless otherwise cited. Lynx depend heavily on snowshoe hares as their primary prey, as evidenced by the nearly complete overlap of their respective distributions in

North America. Lynx tend to occupy habitats where snowshoe hares are most abundant. Even though snowshoe hare densities are typically low in southern boreal forests, they remain the predominant prey species in the lynx diet. The lynx released in Colorado seem to feed heavily on snowshoe hare, although dietary analysis is still ongoing (Shenk 2003). The most important alternative prey species may be the pine squirrel followed by cottontails (*Sylvilagus* spp.), blue grouse (*Dendragapus obscurus*), and a variety of small mammals (Aubry et al. 1999a, Best 2002). Ungulates represent an insignificant proportion of lynx diets, but lynx do occasionally kill young ungulates and scavenge ungulate carcasses (Fitzgerald et al. 1994, Koehler and Aubry 1994).

## 5. Population Dynamics

### *Life History*

Lynx are one of three wild members of the cat family (Felidae) found in the southern Rocky Mountains, the other two being the bobcat and mountain lion (*Puma concolor*, Fitzgerald et al. 1994, Koehler and Aubry 1994, Cook et al. 2000). The lynx is a medium-sized cat weighing from 5-15 kg (Fitzgerald et al. 1994). Males are slightly larger than females, with males averaging 10 kg and females 8.5 kg (Koehler and Aubry 1994). Their paws are large, offering greater support on snow, and their legs are relatively long—also an advantage in deep snow. Lynx are primarily nocturnal and solitary, except females with kittens and when adults come together to breed (Fitzgerald et al. 1994, Koehler and Aubry 1994). Lynx become sexually mature as yearlings, but breeding may be delayed if prey is scarce (Fitzgerald et al. 1994). Breeding occurs in the spring and litters average about 3 young that are raised by the mother (Fitzgerald et al. 1994). Few wild lynx live beyond the age of 6 years, but some individuals may live as long as 11 years (Aubry et al. 1999a). Aubry et al. (1999a) characterized lynx population dynamics in southern boreal forests as being generally similar to those of lynx populations in northern boreal forests during times of snowshoe hare scarcity. Generally, these population characteristics include low yearling pregnancy rates and litter sizes, low overall kitten production, high kitten mortality, and low lynx densities. Koehler and Aubry (1994) attribute the lack of dramatic snowshoe hare cycles and corresponding lynx population cycles in southern boreal forests to additional predators and competitors of snowshoe hares, the presence of alternative prey species, and increased patchiness of suitable habitat in southern boreal forests.

### *Population Density*

Aubry et al. (1999a) noted that lynx population densities in southern boreal forests (2-3 lynx/100 km$^2$) are similar to those in northern forests when snowshoe hare populations are low. They speculated that relatively stable and low snowshoe hare populations in southern boreal forests caused generally low lynx populations that lacked the dramatic population fluctuations of northern populations. In contrast, densities in excess of 35 lynx/100 km$^2$ have been observed in northern boreal forests following peaks in snowshoe hare populations (Koehler and Aubry 1994).

### *Home Range*

Unless otherwise cited, the following information is from Aubry et al. (1999a), who reviewed and summarized studies of lynx home ranges in southern boreal forests. Home ranges of lynx in southern boreal forests tend to be about 1.5 times the size of lynx home ranges in northern forests during periods of low snowshoe hare densities. Average home range sizes for males and females in southern habitats are 151 km$^2$ and 72 km$^2$, respectively. Male home ranges overlapped with those of 1-3 females in most studies. Ridges, major rivers, and major highways may define home range boundaries. Aubry et al. (1999a) advised caution in interpretation of these results because of multiple incongruities among studies.

### *Causes of Death*

Causes of mortality in lynx include trapping or shooting by humans (illegal in the Southern Rocky Mountains), vehicle collisions (especially for translocated lynx), starvation (often significant), and predation (Fitzgerald et al. 1994; Aubry et al. 1999a). Known predators of lynx include gray wolves (*Canis lupus*), mountain lions, wolverines (*Gulo gulo*), and adult male lynxes (Fitzgerald et al. 1994). Because of their low population densities and solitary nature, disease is probably not a major mortality factor in lynx populations (Fitzgerald et al. 1994). The average annual natural mortality rate for adult lynx is about 27%, but kitten mortality may approach 90% when snowshoe hares are at low densities—the normal situation in southern boreal forests (Koehler and Aubry 1994).

### *Population Structure and Viability*

In the Southern Rocky Mountains, habitats suitable for lynx and their primary prey (snowshoe hares and pine squirrels) are high-elevation boreal forests, which tend to be isolated and fragmented in their distribution (Aubry et al. 1999a). Human disturbances have increased isolation and fragmentation of lynx habitat (Seidel et al. 1998, Buskirk et al. 1999). This disjunct spatial arrangement of suitable habitat favors a metapopulation structure for lynx (Buskirk et al. 1999, McKelvey et al. 1999b). Metapopulations are characterized by several subpopulations, each occupying a suitable patch of habitat, that are linked (genetically and demographically) by individuals that move between subpopulations (Levins 1969, cited in Meffe and Carroll 1997).

Patches of lynx habitat vary in their size and quality. Larger, high-quality habitats may support "source" populations whose combined reproductive and survival rates produce excess individuals that become dispersers (Meffe and Carroll 1997). Smaller or low-quality habitats are more likely to support "sink" populations that would eventually go extinct without the influx of immigrants from other subpopulations (Meffe an Carroll 1997). For the metapopulation to persist, the colonization rate of extinction-prone subpopulations must greatly exceed their extinction rate (McKelvey et al. 1999b). If source populations are few, the loss of one or more of these populations could theoretically destabilize the entire metapopulation and lead to its eventual extinction (McKelvey et al. 1999b). McKelvey et al. (1999b) speculated that many southern lynx populations exist near the source-sink threshold, and may change status with small changes in the size or quality of habitat patches.

Connectivity that affords sufficient rates of dispersal among isolated habitats is the key to metapopulation health and persistence (Koehler and Aubry 1994, McKelvey et al. 1999b). Lynx populations in the Southern Rocky Mountains are probably effectively isolated from larger populations to the north by broad expanses of unsuitable lynx habitat (McKelvey et al. 1999b). This isolation increases the importance of maintaining sufficient habitat to support source populations in the Southern Rockies.

Determination of viable population size requires information on age structure, reproductive parameters, and survival rates that is lacking for southern lynx populations (McKelvey et al. 1999b). Thus, reliable minimum population viability analyses for lynx in southern boreal forests are not available, and we must resort to general conservation biology theory to develop minimum population and conservation area recommendations.

Because ideal situations do not exist in nature, the "genetically effective population size" ($N_e$) is always smaller than the actual census size ($N_c$) of the population (Meffe and Carroll 1997). A genetically effective population is generally defined as an ideal, stable population with randomly mating individuals, even sex ratio, equal birth rates among females, and nonoverlapping generations (Meffe and Carroll 1997). The multiple effects often are compounded when they interact (Lacy and Clark 1986). While empirical data for determining $N_e$ of southern lynx populations are not available, general knowledge of lynx population dynamics would suggest an effective population size in the range of $N_e = 0.5$ to $N_c = 1$. We caution, however, that many carnivores have ratios of 3 individuals by census to 1 individual in an effective population, similar to the black-footed ferret (*Mustela nigripes*, Lacy and Clark 1989). Grizzly bears (*Ursus arctos*) have a ratio of 4:1 (Allendorf et al. 1991). Lande (1995) argued that an effective population size of about 5,000 individuals may be necessary to generate sufficient mutations to preserve evolutionary processes. Obviously, larger is better for ensuring viable populations.

Franklin (1980) recommends that populations should not be allowed to fall below an effective population size of 50 ($N_c = 100$). However, a major goal of the Southern Rockies Wildlands Network Vision is to preserve or restore evolutionary processes in natural systems. Genetic variation within and among individuals comprising populations of animals is the currency of evolution. In short, genes that confer advantages to individuals are selected over time (because their carriers are more fit and survive longer) and non-advantageous genes are not selected as often. As a general rule, the preservation of "evolutionarily important amounts of quantitative genetic variation" requires effective population sizes of "at least several hundreds of individuals" (Lande and Barrowclough 1987:119). Franklin (1980) and Frankel and Soulé (1981) recommended an effective population size of 500 for long-term conservation. Following the "precautionary principle", we recommend protected habitat complexes for lynx that are large enough to support an effective population of at least 500 lynx ($N_c = 1,000$).

Using density estimates of 2 lynx/100 km$^2$, and an effective population size of about 500 animals (*i.e.*, a source population), a metapopulation would need interconnected complexes of suitable habitat totaling about 50,000 km$^2$. We recommend two or more such lynx refugia within the Southern Rocky Mountains to increase the probability of lynx population persistence. Ideally, lynx refugia should be comprised of old-growth forest subject to natural disturbance regimes, which create and sustain a dynamic habitat heterogeneity favored by lynx and their primary prey (Buskirk et al. 1999). Any patch of suitable or potentially suitable lynx habitat of at least 1,000 km$^2$ in size has conservation significance for lynx (McKelvey et al. 1999c), especially if similar patches occur within 100 km. Habitat complexes smaller than 50,000 km$^2$ that are greater than 100 km from a source population may require periodic augmentation by translocations of lynx to ensure metapopulation persistence.

#### *Population Status*

As of June 2003, there are somewhere between 64-84 free-ranging adult lynx in southern Colorado, with 16 documented kittens. Up to 130 more lynx are planned to be released in the state over the next four years (Colorado Division of Wildlife 2003). Canada lynx are federally listed as threatened and are protected in Colorado, Utah, and Wyoming. Except for the reintroduced population in Colorado, lynx are considered extirpated from the remainder of the Southern Rockies ecoregion (Shinneman et al. 2000).

## 6. Movements

### *Dispersal*

The following information is from Aubry et al. (1999a), unless otherwise cited. Dispersal is the movement of an organism from a place of residence to its first successful breeding location (Shields 1987, cited in Aubry et al. 1999a). No successful dispersal movements of lynx in southern boreal forest have been reported in the literature. Dispersal distances in excess of 100 km are considered typical for lynx; and dispersal success is a function of the dispersal capability of the species, the size of habitat patches, and the distance between habitat patches. The chances of successful dispersal diminish as population islands decrease in size and the distances separating them increase (McKelvey et al. 1999b). Aubry et al. (1999a) suggested that lynx inhabiting the more fragmented habitats of the southern boreal forests make occasional "exploratory" movements prior to actual dispersal. Such movements of up to 38 km have been observed, after which the lynx returned to their previous home ranges. Similar movements have not been observed in the more contiguous northern boreal forests.

### *Long-distance Movements*

When snowshoe hares are scarce (< 0.5/ha), lynx may abandon their home ranges and move long distances, presumably in search of new territories with more abundant prey (Koehler and Aubry 1994). Long-distance movements exceeding 1,000 km have been reported. Such movements may re-establish lynx in vacant habitats or augment marginal populations near the southern edge of the lynx's range (Koehler and Aubry 1994). However, Ruggiero et al. (1999) noted the lack of hard evidence that long-distance movements actually result in successful dispersal or the augmentation of distant populations.

### *Barriers to Movement*

Obstacles that may impede the movement of lynx across the landscape include paved roads, human developments, large rivers, and large expanses of open or unsuitable habitat (Koehler and Aubry 1994, Aubry et al. 1999a, Ruggiero et al. 1999). Koehler (1990, cited in Aubry et al. 1999a) observed that lynx traveled the edges of meadows, but only crossed openings of less than 100 m in width. However, for some documented lynx movements, successful crossings of busy, paved highways and large rivers had to have occurred (Aubry et al. 1999a).

### *Use of Corridors*

Aubry et al. (1999a) defined adequate travel cover as wooded linkages with 420-640 trees/ha and possibly shrub-dominated habitats during snow-free periods. Aubry et al. (1999b) cautioned that there is no empirical evidence of the use of corridors by lynx.

## 7. Ecology

### *Interspecies Interactions*

The distribution and abundance of the snowshoe hare directly affects the lynx's geographic distribution, habitat selection, foraging behavior, reproductive success, and population density (Koehler and Aubry 1994). Other predators that compete with lynx for snowshoe hares include coyote and bobcat and, to a lesser extent, red fox (*Vulpes vulpes*), great horned owl (*Bubo virginianus*), marten (*Martes americana*), fisher (*Martes pennanti*), and wolverine (Koehler and Aubry 1994). Because of their large paws and long legs, lynx are more effective than other predators in deep snow (Koehler and Aubry 1994). However, maintained roads and snowmobile trails may offer increased access to deep-snow habitats by coyotes and bobcats, increasing competition for scarce food resources during times of high energy demands (Koehler and Aubry 1994, Seidel et al. 1998, Buskirk et al. 1999). Mountain lion predation of lynx has been documented in southern boreal forests, which may add to factors limiting lynx populations (Aubry et al. 1999a). The combination of exploitation competition from bobcats, coyotes, and other mesopredators and interference competition from mountain lions may significantly affect lynx population in southern boreal forests, but sufficient data to demonstrate such an effect are not available (Aubry et al. 1999a, Buskirk et al. 1999). However, Buskirk et al. (1999) noted that the presence of gray wolves will likely reduce overall competition for preferred prey of lynx and they suggest, for example, that the Greater Yellowstone Ecosystem should improve as lynx habitat in response to the re-establishment of wolves in the ecosystem.

### *Ecological Effects*

The ecological effects and relationships of lynx are not clear in southern boreal forests. Because lynx depend upon snowshoe hare and pine squirrel for prey, lynx populations are expected to respond directly to changes in populations of these prey species (Koehler and Aubry 1994, Aubry et al. 1999a). However, lynx are not the only predator of snowshoe hares and pine squirrels. There is no evidence suggesting that lynx exert a regulatory effect on prey populations in southern boreal forests, except perhaps as a member of the larger suite of predators of snowshoe hares, pine squirrels, and other small prey species.

## 8. Management Recommendations

*Our challenge, from the perspective of maintaining lynx and their prey in the context of ecosystem management, is to design management strategies that result in dynamic, sustainable landscapes that approximate the composition of natural systems* (McKelvey et al. 1999c:428).

### *Establish Refugia*

Koehler and Aubry (1994) recognized the importance of protected areas to the persistence of lynx populations in southern boreal forests. Areas designated for lynx conservation should contain a mixture of forest age classes and structural conditions similar to the habitat configuration that would result from natural disturbance regimes (Koehler and Aubry 1994). This complex of habitat types can be estimated from models that mimic large-scale stochastic disturbance regimes such as fire (McKelvey et al. 1999c). Such an approach would provide a continuum of stand ages in a variety of spatial configurations that will benefit lynx, their primary prey species, and other boreal forest-dwelling species (McKelvey et al. 1999c). However, McKelvey et al. (1999c) cautioned that planned or artificial disturbance mechanisms can under-represent old-growth forests, which are important for lynx and take a long time to recover. We recommend the establishment of at least two protected refugia for lynx. Refugia should each contain at least 50,000 $km^2$ of interconnected habitats that collectively provide critical ecological needs of lynx and support a viable source population.

### *Protect Isolated Subpopulations*

The fragmented arrangement of suitable habitat in the Southern Rocky Mountains favors a metapopulation structure for lynx populations (Buskirk et al. 1999, McKelvey et al. 1999b). Metapopulations are comprised of several smaller subpopulations whose individual viability is critical to the long-term persistence of the metapopulation. Management practices should consider the vulnerability of local, isolated lynx populations to extinction. Management considerations should include habitat protection and enhancement, protection or creation of linkages, prohibition of take, and population augmentation through translocation of lynx from source populations.

### *Protect or Restore Linkages*

Because lynx tend to avoid openings and travel in areas with relatively dense woody cover, direct links of forest cover among suitable habitat blocks may be essential to lynx persistence (Koehler 1990, cited in Aubry et al. 1999a, Aubry et al. 1999a). Management practices should result in the protection, restoration, and enhancement of potential movement linkages for lynx.

### *Reintroduce Lynx to Suitable Habitats*

The Southern Rocky Mountains are effectively isolated from potential source populations (McKelvey et al. 1999b). Reintroduction is the only feasible method for restoring lynx populations to this region. One reintroduction project is underway (Colorado Division of Wildlife 2003). Pending completion, monitoring, and assessment of the current project, additional reintroductions should be considered for areas of suitable habitat of sufficient size to support subpopulations of lynx.

### *Reduce or Eliminate Roads*

Roads and recreation trails in otherwise suitable lynx habitat will likely result in human disturbances that detrimentally affect lynx populations, especially during critical times such as the denning season and winter (Aubry et al. 1999a). During winter, roads and trails may facilitate access by competing predators, especially coyotes and bobcats, to deep snow areas, where lynx would otherwise have a competitive advantage (Aubry et al. 1999a). Roads should be eliminated, reduced, or seasonally closed in critical lynx habitat areas.

### *Monitor Lynx Populations*

Because of their rather specific habitat preferences (southern boreal forests) and large area requirements (~2 lynx/100 $km^2$), lynx populations are very sensitive to habitat loss or degradation. Resource managers must implement effective monitoring strategies to detect effects of land management practices on habitat quality and numerical abundance and density of lynx and other biota, especially snowshoe hares and pine squirrels (McKelvey et al. 1999c). Monitoring of hare populations with pellet transects and lynx populations with snow tracking or other effective techniques will likely be critical components of monitoring strategies (McKelvey et al. 1999c). Monitoring plans should have defined criteria for assessing results and include adaptive management provisions for modifying management plans and their implementation if necessary (McKelvey et al. 1999c).

### *Promote Natural Disturbance Regimes*

Fire is an important agent in creating forest diversity. Additionally, factors such as windthrow, disease, and insect infestations create microhabitats for lynx, especially within old-growth forests (McKelvey et al. 1999c). McKelvey et al. (1999c) believed that the dynamic mosaic of habitat types resulting from natural disturbance regimes operating at the landscape scale in southern boreal forests provide high-quality habitat for lynx, their primary prey, and other members of the biotic community. With few exceptions (*e.g.*, threats to human life and property), natural disturbance regimes should be allowed to operate freely in large protected reserves. To the extent practical and feasible, natural disturbance patterns should be mimicked by land use management actions in compatible use areas.

*Multi-jurisdictional Planning*

Conservation planning for broad-ranging carnivores, such as lynx, must be conducted over vast spatial scales and must consider connectivity among subpopulations (Noss et al. 1996). Large areas required for lynx population persistence will most certainly span the jurisdictions of multiple land management agencies and private lands. Establishment of regional, multi-jurisdictional planning authorities or arrangements through appropriate means (*e.g.*, legislative or administrative) should be encouraged and pursued.

*Educate Managers and the Public*

Wildlife managers and policy makers need a thorough understanding of lynx ecology in order to establish appropriate policies and make sound management decisions. In addition, the public needs accurate information and knowledge about lynx to inform their opinions and values and their understanding of appropriate management measures. Knowledge is the key to informed conservation actions and advocacy by both agencies and the public.

*Conduct Meaningful Research*

Knowledge of lynx ecology in southern boreal forests is extremely limited (Koehler and Aubry 1994, Aubry et al. 1999a, b). Research is needed in the following areas: foraging ecology, habitat use, den site characteristics, optimum habitat composition and structure (for lynx and key prey species), and effects of forest management practices and natural disturbance events on the quality and quantity of lynx habitat (Koehler and Aubry 1994).

### 9. Justification

The lynx was selected as a flagship, umbrella, and wilderness quality indicator species.

**Flagship:** Wild predatory cats like the lynx fascinate a wide range of people. Conservation efforts focusing on lynx should generate popular support.

**Umbrella:** Lynx have very large area requirements for population persistence. They require a dynamic mosaic of various boreal forest successional stages mimicking that resulting from natural disturbance regimes. Protection of habitat for viable populations of lynx will, by inclusion, protect the habitats of many other southern boreal forest-dwelling species.

**Wilderness Quality Indicator:** The lynx needs old-growth, unlogged forests subject to dynamic natural disturbance regimes for optimum habitat. These parameters are the essence of "wilderness." The best protection for such forests is formal designation as Wilderness Areas.

This species was not selected for inclusion in the initial SITES modeling because adequate data do not currently exist to model their habitat and dispersal requirements within the Southern Rockies ecoregion. Suitable habitat data based on field observations by the Colorado Division of Wildlife (CDOW) is expected to become available soon, and will be used for future analyses. Once breeding populations of lynx become established within the region, the telemetry data collected by CDOW will become a valuable source of lynx dispersal requirements.

## Pronghorn (*Antilocapra americana*)

### 1. Introduction

Pronghorn (*Antilocapra americana*) are the only living species of the family Antilocapridae (Soulounias 1988, Fitzgerald et al. 1994), thus extinction would present a major loss of a unique genetic branch. Pronghorn once ranged through the Great Plains from southern Saskatchewan and Alberta, Canada to northeastern Durango, Mexico (Kitchen and O'Gara 1982). Their numbers suffered a precipitous decline by the 1920s (about 99%), then they began to recover (Nowak 1991).

Pronghorn occupy shrub and grassland habitats (O'Gara 1978). They are social and form herds. Although the combination of white and buff, along with shadows, creates good camouflage in the open areas they frequent, pronghorn depend on eyesight and speed to avoid predators. Bucks and does are similar in size (about 50 kg.) and reach adult size at 2 years (Kitchen and O'Gara 1982).

### 2. Distribution

*Historic*

Pronghorn evolved in the prairies and deserts of North America and have ranged in historical times from the south-central prairies of Canada through the western grasslands and steppes of the United States, and south to the deserts and plateaus of northern Mexico (Kitchen and O'Gara 1982, Yoakum et al. 1996). They ranged from the California and

Oregon coast on the west to near the Mississippi River on the east (Yoakum et al. 1996). Before European settlement, Nelson (1925 in Yoakum et al. 1996) estimated pronghorn numbers between 30,000,000 and 60,000,000. Between 1550 and 1900 habitat loss from homesteading and agriculture, competition with livestock, year-round hunting (much of it market hunting during the late 1800s), fencing, and railroads reduced that abundance by more than 99%, and they reached a low of 13,000 animals by the 1920s (Kitchen and O'Gara 1982, Yoakum et al. 1996, Byers 1997).

In 1920, hunting restrictions were instituted and refuges were established (Yoakum et al. 1996). During the Dust-Bowl years, many of the semi-arid homesteads were abandoned, and much of the land reverted back to natural vegetation. Through hunting restrictions, conservation, and reintroductions, pronghorn increased to 406,000 by 1976 (Kitchen and O'Gara 1982).

*Current*

There are now close to a million pronghorn (Byers 1997). Although this represents a significant recovery, it is still just 3% (or less) of the pronghorn numbers of pre-settlement times (Kitchen and O'Gara 1982). The total area of suitable habitat has been reduced (Yoakum et al. 1996). Remaining pronghorn are fragmented across 15 states in the U.S. and Mexico and two provinces in Canada (Byers 1997). Wyoming has the largest population.

*Potential*

Various obstacles limit further increases, and the very maintenance of some existing populations is in doubt. Habitat destruction, particularly in winter and fawning areas and travel corridors, has increased recently (Nowak 1991, Yoakum et al. 1996). Habitat destruction and degradation comes from intensified agriculture, urban expansion, mining in historic habitats, fencing, increasing exurban development in grassland habitats, upgrading of roads and/or other human-created obstacles across routes of seasonal movements, resistance by agricultural interests to transplants and range extensions, removal of native vegetation by rangeland rehabilitation projects, and heavy livestock grazing (O'Gara 1978, Yoakum 1986, Yoakum et al. 1996). Approximately 98% of pronghorn share their habitat with domestic livestock (Yoakum and O'Gara 1990).

## 3. Habitat

Pronghorn formerly occupied treeless lands from desert shrub and short-grass plains to mountain grasslands, although occupation may have been intermittent in places due to prolonged drought or severe winter storms (Russell 1964). Pronghorn historically occupied 26 different prairie and shrubland habitats across their range (Kitchen and O'Gara 1982). Key similarities were low, rolling topography, and forage consisting of grasses, forbs, and shrubs (Kitchen and O'Gara 1982). Highest densities occurred on areas that received 25 – 40 cm of rainfall per year (Yoakum et al. 1996).

Grassland habitats seem to hold the best mix of vegetation that pronghorn require for good health (Kitchen and O'Gara 1982). About 68% of North American pronghorn populations live in grasslands, 31% in Great Basin shrub steppe, and 1% in desert. Pronghorn thrive on sub-climax rangelands maintained by fire and seasonal grazing by elk (*Cervus elaphus*) and bison (*Bison bison*, Kitchen and O'Gara 1982). High quality habitat will have no fencing, as fences restrict or prevent movements to food and water or escape from deep snow (Yoakum and O'Gara 1990). Habitat is considered optimal when pronghorn are 1.5 to 6.5 km from water sources (Sundstrom 1968, Kindschy et al. 1982). Of 12,000 pronghorn sighted, 95% were within 6.5 km of water (Sundstrom 1968).

Pronghorn are found from sea level to 3,300 m, but the greatest densities of pronghorn occur from 1,200 to 1,800 m (Yoakum et al. 1996). A mixture of community types is preferred to a monoculture, and the proper percentages, quantities, and distribution of vegetation are important (Yoakum et al. 1996). They rarely feed in one place but keep moving (Yoakum et al. 1996).

Pronghorn are adapted to "sight and flight" behavior and will avoid areas that hinder visibility or their ability to run at full speed (Byers 1997). They are the swiftest land mammal in North America. Herds move at 64 – 72 km/hr with a maximum of 86 km/hr (Kitchen 1974).

Pronghorn may select habitat on a seasonal basis. As "hiders" pronghorn fawns are dependent on cover to protect them from predators and adverse environmental conditions (Byers 1997). Fawns remain hidden and motionless, heads flattened to the ground, for about 10 days, denying predators information on their presence and location (Byers 1997). In winter, high-quality browse must be available above snow-cover, which means more than 40 cm of snow inhibits pronghorn foraging (Yoakum et al. 1996).

Primary factors reducing pronghorn habitat include highways and railroads, substandard fences, water scarcity during times of drought, waters fenced or surrounded by dense vegetation, tree and shrub encroachment from long-term fire suppression and historical grazing abuse, low plant species richness, and human encroachment resulting in habitat loss and fragmentation (O'Gara 1978, Nowak 1991, Yoakum et al. 1996). Encroachment of woody vegetation from long-term fire suppression has reduced species diversity of grasslands, particularly the forbs upon which pronghorn are dependent (Yoakum 1978). As the height of vegetation increases, habitat suitability for pronghorn decreases

(Kitchen and O'Gara 1982, Byers 1997).

### *4. Food Habits*

Pronghorn require an average of 1.0-1.5 kg of air-dry forage per 44 kg animal per day (Zarn 1981). They are selective, opportunistic foragers depending on the availability and palatability of the species present (O'Gara 1978, Yoakum et al. 1996).

Pronghorn take the most palatable and succulent forage available at all seasons. Browse, such as sage (*Artemisia* ssp.), rabbitbrush (*Chrysothamnus {=Ericameria} nauseosus*), and winterfat (*Krascheninnikovia lanata*) is important year round, but especially in the winter, while forbs dominate their diets in spring and summer (Kitchen and O'Gara 1982). Five to ten species of forbs usually comprise the bulk of their diet, and this may include species noxious to livestock (Yoakum 1990). In Colorado, the average diet was 43% browse, 43% forbs, 11% cactus, and 3% grass (Fitzgerald et al. 1994). In winter, the shrubs increased to 90% of the diet (Fitzgerald et al. 1994).

Grasses are important only during their early growth in the spring and fall, rarely comprising more than 10% of their yearlong diet (Yoakum 1990). Fawn survival appeared twice as high where abundant nutritious forbs and grasses were available during late gestation and early lactation (Ellis 1970). Of 21 studies that compared diet selection to availability, overall forbs were the most preferred, then shrubs, with grasses the least preferred (Yoakum 1990). Several studies indicated that prairie dogs (*Cynomys* spp.) enhance prairie for pronghorn by eating grasses and increasing abundance and variety of forbs (King 1955, Koford 1958, Costello 1970, Bonham and Lerwick 1976, Wydeven and Dahlgren 1985, Cid et al. 1991, Whicker and Detling 1993). Lovaas and Bromley (1972) reported the reverse.

Pronghorn require approximately 4 - 6 liters of water per day per adult during hot, dry summers, but if vegetation is succulent 1 liter of water per day may be sufficient (Sundstrom 1968). When the moisture content of succulent forage exceeds 75%, Utah pronghorns do not require free water (Beale and Smith 1970).

### *5. Population Attributes*

#### *Life History*

Northern pronghorn winter in large mixed-sex aggregations, feeding and bedding in close association. Sexes may remain segregated in mild winters. Winter herds can range in size from a few individuals to thousands (Kitchen and O'Gara 1982). The large herds begin to break up in late February or March. Mature males become solitary, while males one to three years old form small bachelor groups; females form small groups often with a single male. In the spring pregnant does seclude themselves to give birth. After the two week hiding phase, they rejoin groups of yearling females and females that have lost fawns to predation (Byers 1997).

Pronghorn are polygamous (Kitchen and O'Gara 1982). The breeding season in northern states is usually from mid-September to early October; in southern populations it extends from July through October (Byers 1997). Does usually breed for the first time at 15 months (Byers 1997). Gestation averages 252 days (more than 8 months, which is longer than most large ungulates), and most yearlings and adults bear twins (O'Gara 1978, Byers 1997).

Yearling males are capable of breeding, but adults prevent most young males from doing so (Kitchen and O'Gara 1982). The most aggressive bucks do most of the breeding. Males may utilize a harem-type or territorial breeding strategy, depending on differences in density and spatial dispersion of the population (O'Gara 1978, Byers 1997). In a harem system, dominant males control and defend females rather than territories. Females select males on the basis of their vigor and ability to hold a harem (Byers 1997). Being sequestered frees females from aggressive and competing bachelor males, and ensures pregnant or lactating does the best rangelands (Byers 1997). In territorial systems, territorial males do almost all the breeding. Bucks prefer territories that allow them to corner does (O'Gara 1978). Males shed their horn sheaths after the rut (O'Gara 1978).

It has been suggested that selection imposed by the numerous predators that co-existed with pronghorn for much of their evolution has influenced their speed and endurance, maternal behavior, grouping tendencies, competition for social rank, and the selection of mates by females (Byers 1997). Twinning may have been an adaptive response to the low probability of a single offspring surviving. Many of these past adaptations have been made irrelevant by the Holocene extinctions of predators (Byers 1997).

#### *Mortality*

Pronghorn rarely live more than 10 years (Byers 1997). Prolonged severe winters are a great cause of mortality in their northern ranges (Yoakum et al. 1996). Over 40 cm of snow may cause winter-kill by covering forage (Yoakum et al. 1996). This is compounded if the forage is of low quality, man-made obstacles such as improper fences prevent movement to areas with less snow cover, and females have not recovered their body condition after reproducing (Kitchen and O'Gara 1982).

Predation on pronghorn is usually not a factor, but it can be under certain circumstances (Byers 1997). The greatest impact of predators is on fawns (Fitzgerald et al. 1994). Adult pronghorn are also vulnerable in deep snowy conditions. Pronghorns probably evolved to escape Pleistocene predators, particularly American cheetahs (*Acinonyx trumani*), which are no longer present (Byers 1997).

Most predation occurs before fawns are 3 weeks old. After the 10-day hiding phase, active fawns are more susceptible to predation. By 3 weeks of age, however, fawns join fawn/doe groups where vulnerability appears to be low (Kitchen and O'Gara 1982). High fawn mortality is a characteristic of most pronghorn populations throughout their range (Byers 1997). Fawn losses of 25-65% are not uncommon, and they can be as high as 80% (Byers 1997). Predation on fawns appears to be a problem primarily when herds are small and fences confine them with experienced predators (Kitchen and O'Gara 1982). Drought and severe winters can also increase predation.

Pronghorn populations with female and male survival rates >0.80 and reasonable fawn recruitment rates will increase (R. Ockenfels pers. comm.). The quantity and quality of habitat appears to be the overriding influence in fawn survival (Ellis 1970). In Colorado, a hunter harvest of 40% caused population declines, whereas a hunter harvest of 20% allowed population growth (Fitzgerald et al. 1994). A ratio of 25-35 bucks to 100 does in a given area is considered sufficient to maintain a viable population (R. Ockenfels pers. comm.). Losses from crippling sustained during hunts may be serious at times.

Other sources of mortality include poaching, road kills, fence entanglement, complications during parturition, starvation, drought, and old age (O'Gara 1978). Pronghorn are sensitive to bluetongue, Epizootic Hemorrhagic Disease (EHD), leptospirosis, anaplasmosis, and parasitic worms (Kitchen and O'Gara 1982, Yoakum et al. 1996). Bluetongue is probably the most serious disease for pronghorn, and cattle are the primary reservoir for this disease (Yoakum et al. 1996). Cattle are chronic carriers of bluetongue, but do not develop clinical or acute symptoms (Yoakum et al. 1996).

## 6. Movements

### *Home Range*

There is great variation in the size of home and seasonal ranges between areas or at different times. Fitzgerald et al. (1994) reported home ranges varied from 165 ha to 2,300 ha. Size depends on habitat quality, history of domestic grazing, herd size, and season.

### *Movements*

The timing and length of seasonal movements vary with altitude, latitude, weather, and rangeland conditions, though pronghorn may have moved longer distances before humans created barriers. Movements are initiated by a need for water or food, natural disturbances, or weather and are often associated with seasonal changes in forage availability or the selection of fawning areas (Kitchen and O'Gara 1982). Summer and winter ranges are differentiated based on snow accumulation, distances between preferred seasonal foraging areas, and sources of drinking water. Natural barriers limiting movements include large bodies of water, rivers, abrupt escarpments, thick shrubs or trees, and deep canyons.

Essentially no migration occurs in areas with sufficient forage, year-round water, and snows of 6' or less (Kitchen and O'Gara 1982). Temporary seasonal movements to lower elevations may occur during periods of deep snow, the animals returning after snows have receded. In areas of high topographic relief with deep winter snows, summer and winter ranges may be as distant as 160 km (O'Gara 1978). One herd in northwestern Wyoming moved over 300 km to a winter range (J. Berger, pers. comm.). Man-made barriers such as fences, highways and railroads prevent movements and thus reduce the carrying capacities of some ranges (Yoakum 1978, Kitchen and O'Gara 1982).

## 7. Management Recommendations

### *Refugia*

Human disturbances (*e.g.* recreation or livestock use) must be unlikely or minimal most of the year in core areas, there should be no humans residing in the area, fences should be removed or constructed as game-standard or better barbed wire fences (smooth bottom strand at least 41-46 cm above ground and numerous locations in which the bottom strand is more than 46 cm, Kindschy et al. 1982, Yoakum et al. 1996).

### *Livestock Grazing*

Livestock may negatively impact pronghorn populations by altering plant structure and species composition, promoting shrub growth at the expense of perennial grasses, reducing fuels, grazing after droughts or fire, and causing erosion and loss of soil fertility (Yoakum et al. 1996). Sheep and goats compete directly with pronghorn for forage (Yoakum et al. 1996). An area with a history of excessive livestock grazing and fire suppression often has enhanced growth of shrubs at the expense of the quantity and diversity of native forbs and grasses, and degraded riparian habitat. Prescribed fire and riparian restoration are primary management tools for enhancing productivity of native forbs on summer ranges of pronghorn (Yoakum et al. 1996).

### *Transplants*

Criteria formulated by Hoover et al. (1959) for selection of transplant sites continue to be used. These criteria specified that there be sufficient continuous rangeland to allow the herd to spread, that a minimum of 100 animals be released, each animal as a rule requiring 2.6 $km^2$, that there be a diversity and abundance of forbs and shrubs, that concurrent use by livestock be evaluated for competition for forage, and that any fences be suitable for pronghorn movements. Public lands should have first preference, then large blocks of private land under one ownership. The least desirable places for transplants were felt to be private lands in small units with many owners. Local acceptance is critical.

Transplanting should involve a feasibility study or management plan to document objectives and procedures, and post-release monitoring of animals and habitat (Yoakum et al. 1996). An effective population size of 500 is considered a minimum for sustaining genetic diversity (Frankel and Soulé 1981).

*Linkages*

A trend with long-term negative consequences is the increasing isolation of populations from each other, preventing interchange and keeping populations small. Natural obstacles, fencing to control livestock, paved highways, utility corridors, subdivisions, urban development, and tree and shrub encroachment all act as barriers to movement that may prevent pronghorn from finding water and fawning sites, and from escaping predators or deep snow. In some areas, fences may determine home ranges.

In the long run, small populations result in genetic deterioration and the risk of local extinction from stochastic processes (Frankel and Soulé 1981). Small populations are more susceptible to extirpation from severe weather, habitat loss, poaching, over-harvest, and negative changes in gene frequency. Polygamous species like pronghorn require larger populations to maintain genetic diversity since only a few males contribute genes in any given year (heterozygosity may be retained by frequent replacement of males which is one rationale for focusing hunts on bucks). Maintaining movement corridors is critical to preventing fragmentation.

### 8. Justification

Recovery and protection of the pronghorn would help achieve the goals of ungulate recovery, restoration of natural fire, and protection and restoration of connectivity (grasslands). Pronghorn were chosen as a flagship and prey species.

**Flagship:** The hunting value of the species has attracted attention for the conservation and restoration of the areas where they are present or will be reintroduced. This contributes to the protection of grasslands (often currently heavily overgrazed) and associated species. Pronghorn represent a unique evolutionary lineage in North America, and are also popular with the American public.

Pronghorn was included in the initial SITES analysis because its suitable habitat covers areas within the ecoregion not otherwise included by those areas suitable for wolves and bears.

## Cutthroat Trout (*Oncorhynchus clarki*)

### 1. Introduction

Of all western trout species in the Southern Rockies, cutthroat (*Oncorhynchus clarki*) have the widest distribution and include 14 different subspecies (Behnke 1992). Their historical range extended from Alaska to northern California and east to central Colorado (Sigler and Sigler 1996, Meehan and Bjornn 1991). Found in rivers, streams, and lakes, they are the only trout species native to Wyoming and other states in the west (Baxter and Simon 1970, McClane 1978).

Cutthroat trout are of special concern because overharvest, habitat loss and degradation (logging, mining, livestock grazing, and irrigation), and introductions of nonnative fish have drastically reduced their numbers (1977 Federal Register Vol. 42 No. 186: 48901-48902, Hickman and Duff 1978, McClane 1978, Warren and Burr 1994). In the latter case, cutthroats hybridize with rainbow trout (*O. mykiss*) and, when sympatric, golden trout (*O. aguabonita*, McClane 1978). In addition, their needs for oxygen, temperature, and water quality tend to be more stringent than other trout (1977 Federal Register Vol. 42 No. 186: 48901-48902). Other trout can thus displace cutthroats by competition and also by direct predation. This review will focus on the subspecies occurring in the Southern Rockies: the Colorado River cutthroat (*O. c. pleuriticus*), the Rio Grande cutthroat (*O. c. virginalis*), and the greenback cutthroat (*O. c. stomias*).

These three subspecies have very similar requirements and life histories. Hence, a general discussion is sufficient in most cases. However, any characteristics and requirements that may vary from statements made in this general discussion will be noted below.

### 2. Colorado River cutthroat trout

*Historical Range*

Historical range descriptions for the Colorado River cutthroat trout are unavailable but the range is thought to include the upper Colorado River basin above the Grand Canyon. This includes the Dolores, Green, San Juan, Yampa, Gunnison, Duchesne, Dirty Devil, and Upper Escalante drainages (Behnke 1992, Young et al. 1996,

Hepworth et al. 2001). Non-native salmonids have been stocked into Colorado River cutthroat range for over 100 years, a practice that continues today.

#### *Current Situation*

The range of Colorado River cutthroat trout has been drastically reduced (Behnke 1992, Young et al. 1996). Colorado River cutthroat may still be found in Colorado, Utah, and Wyoming, but the populations are few, small, and scattered across historical range (Young et al. 1996, Colorado River Cutthroat Trout Task Force 2001). Most remaining populations occupy headwater streams and lakes, but their abundance and status is uncertain (Colorado River Cutthroat Trout Task Force 2001).

Of 152 waters believed to contain remnant populations of Colorado River cutthroat, 70 have been stocked with non-native trout, and 63 of the 70 have been stocked with trout that can hybridize with Colorado River cutthroats (Young et al. 1996). Porcupine Lake, Lake of the Crags, and Lake Diana in Colorado have been repeatedly stocked with non-native cutthroat trout and should no longer be considered part of the Colorado River cutthroat range (Young et al. 1996). Stocking non-native trout is a management practice that continues, and nearly 30% of the waters holding Colorado River cutthroat trout have been stocked with non-native trout recently (Young et al. 1996). Only about 20 waters still support indigenous, genetically pure populations - now isolated by artificial barriers- and these populations may be too small to be viable (Young et al. 1996). Those areas are obviously important to conservation. The Gunnison and Dolores Basins are two such regions (Young et al. 1996).

Artificial barriers are used to protect the isolated populations of pure strain cutthroat. But, those populations are still at risk from illegal stocking above the barrier by anglers (Young et al. 1996). Removing non-natives and reconnecting populations, allowing them to move from rivers to streams or between rivers and lakes, could reduce threat of extinction by reducing the risk of a single environmental disaster (Young et al. 1996).

#### *Legal Status*

The U.S. Fish and Wildlife Service considered the Colorado River cutthroat for listing as a Category 2 species under the Endangered Species Act, but that category was eliminated (Young et al. 1996). It is considered a sensitive species by Region 2 and 4 of the U.S. Forest Service, and it has special status in the states of Colorado, Utah, and Wyoming (Young et al. 1996). Many Wyoming and Colorado populations are protected by harvest closures, catch and release, or catch by artificial lures only.

### 3. Rio Grande cutthroat trout

#### *Historical Range*

Historical range was the Rio Grande drainage, including the Chama, Pecos, and Jemez drainages (Stumpff and Cooper 1996). It is the southernmost subspecies of cutthroat trout (Behnke 1992). After severe overharvest of Rio Grande cutthroats in the 1800s, non-native trout were introduced. The Rio Grande cutthroat evolved without other salmonids, so it was vulnerable to these introductions. Livestock grazing, irrigation, and sedimentation destroyed habitat for the Rio Grande cutthroat (Stumpff and Cooper 1996).

#### *Current Situation*

Presently, there are 39 populations in Colorado, 21 on public land and 18 on private land, and there are 53 populations in New Mexico, 46 on public land and 7 on private land (Stumpff and Cooper 1996). Nearly all of those populations are limited to small, high mountain headwater streams (Stumpff and Cooper 1996). Of the 92 populations, 4 are stable and increasing, 29 are stable and holding, 8 are declining, and 51 are unknown but likely at risk (Stumpff and Cooper 1996). Removing non-native brook trout (*Salvelinus fontinalis*) from a Rio Grande cutthroat stream (West Indian Creek) has helped upgrade the status of that stream from declining to stable (Colorado Division of Wildlife website).

#### *Legal Status*

On February 25, 1998, the Southwest Center for Biological Diversity petitioned the U.S. Fish and Wildlife Service to protect the Rio Grande cutthroat trout as a threatened or endangered species. The U.S. Fish and Wildlife Service ruled that listing was not warranted (1998 Federal Register 63: 49062). The Southwestern Center for Biological Diversity then filed a legal complaint in June of 1999. With additional information on whirling disease in hatchery fish, the U.S. Fish and Wildlife Service initiated a candidate status review in November of 2001.

### 4. Greenback cutthroat trout

#### *Historical Range*

The greenback trout was found largely in Colorado along the east slope of the Rocky Mountains, particularly in the South Platte and Arkansas River drainages (1977 Federal Register Vol 42 No. 186: 48901-48902, 1978 Federal Register Vol 43 No. 75: 16343-16345). It is the most easterly subspecies of the cutthroat trout (Young et al. 2002). It was abundant in the 1800s, and fish from 2 to 4.5 kg were not uncommon (Young et al. 2002). Greenback trout evolved as the only trout in these waters, and they will not coexist with other trout (1977 Federal Register Vol 42 No. 186: 48901-48902). Nonetheless, non-native trout were stocked into greenback waters, while logging, livestock grazing, irrigation, and mining destroyed greenback trout habitat (1977 Federal Register Vol 42 No. 186: 48901-48902). Overharvest, sometimes with explosives, further reduced populations (Young et al. 2002). It was considered extinct in 1930, but a population was discovered in 1957 (Young et al. 2002).

### *Current Situation*

Most recovery efforts focus on starting new populations. By 1999, 44 such populations had been established and 24 others were discovered, mostly in high mountain headwaters (Young et al. 2002). About 20 populations are considered stable, and there is still a shortage of good, low elevation habitat (Young et al. 2002). Many drainages in Colorado contain the parasitic infection whirling disease (*Myxobolus cerebralis*), and scientists predict that all streams will be infected in 60 years (www.waterknowledge.colostate.edu). Whirling disease probably originated in Europe and entered the U.S. from Denmark in 1956. The parasite is transmitted to the wild by hatchery fish. While it is found in the wild in Colorado and Wyoming, it is restricted to hatcheries in New Mexico. In time, however, the parasite will probably infest wild waters of New Mexico as well. There are catch and release fishing programs in selected waters of Rocky Mountain National Park, San Isabel National Forest, Roosevelt National Forest, and Arapaho National Forest.

### *Legal Status*

In 1969, the greenback trout was listed as Endangered under the U.S. Endangered Species Act (1977 Federal Register Vol 42 No. 186: 48901-48902). In 1978, that federal status was changed to Threatened (1978 Federal Register Vol 43 No. 75: 16343-16345). The U.S. Fish and Wildlife Service has turned the greenback program over to the state of Colorado.

## *5. Habitat*

All trout require different, specific habitat types for spawning, juvenile rearing, adult rearing, and overwintering (Behnke 1992, Kershner 1995, U.S. Fish and Wildlife Service 2001b). Cutthroat trout abundance may be limited by availability of suitable habitat for these life stages (Behnke 1992). Optimal cutthroat trout riverine habitat has cold, well-oxygenated waters with a stable temperature regime; there is an approximate 1:1 pool to riffle ratio (Hickman and Raleigh 1982, Raleigh and Duff 1981). In addition, it must have areas of deep, low-velocity water (Raleigh and Duff 1981). The water is typically clear, having a rocky substrate free of fine sediments. Suitable habitat will also have abundant cover, well-vegetated stream banks, and relatively stable stream flow. Lacustrine habitat consists of cold, clear, deep lakes that are usually oligotrophic.

Cutthroat trout are stream spawners and typically require streams with gravel substrate in riffle areas (Hickman and Raleigh 1982). Hence, successful spawning may be limited in high-gradient streams because suitable spawning gravel is carried off with the current (Behnke 1992). However, in lower reaches, high sediment loads may limit successful reproduction by suffocating redds in a blanket of silt.

Cutthroat trout in nursery habitat require low water velocity and protective cover (Behnke 1992). Suitable nursery sites are not likely to be found in high-velocity, high-gradient streams, but rather in small tributaries, along the margins of streams, in side channels, and in spring seeps.

Survival in winter months is dependent upon the availability of areas with low water velocity and sufficient protective cover. Beaver (*Castor canadensis*) ponds and pools with large woody debris and/or boulders are important in these respects (Behnke 1992, Jakober et al. 1998).

## *6. Ecology*

Cutthroat trout migration patterns vary widely depending on locality and subspecies (Hilderbrand 1998, Meehan and Bjornn 1991). However, movement patterns also vary between individuals of the same subspecies. Some individuals remain completely stationary and others frequently move long distances (Hilderbrand 1998).

Life span for cutthroat trout is highly variable between subspecies, but can also vary greatly within subspecies depending environmental factors (Behnke 1992, Gresswell 1995, Gresswell and Varley 1989). For example, life span can be long in cold waters where metabolism is low. Colorado River cutthroats generally do not grow to be more than 200 mm (325 mm for lacustrine forms), and tend to grow slower than other cutthroat subspecies (Utah Department of Natural Resources 1996). For Colorado River cutthroats, sexual maturity is reached between 2 and 4 years of age (Speas et al. 1994, Utah Department of Natural Resources 1996).

Cutthroat trout spawn in the spring or early summer throughout their native range (Meehan and Bjornn 1991). Spawning migration is correlated to fluctuations in water temperature (Gresswell 1995, Varley and Gresswell 1988). Colorado River cutthroat typically begin spawning when peak flows begin to diminish in late spring to early summer and spawning may be cued when temperatures reach 7-8°C (Speas et al. 1994). Fecundity of Colorado River cutthroat is variable, depending on location, life history strategy, and individual size (Utah Department of Natural Resources

1996). One study found the average fecundity of 16 females (mean length 290 mm) to be 667 eggs (Snyder and Tanner 1960).

"All trout are opportunistic feeders that consume a wide range of organisms from among those available in particular habitats" (Behnke 1992). They generally rely on whatever organism is most available at any particular time. Dipterans, Emphemeropterans, Trichopterans, and terrestrial Hymenopterans comprise the main diet of Colorado River cutthroat trout (Speas et al. 1994). Reports regarding whether larger individuals may become piscivorous are conflicting.

## 7. Human Use of Cutthroat Trout Habitat

Human use of cutthroat trout habitat has lead to drastic declines in populations of native cutthroat trout. In fact, all three subspecies have been petitioned for listing under the Endangered Species Act, although none currently have protection under that law.

Activities such as livestock grazing, road building, water diversion and withdrawal, mineral development, timber harvest, angling, and stocking of non-native fish species degrade existing habitat, reduce genetic purity through hybridization, and create direct competition for the cutthroat trout, all causing major decreases in numbers of reproducing individuals (Shepard et al. 1997). Furthermore, any activity that damages habitat favors exotics at the expense of native trout (Duff 1996).

### *Non-native stocking*

Hybridization and competition with exotic rainbow trout, brown trout (*Salmo trutta*), and brook trout pose the greatest threat to cutthroat trout survival (Hickman and Duff 1978). Exotic trout have been stocked in most waters containing native cutthroat trout for over a century; this has largely been an attempt to improve the fishing in lakes and streams where native cutthroat trout populations had been decimated by overfishing (Duff 1996, Young et al. 1996, U.S. Fish and Wildlife Service 2001b). Hybridization between cutthroat and non-native trout can lead to loss of pure populations of cutthroat (Duff 1996), loss of locally adapted populations, loss of physical traits that make the cutthroat different from other species or subspecies, and outbreeding depression (Allendorf and Leary 1988). Any remaining cutthroat trout are generally eliminated from their native streams after introduction of aggressive brook trout (Hilderbrand 1998). Colorado River cutthroat readily hybridize with rainbow trout, and have been widely reported as replaced by brook trout.

### *Habitat Degradation and Fragmentation*

Dams and diversions fragment cutthroat trout habitat, further isolating pure populations and reducing the potential for maintaining a viable metapopulation. This reproductive isolation may result in inbreeding depression and can increase the subspecies' vulnerability to stochastic events (Smith 1998). In addition, dams may de-water channels, blocking movement between summer and winter habitat or for spawning. Water diversions also degrade existing cutthroat trout habitat by reducing stream flows and degrading riparian vegetation (Thurow et al. 1988).

Loss of streamside vegetation from livestock grazing and timber harvest causes bank instability, increased summer temperatures, and increased sedimentation (Gresswell 1995). Because cutthroat trout require relatively clear streams with gravel substrate and minimal fine sediment, substantial changes in timing, duration, or magnitude of sediment transport may have drastic effects (Chamberlin et al. 1991). For example, increased sediment eliminates invertebrate habitat and reduces winter concealment, both vitally important for cutthroat trout (Meehan 1991, Smith 1998).

Increased bank instability causes the stream to become wider and shallower. That reduces the availability of lateral habitat, pools, and undercut banks (Smith 1998). Pools are critical for certain life stages of cutthroat trout, undercut banks provide cover, and lateral habitat can provide needed spawning gravel (Meehan 1991).

Wider, shallower streams also become warmer. Cutthroat trout, with few exceptions, require cold water to persist. Spawning is often triggered by specific temperature fluctuations and modest changes in temperature may preclude spawning migration or affect the time required for eggs to develop and hatch (Chamberlin et al. 1991). In cases where cutthroat trout are able to persist in unusually warm water, their ability to compete with exotic trout species may be lessened, increasing the chances for site extinction (Behnke 1992).

In addition to livestock grazing and logging, some mining techniques also damage stream habitat. Mining operations can strip vegetation, channel streams, and divert water (Meehan 1991). These disruptions cause increased runoff and sedimentation. In addition, poorly operated mines have the potential to contaminate a watershed with heavy metals and toxic chemicals.

Roads directly affect streams by accelerating sediment loading and erosion, increasing runoff, and altering channel morphology (Furniss et al. 1991). The damage to watersheds caused by road construction and use is persistent and not readily reversible. Native cutthroat trout are less likely to have strong populations in streams near roads and are less likely to use streams near roads for rearing and spawning (U.S. Forest Service and U.S. Bureau of Land Management

1997). As discussed in Chapter 2, a high number of roads in the Southern Rockies are near streams.

If fishing is not closely monitored, it can reduce the number of large, breeding cutthroats and favor exotics including brown trout, brook trout, and hybrids (Behnke 1992, Hickman and Duff 1978). When angling occurs in conjunction with a population already depleted by competition from non-native trout, hybridization, dewatering, and other habitat degradation, it can hasten the decline of cutthroat trout (Smith 1998).

### *8. Management Implications*

Persistence of the cutthroat subspecies will depend on: (1) protection from exotic trout species, (2) reestablishing metapopulations, (3) protecting essential habitats, (4) reestablishing migratory corridors, and (5) designating and protecting well-distributed habitat refugia (Young 1996, Kruse 1998). Furthermore, protecting genetically pure populations of native cutthroat trout should be the top priority when planning management actions (Duff 1996, May 1996).

Stocking of non-native fish should cease in all areas where native subspecies persist. In addition, stocking practices should be evaluated for all areas within the subspecies' historical range where restoration may be possible. In most cases, stocking occurs to enhance sport-fishing opportunities (U.S. Fish and Wildlife Service 2001b).

Most cutthroat populations in the Southern Rockies have been relegated to small headwater streams (*i.e.*, the areas least impacted by human activity). Much of this habitat exists in roadless and Wilderness Areas on public land. These areas provide the greatest opportunity for native cutthroat trout conservation (Kruse 1998). Protecting native cutthroat trout requires that activities such as timber harvest, livestock grazing, mining, and road building (or any activity that may cause habitat degradation) be restricted in these important areas (Duff 1996). Wilderness Areas also provide cutthroat trout habitat in large and intact blocks, thus reconnecting headwater streams and main-stem river systems (Van Eimeren 1996, Kershner et al. 1997, Kruse et al. 2000).

Enhancing less-than-ideal cutthroat trout habitat is also imperative to the long-term persistence of the subspecies (May 1996). First and foremost, removing exotic species that hybridize or compete with native cutthroat trout is necessary to enhance habitat (Duff 1996, May 1996). Restoring streamside vegetation should also be a priority (Johnstone 2000). Reduced grazing allows streamside vegetation to grow denser, and streams usually narrow as the dense vegetation encroaches on the stream (Binns and Remmick 1994). Cattle should be permanently removed from core areas for cutthroat trout.

Where logging occurs, large buffer strips along stream banks are necessary to protect cutthroat trout habitat from further degradation. It would be ideal to maintain old growth in cutthroat trout core areas; that would preserve channel integrity (Chamberlin et al. 1991). In areas that have undergone logging, habitat restoration and enhancement is important. Placing large boulders in the stream to stabilize eroding banks (Binns and Remmick 1994) would help to improve habitat. Beaver reintroductions and conservation can also aid in habitat restoration because beaver dams create critical low velocity, relatively deep water habitats (Jakober et al. 1998).

In many situations, barriers (natural or constructed) are the main strategy for ensuring that pure populations of cutthroat trout are not decimated by exotics (May 1996, Young et al. 1996). However, while they may be necessary to remove the threat of hybridization and competition, they must be considered only a temporary solution for populations in danger (Hilderbrand and Kershner 2000). Barriers fragment stream habitat, isolate populations, and limit ability of small populations to persist. Hence, if possible, it would best to avoid building new barriers. If other management actions (*e.g.*, removal of non-natives) can adequately protect genetic purity, it would be best to remove previously constructed barriers. Finally, fishing regulations should be evaluated, and complete fishing closures should be implemented to help streams occupied by pure populations. In waters where sport-fishing is allowed, specific regulations should be implemented (*i.e.*, catch and release only, artificial lures only, or fish limits).

#### *Needed Field Research*

Specifics about the life history and habitat requirements of these subspecies are somewhat limited. More field research is needed to better understand cutthroat trout and how land management activities may affect populations. Specific areas where more information is needed:

- Habitat characteristics and conditions;
- Relationships between habitats (by channel type) and fish numbers and biomass;
- Genetic purity of specific populations;
- Movement, growth, and recruitment – especially in areas where cutthroat trout have been reintroduced to reestablish a population;
- Habitat Suitability Model using the following coverages: channel type, summer temperature ranges, macroinvertebrate numbers, water quality (pH, total dissolved solids, conductivity, CaCO). This would prove highly beneficial to agencies charged with cutthroat trout conservation (Duff, pers. comm. 2002).

### 9. Justification

Cutthroat trout are a flagship species and an indicator species.

**Flagship:** The fishing value of the species has attracted attention for the conservation and restoration of the areas where they are present or will be reintroduced. Native cutthroat trout thus contribute to the restoration and protection of waterways (often heavily degraded) and associated species.

**Habitat Quality Indicator:** Decline of native cutthroat trout because of habitat loss and introduction of exotic species is indicative of more insidious management problems (*e.g.*, the inability to restore waterways to prior levels of complexity and the general lack of long-term, broad-scale vision in fisheries planning).

The subwatersheds supporting populations of cutthroat trout were included in the initial SITES analysis as a way to include important headwaters not covered by the other focal species inputs.

## Gray Wolf (*Canis lupus*)

### 1. Introduction

There is ample evidence to support the role of large carnivores in ecosystem function. Several recent review articles cover the empirical and anecdotal evidence (Terborgh et al. 1999, Estes et al. 2001, Miller et al. 2001, in press a). Macroecological evidence comes from Crête and Manseau (1996), Crête (1999), Oksanen and Oksanen (2000), Schmitz et al. (2000), and Halaj and Wise (2001). All concluded that top-down effects existed under a broader range of habitats and across more systems than previously thought.

Carnivores control prey by direct and indirect methods. Through predation, carnivores directly reduce numbers of prey (Terborgh 1988, Terborgh et al. 1997, 2001, Estes et al. 1998, Schoener and Spiller 1999). Indirect mechanisms cause prey to alter their behavior so that they become less vulnerable (Kotler et al. 1993, Brown et al. 1994, FitzGibbon and Lazarus 1995, Palomares and Delibes 1997, Schmitz 1998, Berger et al. 2001a). They choose different habitats, different food sources, different group sizes, different time of activity, or they reduce the amount of time spent feeding.

By reducing the numerical abundance of a competitively dominant prey species (or by changing its behavior), carnivores erect and enforce ecological boundaries that separate species (at some level) and allow weaker competitors to persist (Estes et al. 2001). If a predator selects from a wide-range of prey species, the presence of the predator may cause all prey species to reduce their respective niches and thus reduce competition among those species. Removing the predator will dissolve the ecological boundaries that check competition. As a result, prey species may compete for limited resources and superior competitors may displace weaker competitors, leading to less diversity through competitive exclusion (see Paine 1966, Terborgh et al. 1997, Henke and Bryant 1999). The impact of carnivores thus extends past the objects of their predation. Because herbivores eat seeds and plants, predation on that group influences the structure of the plant community (Terborgh 1988, Terborgh et al. 1997, 2001, Estes et al. 1998). The plant community, in turn, influences distribution, abundance, and competitive interaction within groups of birds, mammals, and insects.

### 2. Wolves and Ecosystems

Ecosystem health implies that there is a full complement of native species as well as the biological processes associated with these species—*i.e.*, structure and function (Miller et al. in press a). Because carnivores are important in ecological function, their presence over time indicates a healthier system than one from which they are absent (Soulé et al. in press). Granted, ecosystems may continue to exist after species have been lost and natural relationships have been altered, but evidence indicates such systems are impoverished (Terborgh et al. 1999, Estes et al. 2001, Miller et al. 2001, in press a, Soulé et al. in press).

***Wolf Population Dynamics***

To exert an ecological function, gray wolves (*Canis lupus*) must be more than occasionally present in a region (Soulé et al. in press). They must live in numbers (over time) that are sufficient to have an ecological impact. In North America, wolf densities vary widely across regions, but are more or less stable within populations (Miller et al. in press a). Average annual wolf densities do not often exceed about one wolf per 24 $km^2$ and are usually far lower than this (Miller et al. in press a).

Wolf populations are closely linked to population levels of their ungulate prey, with wolf numbers higher when prey biomass is higher (Keith 1983, Fuller 1989). Ungulate biomass was particularly important for pup survival during the first 6 months of life (Fuller 1989). Prey biomass mediates wolf demography through various social factors: pack formation, territorial behavior, exclusive breeding, deferred reproduction, intraspecific aggression, and dispersal (Keith 1983). When ungulate biomass is low, starvation and intraspecific aggression are more common (Mech 1977, Messier 1985). A population of wolves suffering a lack of food may be more vulnerable to disease than one with more food available (Miller et al. in press a). Some studies have reported wolf mortality varying from rates of 2-21% because of disease (Carbyn 1982, Peterson et al. 1984, Fuller 1989, Ballard et al. 1997).

Human-caused mortality is also a major factor affecting wolf density (Miller et al. in press a), in some cases causing up to 69 – 80% of the deaths (Peterson et al. 1984, Ballard et al. 1997). Survival rate of wolves in a semi-protected area of northwest Montana was 0.80 with survival rates of 0.66 for dispersers outside the semi-protected area (Plescher et al. 1997). Fuller (1989) concluded a wolf population would stabilize with an overall rate of annual mortality of 0.35 or rate of human-caused mortality of 0.28. Gasaway et al. (1983), Keith (1983), Peterson et al. (1984), Ballard et al. (1987), and Fuller (1989) found that harvest levels of 20-40% can limit wolf populations, but that the lower rate has a more significant effect in an area with low ungulate biomass (Gasaway et al. 1983).

### *Interactions with Other Carnivores*

Wolves interact with other carnivores, and may change the distribution and abundance of competitors (Paquet 1989, 1991, 1992, Crabtree and Sheldon 1999). In addition to these obvious competitive interactions, wolves also provide a regular supply of carrion, which is exploited by smaller carnivores. Less obvious is how wolves may modify relationships among different carnivores and scavengers (Miller et al. in press a). These relationships can be quite complex in systems with multiple prey and predators.

Wolf reintroduction is a factor that may accent competition among intraguild carnivores (Miller et al. in press a). The gradation of such adjustment may be slower for natural recolonization than it is during reintroductions. During reintroduction, the naiveté of newly coexisting predators probably affects the intensity of interaction dynamics (*sensu* Berger et al. 2001a). The wolves introduced to Yellowstone coexisted with coyotes (*Canis latrans*) in Canada before their transfer. Conversely, coyotes in Yellowstone had no experience with wolves. Educated coyotes do well in the presence of educated gray wolves, but naïve coyotes have fared poorly in Yellowstone, suffering heavy mortality in the early stages (Crabtree and Sheldon 1999). A change in behavior and abundance of coyotes can in turn affect badgers (*Taxidea taxus*), weasels (*Mustela* spp.), rodents, and songbirds.

Kunkel et al. (1999) suggested wolves and mountain lions (*Puma concolor*) may exhibit exploitation and interference competition that affects each other's behavior and dynamics, and that of their prey. That may also be true for interactions between bears (*Ursus* spp.) and wolves.

### *Wolf/Prey Dynamics*

Studies conducted in areas without predators emphasized that ungulates are most influenced by density-dependence and weather (Merrill and Boyce 1991, Singer et al. 1997, Post et al. 1999, Singer and Mack 1999). In contrast, many researchers have reported that wolf predation decreased survival measures or population growth rate of prey (Gauthier and Theberge 1986, Gasaway et al. 1992, Jedrzejewska and Jedrzejewski 1998, Kunkel and Pletscher 1999, Hayes and Harestad 2000). Researchers also found wolf predation on ungulates increased with snow depth (Nelson and Mech 1981, Huggard 1993, Post et al. 1999), indicating predation can interact with weather. While snow is not as deep in the Southern Rockies as it is in the northern areas of these studies, drought may partially act in the same manner.

Wolves are opportunistic predators and have variable prey preferences. The most abundant species typically comprises the bulk of their diet (Huggard 1993, Kunkel 1997, Smith et al. 2000). We assume that elk (*Cervus elaphus*) and deer (*Odocoileus* spp.) would form the main part of their diet in the Southern Rockies (Miller et al. in press a). Wolves generally kill animals that are more vulnerable because of age, condition, or habitat and weather circumstances (Mech 1996a, Kunkel et al. 1999, Kunkel and Pletscher 2001). For example, since 1995, the average age of prey killed by wolves in and around Yellowstone has been 14 while the average age of ungulates killed by hunters is 6 (Smith et al. 2001).

The level that predation affects ungulate populations depends on whether and to what extent that predation is additive or compensatory. In general, compensatory effects are most likely when prey numbers are near carrying capacity (Bartmann et al. 1992, Dusek et al. 1992, White and Bartmann 1998). As prey numbers drop farther below carrying capacity, mortality becomes increasingly more additive (see Turchin et al. 2000). Dusek et al. (1992) reported that hunting mortality by Montana humans was largely additive to other forms of mortality (including predation). There was thus little opportunity for compensatory mortality in the adult segment of the population. Kunkel and Pletscher (1999) reported similar findings for deer and elk populations where the main source of mortality was predation. Singer et al. (1997) reported potential compensation between components to elk calf mortality for Yellowstone's Northern Range, similar to results of Adams et al. (1995) for caribou (*Cervus tarandus*) in Denali, Alaska.

Overall, a lack of carnivores means ungulates have been released from top-down pressure and most populations are likely kept close to carrying capacity of the vegetation (Crête and Daigle 1999). Consistent with the hypothesis of top-down pressures on ecosystems (Oksanen et al. 2001), ungulate biomass in North America is 5 to 7 times higher when wolves are absent than when they are present (Crête 1999). Overabundance of elk causes negative impacts to the ecosystem (Terborgh et al. 1999, Singer and Zeigenfuss 2002).

### 3. Wolf/Livestock Interactions

Conflicts between wolves and livestock have been controversial and complex (Mech 1995, 1999, 2001, Mech et al. 1996, 2000, Phillips and Smith 1998). Even though wolf depredations are relatively uncommon, particularly when compared to other forms of livestock mortality, agricultural interests demand action when wolves kill livestock. Resolution of these conflicts is the most common reason for Mexican wolves (*C. l. baileyi*) to be recaptured for re-release or permanent placement in captivity.

Minimum confirmed livestock losses have annually averaged about 6 cattle, 37 sheep, and 4 dogs in the Yellowstone area and 6 cattle, 24 sheep, and 2 dogs in central Idaho (U.S. Fish and Wildlife Service 2002). In comparison, annual losses from all causes in Montana are 80,000 cattle and 90,000 sheep (Bangs 1998). Since 1995, U. S. Fish and Wildlife Service and Wildlife Services have killed 18 wolves in central Idaho and 26 in the Yellowstone area because of conflicts with livestock. Since 1987, a private compensation fund administered by Defenders of Wildlife has paid livestock producers who experienced confirmed or highly probable wolf-caused losses in Montana, Idaho, and Wyoming about $155,000. This compares to an estimated $45,000,000 in annual losses to all causes for livestock producers in Montana (Bangs 1998).

Most livestock producers have cooperated with wolf recovery because they believe their problems will be addressed in a fair and equitable manner. Financial compensation for livestock losses has helped minimize animosity toward wolves (Fischer 1989, Fischer et al. 1994). Such compensation, however, must be directly linked to the root of the problem and not just buy dead stock (Miller et al. in press a).

Taken as a whole, depredation by wolves would not affect the economy of the Southern Rockies. Agriculture is a small part of the Southern Rocky Mountain economy, and wolves have little effect on that small part. The percentage of contribution made by the farm and ranch sector to the Gross State Economy of Colorado has declined steadily from 1977 (2.7%) to 1999 (0.9%, figures are from the U.S. Department of Commerce, Bureau of Economic Analysis). When the annual percentages are regressed against time, the result is statistically significant at $p \leq 0.05$ with an $r^2 = 0.77$. In other words, the decline is steady and consistent across years. Despite lack of an effect on the agricultural economy, wolf reintroduction would likely have an impact on some individual livestock producers.

The tension between promoting wolf survival and killing wolves to resolve agricultural conflicts complicates wolf recovery. With the exception of lethal control, most approaches for resolving conflicts seem to be ineffective, cost-prohibitive, and/or logistically unwieldy when applied over a large scale (Cluff and Murray 1995, Mech et al. 1996). The conundrum, of course, is that resolving issues by killing wolves reinforces the negative attitudes and values toward wolves (*i.e.*, wolves are bad and livestock good). And, wolves affect ecosystems through predation. If wolves are to reclaim a functional role in ecosystem regulation, we must allow them to act like wolves.

### 4. Wolf and Human Interactions

The following is based on information compiled by the International Wolf Center (www.wolf.org), a non-profit education organization that focuses on the wolf. Much of the information below is from Bishop (1998), Mech (1990, 1996b, 1998), Route (1998), and Phillips et al. (in press).

Persecution of wolves by humans probably has made wolves wary of humans. In Minnesota's Superior National Forest, there have been some 19,000,000 visitor-days without any wolf attacks on humans. Millions of visitor-days are recorded at parks and wilderness areas in Canada and Alaska as well without incident. Nonetheless, like bears and mountain lions, wolves are instinctive predators that should be respected and allowed to be wild.

In rare cases, wolves have become fearless of humans. The result has lead to serious injury, and in some countries, even death. During 1996 and 1997, one or more wolves in India attacked 75 children, some fatally. Additional incidents seem to happen because of mistaken identities, defensive reactions, or a person getting between a wolf and a dog it was attacking.

Many oppose wolf recovery on the belief that it will lead to significant changes in land use. Opponents predict that the federal government will close vast areas of public land to promote wolf conservation. In 1978, about 25,000 km$^2$ of public land in Minnesota (11% of the state) was designated as critical habitat for the gray wolf (Nowak 1978). This designation was supported by local and federal land management agencies including the Forest Service and National Park Service.

Critical habitat imposes the "substantive duty that federal agencies actions not modify or destroy critical habitat" and serves "to guide federal agencies in fulfilling their obligations under section 7 of the Act" (*i.e.*, Interagency Cooperation, Bean and Rowland 1997: 202). Critical habitat in Minnesota does not impose restrictions on the movement or activities of private citizens or state agencies (unless an Environmental Impact Statement is needed as per the National Environmental Policy Act). It is important to note that up to the present, there have been no provisions in federal or state wolf management plans for restricting activities on private land to promote wolf recovery except for potential use of M44 cyanide devices (used to kill coyotes) in areas occupied by wolves.

### *5. Human Attitudes Toward Wolves*

Much of this section is summarized from Miller et al. (in press b). The attitudes people hold may be the most important factor influencing the success of wolf recovery. These attitudes can be affected by level of knowledge, human/animal relationships, personal experience with the species, real and perceived impacts on economies or lifestyles, and the species' economic or cultural value (Reading 1993, Kellert 1996). When opinions about a species are highly polarized, the challenges faced by a recovery program are heightened (Chaiken and Stangor 1987, Reading 1993, Kellert 1996).

Challenges become particularly difficult if one of the opposing interest groups, or stakeholders, is powerful, wealthy, and influential. Our political system is a representative democracy where, although one person represents one vote, all votes are not equal. While Representatives to the House are elected from districts based on population numbers, people residing in rural regions have a *defacto* stronger representation in the Senate. Senators from rural regions represent fewer constituents than do senators from urban regions.

In addition to representative democracy, we also have a market economy. While one person may have one vote in an election, in the market economy, more dollars equals more votes (Korten 1995). As such, there is often a dynamic tension within our political system, and when special interests dominate the democratic process, it has been called the "tyranny of the minority" (Terborgh 1999).

Some level of federal protection for a wildlife species often elevates polarization, especially if certain stakeholders perceive economic or political consequences. Thus, we must also consider opinions and attitudes toward the Endangered Species Act, state *vs.* federal control, decision-making power and goals of stakeholders, issues of wilderness, and traditional control over public and private land. Indeed, reintroduction of wolves, and their legal protection, infers a loss of political power to ranchers, loggers, and miners, who once undisputedly held that power throughout the West; they believe that power is transferred to urban vacationers and conservationists (Reading et al. 1994). In many ways, wolves, endangered species, and proposals for Wilderness Areas are "straw men" for the real issue—who controls the land.

It is also important to consider the attitudes of people within the managing agencies. After all, they will be making the decisions dictating the future of the species. We often assume that everyone in a managing body has the same values toward species recovery. That is unlikely. Even if all members of the group share the same values, they may not hold those values in the same order of preference. A member representing a non-governmental conservation organization may place wolf recovery at the top of his/her list of goals, whereas a representative of an agency with a multi-use mandate may have to balance wolf recovery against other goals—some of which may even compete with wolf recovery. Fritts et al. (1995) very aptly described the perspectives of people from the organizations involved with wolf recovery in Montana, Wyoming, and Idaho.

As for the attitudes of the general public toward wolves in the Southern Rockies, rural residents of Colorado tend to support traditional forms of wildlife management (*e.g.*, trapping and hunting), whereas urban residents are more likely to oppose these practices and support animal rights (Kellert 1984, Manfredo et al. 1993). A positive correlation was found between level of education and likelihood of involvement in animal activism. Some evidence suggests that long-time residents of a state or area are more likely to support traditional forms of wildlife management than are newer residents (Zinn and Andelt 1999).

Manfredo et al. (1993) found that the public generally supported the idea of wolf reintroduction, and 71% indicated they would vote for reintroducing wolves. East slope residents were more supportive than west slope residents, with 74% *vs* 65%, respectively, saying they would vote "Yes". Most people considered wolf reintroduction at least as important as protecting several other endangered or threatened species in the state.

In a recent poll of Colorado, New Mexico, and Arizona residents, Meadow (2001) found that 64% of respondents favored wolf restoration, while 31% opposed it (Colorado and Arizona reported 68% in favor and 28% opposed, while New Mexico reported 59% in favor and 38% opposed). An informal poll of people visiting the Denver Zoo (N = 6,000) produced a similar, and slightly more favorable response.

In the Meadow (2001) study, people registered as Democrats showed the highest support (80%), followed by Independents (68%) and Republicans (56%). There was little difference in levels of support between women (67%) and men (66%) or between hunters (61%) and non-hunters (70%, Meadow 2001). Voters wanted restoration and management based on science, and wildlife biologists were a professional group highly trusted by the public (Meadow 2001). There was also strong support for large, interconnected lands that were managed for wildlife (Meadow 2001).

Colorado residents favored wolf reintroduction because it would preserve the wolf, balance deer and elk populations, increase understanding of the importance of wilderness, control rodent populations, and restore the natural environment (Manfredo et al. 1993, Pate et al. 1996, Meadow 2001).

Those opposed thought ranchers would lose money, wolves would attack humans, pets, and livestock, and wolves would reduce deer and elk populations (Manfredo et al. 1993, Pate et al. 1996). Both groups thought that ranchers would shoot wolves (Manfredo et al. 1993).

The results of these surveys demonstrate that there is general public support for wolf reintroduction in the Southern Rockies, increasing the likelihood that such an undertaking might be feasible, at least from a public opinion perspective (see Miller et al. in press b). In addition, the demographics and economics of the region – increasingly urban, well-educated people not connected to livelihoods involving livestock – continue to move in directions that would seem to increase support for wolf reintroduction over time. Still, a relatively large proportion of the public opposes reintroduction and could present a formidable challenge to a reintroduction project, especially if that group of people is strongly opposed and politically powerful. Strong, vocal, and active opposition by minority stakeholders has effectively prevented, delayed, or greatly complicated reintroduction programs in the past (Reading and Clark 1996).

The wolf recovery program must address the hostility and antagonism of strong opponents to wolf reintroduction, while simultaneously maintaining the support and addressing the almost diametrically opposite concerns of conservationists and the larger, general population (Miller et al. in press b). Public relations and education programs have been successful in developing support for some reintroduction programs, but such programs usually worked with largely uninformed publics who had relatively indifferent attitudes. Programs directed at *changing* strongly held attitudes and values are rarely successful (Chaiken and Stangor 1987). The data suggest that beliefs toward, and the symbolism of, wolves (based on associated cultures, perceptions and values) may be the most important factors influencing attitudes, as very few people now have personal experience with the species (Miller et al. in press b). With myth playing such a strong role, it will be difficult to reach consensus.

Effective persuasion requires that people both receive and acquiesce to a persuasive message (Olson and Zanna 1993). Receptivity depends on several factors, including motivation, the identity of the messenger, the strength and frequency of the message, the clarity of the message, and the state of the recipient (Chaiken and Stangor 1987, Petty et al. 1997). Peer pressure can play a large role in maintaining or changing values, attitudes, and behaviors (Chaiken and Stangor 1987, Tessler and Shaffer 1990). In addition, changes are more likely to occur when alternative choices are provided that invalidate current values (Tessler and Shaffer 1990, Petty et al. 1997). Thus economic incentives and enforcement both have value in implementation.

### 6. General Projections for a Southern Rockies with Wolves

Wolf survival in Southern Rocky Mountain systems that are outside of protected areas will probably be similar to northern systems outside protected areas, where wolves are lightly exploited (via harvest or control) due to the inevitable conflicts that occur between wolves and livestock and the subsequent killing of wolves (Phillips et al. in press). To date, every pack of wolves that has established itself outside of Yellowstone Park has had conflicts with livestock; offenders were eliminated (U.S. Fish and Wildlife Service 2002). Pletscher et al. (1997) reported lower survival rates for wolves moving outside a semi-protected area of northwest Montana.

The addition of wolves to this system likely will exacerbate the impacts of other predators unless interference competition is significant—which it may be between coyotes and wolves if both are preying on deer (see Ballard et al. 2001, Crabtree and Sheldon 1999). The impact of wolves on some ungulate populations may be locally significant. This may be particularly true when predation is combined with environmental swings (Miller et al. in press a). If that is the case, managers should be ready to eliminate or greatly reduce human harvest of females until the pendulum swings the other direction. However, in systems where elk are the dominant ungulate, we would expect reduced impacts from wolves on mule deer (*Odocoileus hemionus*). Indeed, mule deer may be somewhat released from competition with elk if wolves preferentially choose to prey on elk (as they do in Yellowstone, Smith et al. 2000) and that predation reduces elk numbers.

We predict that impacts of wolf reintroduction on large elk herds (>5,000 animals) will be less noticeable for a longer period of time than impacts on smaller herds (Miller et al. in press a). Wolves have not yet significantly affected population dynamics of the northern Yellowstone or Jackson elk herds. Impacts to smaller herds in multi-prey environments have been more significant. Wolves did have a noticeable impact on a small herd in northwest Montana, which prompted managers to stop the female harvest (Kunkel and Pletscher 1999), and on the small elk population in northern British Columbia (Bergerud and Elliot 1998). Decker et al. (1995) reported that following years of high wolf numbers (>4 wolves/1000 $km^2$) within areas of Banff National Park, small elk herds (<1,000) declined, but they increased following years of low wolf numbers (2-4 wolves/1,000 $km^2$). Hebblewhite (2000) reported similar results when comparing impacts of wolves whose densities differed spatially in the park. For small herds we predict the need to eliminate female harvest when wolves are present in high density (Phillips et al. in press).

In general, however, Colorado has higher elk numbers

than any other state in the U.S.—305,000 in 2002 (Colorado Division of Wildlife unpublished data, Meyers 2002). The Trincheria, White River, and San Juan herds are large and exist in areas important to wolves. Because of the high and growing elk numbers throughout the region, a combination of factors working simultaneously may be required to pull large populations of elk down to low-density equilibria. In addition, it may be that wolves and elk would exist in a multi-equilibrial state.

Survival rates for adult female elk in Colorado were 0.78, and almost all of that mortality resulted from hunting (Freddy 1987). In North America where hunting is allowed, it is the major source of mortality to elk populations with predation being secondary (Ballard et al. 2000). The survival rate for adult female elk in Colorado is lower than that for female elk in northwest Montana, and yet the elk population of Colorado is still growing. Hunting has been unable to hold elk numbers in Colorado at predicted carrying capacity of the range, and it appears that there are plenty of elk to support both hunters and wolves in the Southern Rockies. Given that it will take wolves much longer to reduce such a large herd, and that the present level of wolves leaves many unused elk tags each year, should not affect elk harvest for human hunters in the Southern Rockies anytime soon. That is of course, unless elk abundance exceeds carrying capacity of the range by an amount large enough to cause a crash in elk numbers, or chronic wasting disease decimates the Southern Rockies elk herd.

The mean annual survival rate for adult female mule deer in Colorado, Idaho, and Montana was 0.85 (Unsworth et al. 1999). This rate is higher than the survival rate for white-tailed deer (*Odocoileus virginianus*) in northwestern Montana and may provide room for compensation with wolf restoration. If, however, deer recruitment is lowered by the presence of wolves, or by changes in climate, productivity, *etc.*, then reductions in female deer harvest will be likely. Wide year-to-year variation in environmental conditions is characteristic of the Southern Rockies, and these conditions are the primary source of variation in estimates of over-winter fawn survival but have little affect on adult survival rates (Unsworth et al. 1999).

In summary, we predict that over the short term, there are plenty of ungulates for wolves in the Southern Rockies. Indeed, elk numbers in Colorado are higher than predicted carrying capacity of the range and the Division of Wildlife has a goal of reducing elk numbers through hunting, yet hunting alone has so far not been effective at reaching the goal.

In the long-term, wolf restoration in the Southern Rockies should cause deer and elk biomass to decline significantly. That is in line with goals by the Division of Wildlife, and the negative effects of high ungulate numbers on other flora and fauna are known (Singer and Zeigenfuss 2002). At present the number of people hunting is declining across the nation, but over the long-term it is likely that ungulate numbers would decline in the presence of wolves more than the number of hunting days or tags sold. Thus, environmental fluctuations plus lower ungulate recruitment in the face of wolves may mean that female harvests may be reduced or eliminated for periods of time in at least the core of wolf recovery areas. Depending on objectives, managers should be prepared to quickly reduce hunting pressure on cervids to prevent potentially long-term low equilibria for prey in such areas (Fuller 1990, Gasaway et al. 1992).

### *7. Recovery of wolves*

For discussion of wolf recovery, please see Phillips and Miller (in press). Here we will only touch on whether recovery represents ecological function or taxonomic representation.

Wolf recovery in the Southern Rockies could be especially significant when considered against a continental perspective. Because the ecoregion is nearly equidistant from the Northern Rockies and the Blue Range Wolf Recovery Area it is possible that a Southern Rockies population, through the production and movement of dispersers, would contribute to the establishment and maintenance of a metapopulation of wolves extending from the Arctic to Mexico (Carroll et al. 2003, in press a, Phillips and Miller in press). There may be no other region in the world where large carnivore conservation can be connected to a landscape of such significant ecological proportions. Currently several non-governmental conservation organizations are actively advocating the wolf's return to the Southern Rockies. Public opinion surveys indicate strong support for the idea (Manfredo et al. 1994, Pate et al. 1996, Meadows 2001).

The U.S. Fish and Wildlife Service is currently planning to release new recovery objectives for wolves that are based on distinct population segments. The numerical goals for the Western distinct population segment were 30 pairs of wolves breeding three consecutive years in the Montana, Idaho, and Wyoming area. The new proposal increases the area considerably—by including parts or all of 6 more states—but maintains the same numerical goals. The U.S. Fish and Wildlife Service (1994) considered the recovery goals for the 1987 plan to be conservative and minimal. Yet, now they have enlarged the area while leaving the numerical goals unchanged. It seems intuitive that if the area goal for recovery changes, the numerical goal must change as well (Phillips and Miller in press).

Thus, one result of the new criteria by the U.S. Fish and Wildlife Service may be delisting of wolves in areas where they do not exist. That could have severe repercussions to proposed reintroductions (Phillips and Miller in press). The welfare of wolves would be left to the states, and Colorado has a state law requiring that any reintroduction of a native species must be approved by the state legislature. That event is unlikely for wolves.

Finally, if the pending proposal by the U.S. Fish and Wildlife Service delists wolves in the Southern Rockies, it would also appear to violate previous statements from the Service. Refsnider (2000: 43476) wrote, "provisions of the Act are not needed when these four conditions jointly exist". The four conditions that must jointly exist are: *(1) wolves currently do not occur, (2) wolves are unlikely to arrive on their own, (3) wolf restoration is not potentially feasible, and (4) wolf restoration is not needed to achieve recovery.*

Condition # 3 alone should prevent wolves from being delisted in the Southern Rockies. The Southern Rockies and Utah have the largest unoccupied wolf habitat in the western U.S. (Carroll et al. 2001, 2003, in press a). Indeed, Refsnider (2000) admits that Colorado has habitat for wolves and that a viable number of wolves could exist in Colorado, even though he also says on the same pages that the Service has no plans to initiate wolf recovery in Colorado (Phillips and Miller in press). Finally, the U.S. 9th Circuit Court of Appeals (2001) ruled that recovery should consider a significant portion of historic range, at least where suitable habitat exists.

Disagreements over the specifics of the Fish and Wildlife Service's (FWS) proposal are to be expected, and they center on the federal government's responsibilities under the Endangered Species Act. Unfortunately, the Act does not define the term "recovery". Indeed, the Act provides no clear answer to the question, What is recovery? On this matter FWS policy states: "The goal of this process [recovery] is to restore listed species to a point where they are secure, self-sustaining components of their ecosystem and, thus, to allow delisting" (U.S. Fish and Wildlife Service 1996: 2).

FWS seems to promote a minimal definition of wolf recovery—one that merely maintains a taxonomic representation in the lower 48 states of the U.S. Many biologists, however, believe that the goal of recovery should be more than taxonomic. It does not suffice that some species can be maintained, but at numbers lower than necessary to exert their historic ecosystem function. Thus, there is a call to establish ecologically functional densities of the species over significant portions of suitable habitat within the species' historic range (Tear et al. 1993, Rohlf 1991, Shaffer and Stein 2000, Soulé et al. in press). Such a goal seems consistent with the Endangered Species Act (U.S. Ninth Circuit Court of Appeals 2001).

In each case that FWS delisted a species based on recovery, that delisting was justified by evidence that the species was distributed throughout its former range, at numbers near to original abundance, and faced no foreseeable threats (Phillips and Miller in press). In two instances before delisting, the Service considered abundant and well-distributed distinct populations of endangered species separately from other populations of the same species that were faring poorly (Phillips and Miller in press). Thus, FWS removed the southeastern population of brown pelican (*Pelecanus occidentalis*) from the endangered species list because in the southeast pelican nesting populations were at or above known historical levels, while retaining the species' endangered status throughout the remainder of its range where threats still persisted (Jacobs 1985). The same was done for distinct population segments of the gray whale (*Eschrichtius robustus*, 1994 Federal Register 59: 31094-31095).

On April 1 2003, FWS released the final ruling for reclassification of the gray wolf. Wolves remain endangered south of I-70, an area which is now part of an expanded SW Distinct Population Segment (DPS). Areas north of I-70 have been added to the Northern Rockies DPS and are slated for downlisting.

Delisting under present conditions does not represent recovery of ecological function for the wolf, and it appears to contradict the above examples, given that wolves only occupy about 5% of the species' historic range. Significant portions of the former range, such as the Southern Rockies and the state of Utah, have biological conditions that could accommodate viable populations of wolves if the U.S. Fish and Wildlife Service chose to undertake recovery in those areas.

## 8. Justification

Protecting viable populations of gray wolves in the Southern Rockies would restore an essential link toward ecosystem health of the area. The wolf was selected as an umbrella, keystone, and flagship species.

**Umbrella:** Wolves require large areas for population persistence. Protecting sufficient area and habitat for viable populations of wolves will, by inclusion, protect the habitats of many other species.

**Flagship:** The gray wolf is a charismatic animal with considerable public appeal. Conservation programs that benefit wolves should readily capture the public's attention and interest.

**Keystone:** Wolves play a significant role in top-down ecosystem regulation. Their presence both represents and maintains a healthy ecosystem.

Gray wolves are included as an input into the initial SITES analysis because wide-ranging carnivores are a primary focus of this network design and their inclusion covers a large portion of the native biodiversity within the ecoregion.

# APPENDIX 2: NETWORK UNITS

Unit classes were assigned to Rocky Mountain Region 2 U. S. Forest Service prescriptions that have been used in Citizens' Management Alternatives. Additional codes were assigned for non-forest management areas. *MGT_P=Management Prescription code **W&SR=Wild and Scenic RIver, ***WND=Wildlands Network Design.

**Table A2.1 Cross-reference of Vision unit classifications to U.S. Forest Service management prescriptions.**

| MGT_P* | UNIT CODE | UNIT CLASS | UNIT SUBCLASS | VISION DEFINITION | USFS DESCRIPTION | SOURCE |
|---|---|---|---|---|---|---|
| 1.41 | CA | Core Wild Area | Core Agency | Non wilderness protected public lands areas | Core areas | USFS |
| 2.10 | CA | Core Wild Areas | Core Agency | Non wilderness protected public lands areas | Special Interest Areas - Minimal Use and | USFS |
| 2.20 | CA | Core Wild Areas | Core Agency | Non wilderness protected public lands areas | Interpretation Research Natural Areas | USFS |
| 100.10 | CA | Core Wild Areas | Core Agency | National Parks and Monument | National Parks and Monuments | WND*** |
| 100.20 | CA | Core Wild Areas | Core Agency | National Wildlife Refuges | National Wildlife Refuges | WND |
| 100.30 | CA | Core Wild Area | Core Agency | State Parks managed for high protection | State Parks managed for high protection | WND |
| 100.40 | CA | Core Wild Areas | Core Agency | State Wildlife Areas managed for high protection | State Wildlife Areas managed for high protection | WND |
| 100.50 | CP | Core Wild Area | Core Private | Private Reserves and Conservation Ranches | Private Reserves and Conservation Ranches | WND |
| 100.60 | CPW | Core Wild Areas | Core Private Wilderness | Other Private Cores include private land managed for wilderness | Other Private Cores include private land managed for wilderness | WND |
| 1.10 | CW | Core Wild Areas | Federal Core Wilderness | Existing Wilderness | Wilderness | USFS |
| 1.11 | CW | Core Wild Areas | Federal Core Wilderness | Existing Wilderness | Wilderness, pristine | USFS |
| 1.12 | CW | Core Wild Areas | Federal Core Wilderness | Existing Wilderness | Wilderness, primitive | USFS |
| 1.13 | CW | Core Wild Areas | Federal Core Wilderness | Existing Wilderness | Wilderness, semi-primitive | USFS |
| 1.20 | CW | Core Wild Areas | Federal Core Wilderness | Recommended for Wilderness | Recommended for Wilderness | USFS |
| 1.20 | CW | Core Wild Areas | Federal Core Wilderness | Proposed Wilderness | Colorado Citizens BLM Wilderness Proposal | USFS |
| 200.20 | LD | Landscape Linkage | Dispersal Linkage | Areas of federal, state, private, or mixed land for dispersing animals | Non-resident dispersal linkages, public or private lands | WND |
| 1.50 | LR | Landscape Linkage | Riparian Linkage | Rivers protected for continuous aquatic habitat and riparian species | W&SR**, Wild classification, designated and eligible | USFS |
| 3.40 | LR | Landscape Linkage | Riparian Linkage | Rivers protected for continuous aquatic habitat and riparian species | W&SR, Scenic classification, designated and eligible | USFS |
| 4.40 | LR | Landscape Linkage | Landscape Linkage | Rivers protected for continuous aquatic habitat and riparian species | W&SR, Recreation classification, designated and eligible | USFS |

| MGT_P | UNIT CODE | UNIT CLASS | UNIT SUBCLASS | VISION DEFINITION | DESCRIPTION | SOURCE |
|---|---|---|---|---|---|---|
| 200.10 | LR | Landscape Linkage | Riparian Linkage | Riparian linkages are found along rivers, including W&SR | Riparian linkages, other-public or private lands | WND |
| 3.55 | LW | Landscape Linkage | Wildlife Movement Linkage | Linkages for general movements of animals | Corridors connecting core areas | USFS |
| 200.00 | LW | Landscape Linkage | Wildlife Movement Linkage | These provide terrestrial linkages for wildlife | Wildlife Movement linkages, other, public or private lands | WND |
| 401.00 | SAP | Study Areas | Assess for conservation | Private lands that need more research | Assess for nonwilderness conservation protection | WND |
| 400.00 | SAW | Study Areas | Study Areas wilderness | Public land areas that need additional fieldwork | Study to assess wilderness status | |
| 3.31 | UH | Compatible Use Lands | High Use Compatible-use Lands | OHV recreation areas, developed ski areas, water based sports areas | Year round motorized use on primitive roads and trails | WND |
| 3.33 | UH | Compatible Use Lands | High Use Compatible-use Lands | OHV recreation areas, developed ski areas, water based sports areas | Motorized recreation | USFS |
| 4.32 | UH | Compatible Use Lands | High Use Compatible-use Lands | OHV recreation areas, developed ski areas, water based sports areas | Dispersed recreation, high use around developed campgrounds and water bodies. | USFS |
| 5.13 | UH | Compatible Use Lands | High Use Compatible-use Lands | Logging | Forest products - fuel wood, logging | USFS |
| 7.10 | UH | Compatible Use Lands | High Use Compatible-use Lands | Residential- forest intermix | Residential- forest intermix | USFS |
| 8.21 | UH | Compatible Use Lands | High Use Compatible-use Lands | Developed recreation complexes | Developed recreation complexes | USFS |
| 8.22 | UH | Compatible Use Lands | High Use Compatible-use Lands | OHV recreation areas, developed ski areas, water based sports areas | Ski resorts | USFS |
| 8.25 | UH | Compatible Use Lands | High Use Compatible-use Lands | OHV recreation areas, developed ski areas, water based sports areas | Ski resorts | USFS |
| 300.20 | UH | Compatible Use Lands | High Use Compatible-use Lands | Other high use lands not covered in other categories | Other high use lands not covered in other categories | USFS |
| 1.31 | UL | Compatible Use Lands | Low Use Compatible-use Lands | Road 0.5 mi/mi$^2$, low-intensity uses | Backcountry recreation, non-motorized | WND |
| 1.32 | UL | Compatible Use Lands | Low Use Compatible-use Lands | Road 0.5 mi/mi$^2$, low-intensity uses | Backcountry recreation withlimited winter motorized use | USFS |
| 3.21 | UL | Compatible Use Lands | Low Use Compatible-use Lands | Road 0.5 mi/mi$^2$, low-intensity uses | Limited use, corresponds to buffer in Noss model | USFS |
| 3.22 | UL | Compatible Use Lands | Low Use Compatible-use | Road 0.5 mi/mi$^2$, low-intensity uses | Spruce restoration | USFS |

| MGT_P | UNIT CODE | UNIT CLASS | UNIT SUBCLASS | VISION DEFINITION | DESCRIPTION | SOURCE |
|---|---|---|---|---|---|---|
| 3.32 | UL | Compatible Use Lands | Low Use Compatible-use Lands | Road 0.5 mi/mi$^2$, low-intensity uses | Backcountry non-motorized with winter motorized - snowmobiles | USFS |
| 3.50 | UL | Compatible Use Lands | Low Use Compatible-use Lands | Road 0.5 mi/mi$^2$, low-intensity uses | Forested flora and fauna. (Management is more protective than the 5.4 FFF) | USFS |
| 3.51 | UL | Compatible Use Lands | Low Use Compatible-use Lands | Road 0.5 mi/mi$^2$, low-intensity uses | Sage grouse recovery | USFS |
| 3.54 | UL | Compatible Use Lands | Low Use Compatible-use Lands | Road 0.5 mi/mi$^2$, low-intensity uses | Special Wildlife Area - Sheep Mtn. | USFS |
| 3.58 | UL | Compatible Use Lands | Low Use Compatible-use Lands | Road 0.5 mi/mi$^2$, low-intensity uses | Deer & Elk Winter Range - Limited Management | USFS |
| 5.41 | UL | Compatible Use Lands | Low Use Compatible-use Lands | Road 0.5 mi/mi$^2$, low-intensity uses | Deer and elk winter range | USFS |
| 5.42 | UL | Compatible Use Lands | Low Use Compatible-use Lands | Road 0.5 mi/mi$^2$, low-intensity uses | Bighorn sheep | USFS |
| 5.43 | UL | Compatible Use Lands | Low Use Compatible-use Lands | Road 0.5 mi/mi$^2$, low-intensity uses | Elk habitat | USFS |
| 5.45 | UL | Compatible Use Lands | Low Use Compatible-use Lands | Road 0.5 mi/mi$^2$, low-intensity uses | Forest carnivores | USFS |
| 6.40 | UL | Compatible Use Lands | Low Use Compatible-use Lands | Road 0.5 mi/mi$^2$, low-intensity uses | Grazing Shortgrass prairie, mid-composition, high structure | USFS |
| 300.00 | UL | Compatible Use Lands | Low Use Compatible-use Lands | Road 0.5 mi/mi$^2$, low-intensity uses | Other low use lands not covered in other categories | WND |
| 3.10 | UM | Compatible Use Lands | Low Use Compatible-use Lands | Road 1 mi/mi$^2$, more intensive use | Special Interest Areas - Emphasis on use and interpretation | USFS |
| 4.20 | UM | Compatible Use Lands | Moderate Use Compatible-use Lands | Road 1 mi/mi$^2$, more intensive use | Scenery | USFS |
| 4.21 | UM | Compatible Use Lands | Moderate Use Compatible-use Lands | Road 1 mi/mi$^2$, more intensive use | Scenic Byway | USFS |
| 4.30 | UM | Compatible Use Lands | Moderate Use Compatible-use Lands | Road 1 mi/mi$^2$, more intensive use | Dispersed recreation Undeveloped recreation in natural appearing landscapes. | USFS |
| 5.11 | UM | Compatible Use Lands | Moderate Use Compatible-use Lands | Road 1 mi/mi$^2$, more intensive use | Range and forest | USFS |
| 5.12 | UM | Compatible Use Lands | Moderate Use Compatible-use Lands | Road 1 mi/mi$^2$, more intensive use | Livestock Grazing | USFS |

| MGT_P | UNIT CODE | UNIT CLASS | UNIT SUBCLASS | VISION DEFINITION | DESCRIPTION | SOURCE |
|---|---|---|---|---|---|---|
| 5.31 | UM | Compatible Use Lands | Moderate Use Compatible-use Lands | Road 1 mi/mi$^2$, more intensive use | Experimental forest | USFS |
| 5.40 | UM | Compatible Use Lands | Moderate Use Compatible-use Lands | Road 1 mi/mi$^2$, more intensive use | Forested flora and fauna habitats | USFS |
| 6.60 | UM | Compatible Use Lands | Moderate Use Compatible-use Lands | Road 1 mi/mi$^2$, more intensive use | Grazing mid composition, low structure | USFS |
| 300.10 | UM | Compatible Use Lands | Moderate Use Compatible-use Lands | Road 1 mi/mi$^2$, more intensive use | Other moderate use lands not covered in other categories | WND |
| 300.20 | UP | Compatible Use Lands | Private Compatible-Use Lands | Private lands voluntarily managed to protect wildlife | Private lands managed for conservation values, but not wilderness | WND |
| 0.00 | | Private | Private | Private/other | Private, no conservation status recommended by local group | SJ |
| 0.01 | | Private | Private | Private/other | Private/non-forest, no conservation status recommended by local group | WR |

# APPENDIX 3:

# TARGETS AND OTHER ELEMENTS COVERED BY THE WILDLANDS NETWORK DESIGN

Note in the following tables:

**Current Protected** areas are defined as federal Wilderness Areas, National Park Service lands, and other congressionally protected areas managed as wilderness, as of May 2003. This includes the recently expanded boundary of the Great Sand Dunes National Monument and Preserve.

**Core Areas** include those areas defined in this Vision as Core Agency, Core Private, and Core Wilderness (both designated and proposed).

**Table A3.1 Proportion of The Nature Conservancy's targets for the ecoregion included in the Network Design.**

| Taxonomic/ Ecological Group | Current Protected | Core Wilderness | Core Areas | Entire Vision | TNC Portfolio |
|---|---|---|---|---|---|
| Amphibians | 9% | 16% | 26% | 96% | 98% |
| Birds | 7% | 13% | 16% | 69% | 94% |
| Ecological Systems | 17% | 36% | 36% | 72% | 69% |
| Fish | 15% | 22% | 28% | 80% | 97% |
| Invertebrates | 40% | 41% | 50% | 73% | 97% |
| Mammals | 36% | 56% | 65% | 90% | 99% |
| Mollusks | 20% | 30% | 30% | 80% | 100% |
| Plant Communities | 18% | 27% | 38% | 68% | 98% |
| Plants | 13% | 23% | 30% | 71% | 97% |
| Reptiles | 0% | 8% | 17% | 50% | 92% |

See Chapter 8 for more information.

Bighorn sheep (*Ovis canadensis*)

**Table A3.2 Inclusion of Element occurrences for Colorado** (Colorado Natural Heritage Program low resolution data, Colorado portion of the Network Design only).

| Group | Scientific Name | Common Name | Total Available | Current | CW | CA | VISION | TNC | G-rank | S-rank |
|---|---|---|---|---|---|---|---|---|---|---|
| Amphibians | *Bufo boreas pop 1* | Boreal Toad *(Southern Rocky Mountain Population)* | 185 | 39% | 64% | 68% | 87% | 84% | T1 | S1 |
| Amphibians | *Hyla arenicolor* | Canyon Treefrog | 3 | | 67% | 67% | 100% | 67% | G5 | S2 |
| Amphibians | *Rana sylvatica* | Wood Frog | 83 | 42% | 58% | 64% | 87% | 89% | G5 | S3 |
| Amphibians | *Spea intermontana* | Great Basin Spadefoot | 8 | | | | 25% | 50% | G5 | S3 |
| Birds | *Aegolius funereus* | Boreal Owl | 90 | 14% | 32% | 43% | 98% | 59% | G5 | S2 |
| Birds | *Amphispiza belli* | Sage Sparrow | 12 | 8% | 8% | 17% | 58% | 75% | G5 | S3B,SZN |
| Birds | *Asio flammeus* | Short-Eared Owl | 6 | | | | 50% | 83% | G5 | S2B,SZN |
| Birds | *Bucephala islandica* | Barrow's Goldeneye | 11 | 73% | 82% | 82% | 82% | 100% | G5 | S2B,SZN |
| Birds | *Buteo regalis* | Ferruginous Hawk | 5 | 20% | | 20% | 60% | 80% | G4 | S3B,S4N |
| Birds | *Catoptrophorus semipalmatus* | Willet | 2 | | | | | 50% | G5 | S1B,SZN |
| Birds | *Centrocercus minimus* | Gunnison Sage Grouse | 12 | 8% | 25% | 33% | 67% | 83% | G1 | S1 |
| Birds | *Centrocercus urophasianus* | Sage Grouse | 1 | | 100% | 100% | 100% | 100% | G4 | S4 |
| Birds | *Charadrius alexandrinus nivosus* | Western Snowy Plover | 6 | 17% | | 17% | 100% | 100% | T3 | S1B,SZN |
| Birds | *Charadrius montanus* | Mountain Plover | 14 | | | | 50% | 100% | G2 | S2B,SZN |
| Birds | *Cypseloides niger* | Black Swift | 21 | 38% | 62% | 67% | 90% | 86% | G4 | S3B |
| Birds | *Dendroica graciae* | Grace's Warbler | 5 | | | | 40% | 20% | G5 | S3B,SZN |
| Birds | *Egretta thula* | Snowy Egret | 4 | | | | 50% | 75% | G5 | S2B,SZN |
| Birds | *Empidonax traillii extimus* | Southwestern Willow Flycatcher | 4 | | | | 75% | 100% | T1 | SR |
| Birds | *Falco peregrinus anatum* | American Peregrine Falcon | 54 | 41% | 72% | 80% | 98% | 94% | T3 | S2B,SZN |
| Birds | *Grus Americana* | Whooping Crane | 1 | | | | | | G1 | SAN |
| Birds | *Grus canadensis tabida* | Greater Sandhill Crane | 96 | 2% | 23% | 34% | 79% | 90% | T4 | S2B,S4N |
| Birds | *Haliaeetus leucocephalus* | Bald Eagle | 56 | 11% | 30% | 46% | 88% | 88% | G4 | S1B,S3N |
| Birds | *Himantopus mexicanus* | Black-Necked Stilt | 2 | | | | | 50% | G5 | S3B,SZN |
| Birds | *Leucosticte australis* | Brown-Capped Rosy-Finch | 1 | 100% | 100% | 100% | 100% | 100% | G4 | S3B,S4N |
| Birds | *Numenius americanus* | Long-Billed Curlew | 2 | 50% | | 50% | 50% | 100% | G5 | S2B,SZN |
| Birds | *Pelecanus erythrorhynchos* | American White Pelican | 4 | | | | 50% | 75% | G3 | S1B,SZN |
| Birds | *Plegadis chihi* | White-Faced Ibis | 9 | 11% | 11% | 22% | 67% | 89% | G5 | S2B,SZN |
| Birds | *Seiurus aurocapillus* | Ovenbird | 5 | | | | 60% | 80% | G5 | S2B,SZN |
| Birds | *Sterna forsteri* | Forster's Tern | 3 | | | | 67% | 100% | G5 | S2B,S4N |
| Birds | *Strix occidentalis lucida* | Mexican Spotted Owl | 28 | 4% | 79% | 82% | 100% | 93% | T3 | S1B,SUN |
| Birds | *Tympanuchus phasianellus columbianus* | Columbian Sharp-Tailed Grouse | 31 | | 3% | 6% | 68% | 39% | T3 | S2 |
| Birds | *Vireo vicinior* | Gray Vireo | 5 | | 20% | 20% | 60% | 60% | G4 | S2B,SZN |
| Fish | *Catostomus plebeius* | Rio Grande Sucker | 3 | | | | 33% | 67% | G3 | S1 |

| Group | Scientific Name | Common Name | Total Available | Current | CW | CA | VISION | TNC | G-rank | S-rank |
|---|---|---|---|---|---|---|---|---|---|---|
| Fish | *Gila cypha* | Humpback Chub | 1 | | 100% | 100% | 100% | | G1 | S1 |
| Fish | *Gila Pandora* | Rio Grande Chub | 25 | 8% | 8% | 20% | 52% | 88% | G3 | S1? |
| Fish | *Gila robusta* | Roundtail Chub | 18 | | 6% | 11% | 72% | 50% | G3 | S2 |
| Fish | *Oncorhynchus clarki pleuriticus* | Colorado River Cutthroat Trout | 157 | 57% | 80% | 87% | 100% | 97% | T3 | S3 |
| Fish | *Oncorhynchus clarki stomias* | Greenback Cutthroat Trout | 69 | 84% | 91% | 96% | 99% | 100% | T2 | S2 |
| Fish | *Oncorhynchus clarki virginalis* | Rio Grande Cutthroat Trout | 82 | 56% | 62% | 83% | 93% | 96% | T3 | S3 |
| Fish | *Prosopium williamsoni* | Mountain Whitefish | 1 | | | | 100% | 100% | G5 | S3 |
| Fish | *Ptychocheilus lucius* | Colorado Pikeminnow | 10 | | 60% | 60% | 70% | 50% | G1 | S1 |
| Fish | *Xyrauchen texanus* | Razorback Sucker | 4 | | | | 50% | 50% | G1 | S1 |
| Insects | *Agapema homogena* | Agapema Homogena | 1 | | | | | 100% | G4 | S2 |
| Insects | *Amblyderus werneri* | Great Sand Dunes Anthicid Beetle | 1 | 100% | 100% | 100% | 100% | 100% | G1 | S1 |
| Insects | *Amblyscirtes simius* | Simius Roadside Skipper | 2 | | | | | 100% | G4 | S3 |
| Insects | *Atrytone arogos* | Arogos Skipper | 13 | | | | | 92% | G3 | S2 |
| Insects | *Atrytonopsis hianna* | Dusted Skipper | 1 | | | | | 100% | G4 | S2 |
| Insects | *Boloria improba acrocnema* | Uncompahgre Fritillary | 23 | 83% | 100% | 100% | 100% | 100% | T1 | S1 |
| Insects | *Callophrys mossii schryveri* | Moss's Elfin | 12 | | | | | 75% | T3 | S2S3 |
| Insects | *Celastrina humulus* | Hops Feeding Azure | 9 | | | | 11% | 100% | G2 | S2 |
| Insects | *Cicindela nebraskana* | A Tiger Beetle | 1 | | | | | 100% | G4 | S1? |
| Insects | *Cicindela theatina* | San Luis Dunes Tiger Beetle | 1 | 100% | 100% | 100% | 100% | 100% | G1 | S1 |
| Insects | *Coloradia luski* | A Buckmoth | 1 | 100% | 100% | 100% | 100% | 100% | G4 | S1? |
| Insects | *Daihinibaenetes giganteus* | Giant Sand Treader Cricket | 1 | 100% | 100% | 100% | 100% | 100% | G? | S1 |
| Insects | *Doa ampla* | A Moth | 1 | | | | | 100% | G? | S1 |
| Insects | *Erebia pawlowskii* | Theano Alpine | 13 | 85% | 85% | 85% | 85% | 77% | G5 | S3 |
| Insects | *Erynnis martialis* | Mottled Dusky Wing | 18 | | | | | 100% | G3 | S2S3 |
| Insects | *Euphilotes spaldingi* | Spalding's Blue | 1 | | | | 100% | 100% | G3 | S2S3 |
| Insects | *Euphyes bimacula* | Two-Spotted Skipper | 1 | | | | | 100% | G4 | S2 |
| Insects | *Grammia* sp 1 | A Tiger Moth | 1 | | | | | 100% | G? | S? |
| Insects | *Hesperia leonardus montana* | Pawnee Montane Skipper | 6 | | 83% | 100% | 100% | 100% | T1 | S1 |
| Insects | *Hesperia ottoe* | Ottoe Skipper | 9 | | | | | 100% | G3 | S2 |
| Insects | *Hyles gallii* | Galium Sphinx Moth | 1 | | | | 100% | 100% | G5 | S3? |
| Insects | *Libellula nodisticta* | Hoary Skimmer | 1 | | | | | 100% | G4 | S1 |
| Insects | *Lycaeides idas sublivens* | Dark Blue | 4 | 25% | 100% | 100% | 100% | 75% | T3 | S2S3 |
| Insects | *Ochlodes yuma* | Yuma Skipper | 2 | | 50% | 50% | 100% | 50% | G5 | S2S3 |
| Insects | *Oeneis alberta* | Alberta Arctic | 3 | | 33% | 33% | 67% | 100% | G4 | S3 |
| Insects | *Oeneis bore* | White-Veined Arctic | 4 | 100% | 100% | 100% | 100% | 100% | G5 | S3 |
| Insects | *Oeneis jutta reducta* | Rocky Mountain Arctic Jutta | 4 | 100% | 100% | 100% | 100% | 100% | T4 | S1 |
| Insects | *Oeneis polixenes* | Polixenes Arctic | 6 | 50% | 100% | 100% | 100% | 100% | G5 | S3 |

| Group | Scientific Name | Common Name | Total Available | Current | CW | CA | VISION | TNC | G-rank | S-rank |
|---|---|---|---|---|---|---|---|---|---|---|
| Insects | *Pachysphinx modesta* | Modest Sphinx Moth | 1 | | | | 100% | 100% | G4 | S3? |
| Insects | *Polites origenes* | Cross-Line Skipper | 8 | | | | | 100% | G5 | S3 |
| Insects | *Polites rhesus* | Rhesus Skipper | 3 | | | | 67% | 100% | G4 | S2S3 |
| Insects | *Pyrgus xanthus* | Xanthus Skipper | 1 | | | | 100% | 100% | G3 | S3 |
| Insects | *Satyrodes eurydice fumosa* | Smoky Eyed Brown Butterfly | 1 | | | | | 100% | T3 | S1 |
| Insects | *Schinia avemensis* | Schinia Avemensis | 1 | 100% | 100% | 100% | 100% | 100% | G? | S? |
| Insects | *Speyeria idalia* | Regal Fritillary | 2 | | | | | 100% | G3 | S1 |
| Insects | *Speyeria nokomis nokomis* | Great Basin Silverspot Butterfly | 5 | | 40% | 60% | 100% | 100% | T1 | S1 |
| Mammals | *Conepatus leuconotus* | Common Hog-Nosed Skunk | 2 | | 100% | 100% | 150% | 100% | G4 | S1 |
| Mammals | *Cynomys gunnisoni* | Gunnison's Prairie Dog | 19 | | 5% | 5% | 42% | 100% | G5 | S5 |
| Mammals | *Euderma maculatum* | Spotted Bat | 2 | 50% | 100% | 100% | 100% | 50% | G4 | S2 |
| Mammals | *Gulo gulo* | Wolverine | 35 | 63% | 86% | 91% | 97% | 97% | G4 | S1 |
| Mammals | *Lynx canadensis* | Lynx | 26 | 50% | 65% | 69% | 100% | 100% | G5 | S1 |
| Mammals | *Mustela nigripes* | Black-footed Ferret | 8 | | 25% | 25% | 50% | 63% | G1 | S1 |
| Mammals | *Myotis thysanodes* | Fringed Myotis | 1 | | | | 200% | | G4 | S3 |
| Mammals | *Perognathus flavescens relictus* | Plains Pocket Mouse subsp. | 6 | 100% | 83% | 100% | 100% | 100% | T2 | S2 |
| Mammals | *Perognathus flavus sanluisi* | Silky Pocket Mouse subsp. | 14 | 14% | 7% | 21% | 79% | 93% | T3 | S3 |
| Mammals | *Plecotus townsendii pallescens* | Townsend's Big-Eared Bat subsp. | 35 | | 9% | 14% | 66% | 57% | T4 | S2 |
| Mammals | *Sorex hoyi montanus* | Pygmy Shrew | 7 | 57% | 57% | 86% | 100% | 100% | T2 | S2 |
| Mammals | *Sorex nanus* | Dwarf Shrew | 7 | 14% | 14% | 29% | 43% | 29% | G4 | S2 |
| Mammals | *Spermophilus tridecemlineatus blanca* | Thirteen-lined Ground Squirrel subsp. | 6 | | | | 100% | 100% | T3 | S3 |
| Mammals | *Thomomys bottae pervagus* | Botta's Pocket Gopher subsp. | 2 | | | | 50% | 100% | T3 | S3 |
| Mammals | *Thomomys talpoides agrestis* | Northern Pocket Gopher subsp. | 4 | 25% | | 50% | 50% | 75% | T3 | S3 |
| Mammals | *Zapus hudsonius preblei* | Meadow Jumping Mouse subsp. | 78 | 8% | 12% | 26% | 92% | 96% | T2 | S1 |
| Mollusks | *Acroloxus coloradensis* | Rocky Mountain Capshell | 4 | 75% | 75% | 75% | 75% | 100% | G1 | S1 |
| Mollusks | *Ferrissia walkeri* | Cloche Ancylid | 2 | | | | | 100% | G4 | S3 |
| Mollusks | *Lymnaea stagnalis* | Swampy Lymnaea | 2 | | 50% | 50% | 100% | 50% | G5 | S2 |
| Mollusks | *Physa cupreonitens* | Hot Springs Physa | 1 | | | | 100% | 100% | G2 | S2 |
| Mollusks | *Physa skinneri* | Glass Physa | 1 | | | | | 100% | G5 | S2 |
| Mollusks | *Promenetus exacuous* | Sharp Sprite | 7 | | 14% | 14% | 29% | 71% | G5 | S2 |
| Mollusks | *Promenetus umbilicatellus* | Umbilicate Sprite | 11 | 9% | 9% | 18% | 36% | 82% | G4 | S3 |
| Mollusks | *Valvata sincera* | Mossy Valvata | 9 | 11% | 56% | 78% | 89% | 100% | G5 | S3 |
| Natural Communities | *Abies concolor/mahonia repens* | Mixed Montane Forests | 2 | | 50% | 50% | 50% | 100% | G5 | S4 |
| Natural Communities | *Abies lasiocarpa/trautvetteria caroliniensis* | Subalpine Fir/Carolina Tasselrue | 2 | 100% | 100% | 100% | 100% | 50% | G3 | S2? |
| Natural Communities | *Abies lasiocarpa/vaccinium myrtillus* | Subalpine Forests | 9 | 22% | 78% | 78% | 100% | 78% | G5 | S5 |

| Group | *Scientific Name* | Common Name | Total Available | Current | CW | CA | VISION | TNC | G-rank | S-rank |
|---|---|---|---|---|---|---|---|---|---|---|
| Natural Communities | *Abies lasiocarpa/vaccinium myrtillus Abies lasiocarpa-picea engelmannii/ribes spp.* | Coniferous Wetland Forests | 3 | 67% | 100% | 100% | 100% | 67% | G5 | S3 |
| Natural Communities | *Acer negundo/cornus sericea* | Montane Riparian Deciduous Forest | 4 | 50% | 50% | 75% | 100% | 75% | G3 | S2 |
| Natural Communities | *Acer negundo-juniperus scopulorum/salix exigua* | *Acer negundo-Juniperus Scopulorum/Salix exigua* | 1 | | | | 100% | 100% | GU | SU |
| Natural Communities | *Acer negundo-populus angustifolia/cornus sericea* | Narrowleaf Cottonwood Riparian Forests | 62 | | | | 74% | 81% | G2 | S2 |
| Natural Communities | *Alnus incana/equisetum arvense* | Montane Riparian Shrublands | 3 | 67% | 67% | 67% | 67% | 100% | G3 | S3 |
| Natural Communities | *Alnus incana/mesic forb* | Thinleaf Alder/Mesic Forb Riparian Shrubland | 32 | 13% | 41% | 50% | 88% | 91% | G3 | S3 |
| Natural Communities | *Alnus incana/mesic graminoid* | Montane Riparian Shrubland | 36 | 3% | 42% | 53% | 86% | 86% | G5 | S3 |
| Natural Communities | *Alnus incana-cornus sericea* | Thinleaf Alder-Red-Oiser Dogwood Riparian Shrubland | 17 | 18% | 53% | 59% | 94% | 71% | G3 | S3 |
| Natural Communities | *Alnus incana-mixed salix species* | Thinleaf Alder-Mixed Willow Species | 17 | 12% | 41% | 65% | 76% | 82% | G3 | S3 |
| Natural Communities | *Andropogon gerardii-schizachyrium scoparium* | Xeric Tallgrass Prairies | 8 | | | | 13% | 100% | G2 | S2 |
| Natural Communities | *Aquilegia micrantha-mimulus eastwoodiae* | Hanging Gardens | 2 | | 50% | 50% | 100% | 100% | G2 | S2S3 |
| Natural Communities | *Arctostaphylos patula* | Montane Shrublands | 1 | | 100% | 100% | 100% | 100% | G3 | S2 |
| Natural Communitie | *Artemisia tridentata ssp. Vaseyana /leucopoa kingii* | Western Slope Sagebrush Shrublands | 18 | 6% | 33% | 39% | 78% | 94% | G3 | S1S2 |
| Natural Communities | *Artemisia tridentata ssp. Vaseyana/ pascopyrum smithii* | Sagebrush Bottomland Shrublands | 10 | | | | 50% | 60% | G3 | S1S2 |
| Natural Communities | *Artemisia tridentata ssp. Vaseyana/symphori-carpos oreophilus/agropyron trachycaulum* | West Slope Sagebrush Shrubland | 3 | | 100% | 100% | 100% | 100% | G3 | S3S4 |
| Natural Communities | *Artemisia tridentata ssp. Wyomingensis/pseudoroegneria spicata* | Xeric Sagebrush Shrublands | 10 | 10% | 10% | 30% | 60% | 90% | G4 | S3? |
| Natural Communities | *Artemisia tridentata ssp. Wyomingensis-purshia tridentata/pseudoroegneria spicata shrubland* | Mesic Sagebrush Shrublands | 1 | | | | | 100% | G3 | S2S3 |
| Natural Communities | *Artemisia tripartita/festuca idahoensis* | Mixed Foothill Shrublands | 35 | | | | | 89% | G3 | S1? |
| Natural Communities | *Artrt/agsm phase sarcobatus vermiculatus* | Saline Bottomland Shrublands | 9 | | | | | 44% | G3 | S3 |
| Natural Communities | *Atriplex canescens/bouteloua gracilis* | Shortgrass Prairies | 1 | | | | 100% | | G3 | S3 |

| Group | *Scientific Name* | Common Name | Total Available | Current | CW | CA | VISION | TNC | G-rank | S-rank |
|---|---|---|---|---|---|---|---|---|---|---|
| Natural Communities | *Atriplex confertifolia/leymus salinus* | Cold Desert Shrublands | 13 | | 23% | 38% | 100% | | G4 | S3 |
| Natural Communities | *Atriplex corrugata shale barren* | Alkali Mat Saltbush Shrublands | 2 | | | | | | G5 | S2? |
| Natural Communities | *Betula glandulosa/mesic forb-mesic graminoid* | Subalpine Riparian Shrubland | 12 | 17% | 33% | 33% | 75% | 92% | G3 | S3 |
| Natural Communities | *Betula glandulosa/sphagnum spp.* | Dwarf Birch/Sphagnum Shrubland | 1 | | 100% | 100% | 100% | 100% | GU | SU |
| Natural Communities | *Betula occidentalis/mesic forb* | Foothills Riparian Shrubland | 41 | 2% | | | | 85% | G4 | S2 |
| Natural Communities | *Betula occidentalis/mesic graminoid* | Lower Montane Riparian Shrublands | 10 | | 30% | 50% | 100% | 90% | G3 | S2 |
| Natural Communities | *Calamagrostis stricta* | Slimstem Reedgrass | 2 | | 50% | 50% | 100% | 100% | GU | S1? |
| Natural Communities | *Carex aquatilis-carex utriculata* | Montane Wet Meadows | 59 | 15% | 64% | 71% | 86% | 95% | G4 | S4 |
| Natural Communities | *Carex aquatilis-carex utriculata perched wetland* | Wet Meadow-Perched Wetland | 1 | 100% | 100% | 100 | 100% | 100% | G3 | S3 |
| Natural Communities | *Carex diandra* | Quaking Fen | 1 | | | | | 100% | GU | SU |
| Natural Communities | *Carex foenea* | Montane Riparian Meadow | 1 | | | | 100% | 100% | GU | S1? |
| Natural Communities | *Carex lasiocarpa* | Montane Wetland | 3 | | | | 100% | 100% | G4 | S1 |
| Natural Communities | *Carex nebrascensis* | Wet Meadows | 5 | | 20% | 20% | 20% | 40% | G4 | S3 |
| Natural Communities | *Carex nebrascensis-slope* | Nebraska Sedge-Slope Wetland | 1 | | | | 100% | 100% | GU | S2? |
| Natural Communities | *Carex praegracilis* | Clustered Sedge Wetland | 3 | | | | 67% | 100% | G3 | S2 |
| Natural Communities | *Carex simulata* | Wet Meadow | 10 | 20% | 20% | 30% | 80% | 100% | G4 | S3 |
| Natural Communities | *Carex utriculata* | Beaked Sedge Montane Wet Meacows | 35 | 26% | 46% | 60% | 80% | 91% | G5 | S4 |
| Natural Communities | *Carex utriculata perched wetland* | Beaked Sedge Perched Wetland | 2 | | 50% | 50% | 100% | 100% | G3 | S3 |
| Natural Communities | *Catabrosa aquatica-mimulus spp.* | Spring Wetland | 1 | 100% | | 100% | 100% | 100% | GU | S3 |
| Natural Communities | *Cercocarpus montanus/pseudoroegneria spicata* | Mixed Mountain Shrublands | 28 | | | | | | G4 | S3 |
| Natural Communities | *Cercocarpus montanus/stipa* | Foothills ShrublandE | 6 | | | | | | G2 | S2S3 |

| Group | *Scientific Name* | Common Name | Total Available | Current | CW | CA | VISION | TNC | G-rank | S-rank |
|---|---|---|---|---|---|---|---|---|---|---|
| Natural Communities | *Corylus cornuta* | Lower Montane Forests | 30 | 3% | 27% | 40% | 80% | 90% | G3 | S1 |
| Natural Communities | *Deschampsia cespitosa* | Mesic Alpine Meadow | 3 | 100% | 100% | 100% | 100% | 100% | G4 | S4 |
| Natural Communities | *Deschampsia cespitosa-geum rossii* | Mesic Alpine Meadow | 5 | 40% | 80% | 80% | 80% | 100% | G5 | S5 |
| Natural Communities | *Distichlis spicata* | Salt Meadows | 7 | | | | 29% | 57% | G5 | S3 |
| Natural Communities | *Eleocharis palustris* | Emergent Wetland | 8 | 25% | 38% | 50% | 63% | 88% | G5 | S4 |
| Natural Communities | *Eleocharis quinqueflora* | Alpine Wetlands | 24 | 25% | 83% | 83% | 100% | 96% | G4 | S3S4 |
| Natural Communities | *Eleocharis quinqueflora-triglochin spp.* | Alkaline Spring Wetland | 6 | | | 17% | 67% | 100% | GU | S2 |
| Natural Communities | *Geum rossii/trifolium spp.* | Alpine Meadows | 6 | 50% | 100% | 100% | 100% | 100% | G3 | S3S4 |
| Natural Communities | *Glyceria borealis* | Montane Emergent Wetland | 2 | | | 50% | 100% | 100% | G3 | S3 |
| Natural Communities | *Hippuris vulgaris* | Hippuris vulgaris | 2 | | 50% | 50% | 100% | 100% | G5 | S4 |
| Natural Communities | *Juncus balticus var. Montanus* | Western Slope Wet Meadows | 4 | 25% | 50% | 50% | 75% | 100% | G5 | S5 |
| Natural Communities | *Juniperus osteosperma/artemisia nova/ rock woodland* | Utah Juniper/Black Sagebrush/ Rock Woodlands | 1 | | | | 100% | 100% | G5 | SU |
| Natural Communities | *Juniperus osteosperma/coleogyne ramosissima* | West Slope Juniper Woodland | 1 | | 100% | 100% | 100% | 100% | GU | SU |
| Natural Communities | *Juniperus osteosperma/stipa comata* | *Juniperus osteosperma/Stipa comata* | 7 | | 29% | 29% | 100% | 43% | G2 | S2? |
| Natural Communities | *Juniperus scopulorum/cercocarpus mon- tanus* | Foothills Piñon-Juniper Woodlands/Scarp Woodlands | 6 | 50% | 50% | 50% | 50% | 100% | G2 | S2 |
| Natural Communities | *Juniperus scopulorum/cornus sericea* | Riparian Woodland | 4 | | 75% | 75% | 100% | 100% | G4 | S2 |
| Natural Communities | *Juniperus scopulorum/pseudoroegneria spicata* | Xeric Western Slope Pinyon-Juniper Woodlands | 7 | | | | | 86% | G4 | S2S3 |
| Natural Communities | *Juniperus scopulorum/purshia tridentata* | Foothills Piñon-Juniper Woodlands | 6 | 50% | 50% | 67% | 100% | 83% | G2 | S2 |
| Natural Communities | *Juniperus scopulorum-quercus gambelli* | *Juniperus scopulorum-Quercus gambelli* | 1 | | | | 100% | 100% | GU | SU |
| Natural Communities | *Kobresia myosuroides-thalictrum alpinum* | Extreme Rich Fen | 15 | 7% | 7% | 13% | 40% | 100% | G1 | S1 |
| Natural Communities | *Kobresia simpliciuscula-scirpus pumilus* | Extreme Rich Fen | 11 | | | 9% | 64% | 100% | G2 | S1 |

| Group | Scientific Name | Common Name | Total Available | Current | CW | CA | VISION | TNC | G-rank | S-rank |
|---|---|---|---|---|---|---|---|---|---|---|
| Natural Communities | *Krascheninnikovia lanata/stipa comata* | Winter Fat/Needle-And-Thread Dwarf Shrubland | 1 | | | | 100% | | G3 | SU |
| Natural Communities | *Muhlenbergia montana-danthonia parryi* | Montane Grasslands | 73 | 1% | 26% | 47% | 81% | 82% | G3 | S2? |
| Natural Communities | *Myriophyllum exalbescens wetland* | Western Slope Floating/Submergent Palustrine Wetlands | 1 | | | | 100% | 100% | GU | SU |
| Natural Communities | *Nuphar luteum ssp polysepalum* | Western Slope Floating/Submergent Palustrine Wetlands | 5 | 60% | 60% | 60% | 100% | 100% | G5 | S3 |
| Natural Communities | *Oryzopsis hymenoides-psoralidium lanceolatum* | *Oryzopsis hymenoides-Psoralidium lanceolatum* | 1 | 100% | 100% | 100% | 50% | 100% | G3 | S1 |
| Natural Communities | *Paronychia pulvinata-silene acaulis var subacaulis* | Alpine Fellfields | 2 | 100% | 100% | 100% | | 100% | G5 | S5 |
| Natural Communities | *Phragmites australis* | Western Slope Marshes | 6 | | | | | 50% | G5 | S3 |
| Natural Communities | *Picea engelmannii/trifolium dasyphyllum* | Timberline Forests | 1 | | | | | 100% | G2 | S2 |
| Natural Communities | *Picea pungens/alnus incana-corylus cornuta* | Foothills Riparian Forest | 1 | | | | | 100% | GU | SU |
| Natural Communities | *Pinus aristata/trifolium dasyphyllum* | Upper Montane Woodlands | 17 | 35% | 65% | 76% | 94% | 88% | G2 | S3 |
| Natural Communities | *Pinus aristata/vaccinum myrtillus* | Montane Woodlands | 17 | | 65% | 65% | 94% | 88% | GU | SU |
| Natural Communities | *Pinus contorta/shepherdia canadensis* | Persistent Lodgepole Pine Forests | 2 | 50% | 50% | 50% | 100% | 50% | G3 | S3S4 |
| Natural Communities | *Pinus contorta/vaccinium scoparium* | Seral Lodgepole Pine Forests | 1 | 100% | 100% | 100 | 100% | 100% | G5 | S4 |
| Natural Communities | *Pinus edulis/cercocarpus montanus* | Mesic Western Slope Pinyon-Juniper Woodlands | 8 | | | | | 75% | G5 | S4 |
| Natural Communities | *Pinus edulis/stipa scribneri* | Two-Needle Pinyon/Scribner's Needle Grass | 4 | | | 50% | 125% | 100% | G3 | S2 |
| Natural Communities | *Pinus flexilis/leucopoa kingii* | Lower Montane Woodlands | 12 | | | | 92% | 83% | G3 | S3 |
| Natural Communities | *Pinus ponderosa/alnus incana* | Ponderosa Pine/Thin Leaf Alder | 2 | | | 50% | 100% | 100% | G2 | S2 |
| Natural Communities | *Pinus ponderosa/cercocarpus montanus/andropogon gerardii* | Foothills Ponderosa Pine Scrub Woodlands | 14 | | | 7% | | 57% | G2 | S2? |
| Natural Communities | *Pinus ponderosa/leucopoa kingii* | Foothills Ponderosa Pine Savannas | 18 | 17% | | 56% | 67% | 94% | G3 | S3 |
| Natural Communities | *Populus angustifolia sand dune forest* | Populus Angustifolia Sand Dune Forest | 1 | 100% | | 100% | 100% | 100% | G1 | S1 |
| Natural Communities | *Populus angustifolia/cornus sericea* | Cottonwood Riparian Forest | 35 | 6% | 29% | 43% | 74% | 74% | G4 | S3 |

| Scientific Name | Common Name | Total Available | Current | CW | CA | VISION | TNC | G-rank | S-rank |
|---|---|---|---|---|---|---|---|---|---|
| *Populus angustifolia/mixed salix species* | Narrowleaf Cottonwood/Mixed Willows Montane Riparian Forest | 2 | | | 100% | 100% | 100% | G3 | S3 |
| *Populus angustifolia/prunus virginiana* | Narrowleaf Cottonwood/ Common Chokecherry | 4 | | 25% | 100% | 75% | 100% | G2 | S1 |
| *Populus angustifolia/rhus trilobata* | Narrowleaf Cottonwood/Skunkbrush | 25 | 16% | 20% | | 96% | 84% | G3 | S3 |
| *Populus angustifolia/salix drummondiana-acer glabrum* | *Populus angustifolia/Salix drummondiana-Acer glabrum* | 1 | 100% | 100% | 67% | 100% | 100% | G2 | S1? |
| *Populus angustifolia/salix irrorata* | Foothills Riparian Woodland | 3 | | | | 33% | 67% | G2 | S2 |
| *Populus angustifolia/salix lucida var. Caudata* | *Populus angustifolia/Salix lucida Var. caudata* | 1 | 100% | | 8% | 100% | 100% | G1 | S1 |
| *Populus angustifolia/symphoricarpos spp.* | Narrowleaf Cottonwood/Snowberry Montane Riparian Forest | 3 | 33% | 67% | | 100% | 100% | G2 | S3 |
| *Populus angustifolia-juniperus scopulorum* | Montane Riparian Forest | 150 | 2% | | | | 87% | G2 | S2 |
| *Populus balsamifera* | Montane Riparian Woodland | 15 | 13% | 67% | 73% | 87% | 73% | GU | SU |
| *Populus deltoides ssp. Monilifera-(salix amygdaloides)/salix exigua* | Plains Cottonwood Riparian Woodland | 2 | | | 60% | 50% | 100% | G4 | S3 |
| *Populus deltoides ssp. Wislizeni/rhus triloba-ta* | Fremont's Cottonwood Riparian Forests | 26 | | 8% | | 19% | 85% | G2 | S2 |
| *Populus deltoides ssp. Wislizeni/salix exigua* | Fremont's Cottonwood Riparian Forests | 2 | | | 17% | 100% | 100% | GU | S1S2 |
| *Populus tremuloides-(pinus ponderosa)/ danthonia parryi* | Montane Forest | 1 | | | 100% | | 100% | G3 | S3S4 |
| *Populus tremuloides/acer glabrum* | Montane Riparian Forests | 245 | 0% | 66% | 100% | 92% | 85% | G2 | S1S2 |
| *Populus tremuloides/betula occidentalis* | *Populus tremuloides/Betula occidentalis* | 5 | 60% | 60% | 44% | 80% | 100% | G3 | S2 |
| *Populus tremuloides/ceanothus velutinus* | Aspen Forests | 3 | | | 50% | 100% | 100% | G2 | S2S3 |
| *Populus tremuloides/pteridium aquilinum* | Aspen Wetland Forests | 6 | | 17% | | 83% | | G4 | S3S4 |
| *Populus tremuloides/salix drummondiana* | *Populus tremuloides/Salix drummondiana* | 1 | 100% | 100% | | 100% | 100% | GU | SU |
| *Populus tremuloides/shepherdia canadensis* | Persistent Aspen Forests | 2 | 50% | 100% | | 100% | 100% | G3 | S3? |
| *Populus tremuloides/tall forbs* | Montane Aspen Forest | 9 | 44% | 44% | | 100% | 89% | G5 | S5 |
| *Potamogeton natans* | Montane Floating/Submergent Weltand | 2 | | | 50% | 100% | 100% | G5 | S1 |

| Group | *Scientific Name* | Common Name | Total Available | Current | CW | CA | VISION | TNC | G-rank | S-rank |
|---|---|---|---|---|---|---|---|---|---|---|
| Natural Communities | *Pseudoroegneria spicata-bouteloua gracilis* | Western Slope Grasslands | 9 | | | | | 100% | G4 | S1 |
| Natural Communities | *Pseudotsuga menziesii/cornus sericea* | Lower Montane Riparian Forests | 7 | 14% | 57% | 71% | 71% | 100% | G4 | S2 |
| Natural Communities | *Pseudotsuga menziesii/mahonia repens* | Douglas Fir/Creeping Oregon-Grape | 1 | | 100% | 100% | 100% | 100% | G5 | S1? |
| Natural Communities | *Pseudotsuga menziesii/symphoricarpos oreophilus* | Western Slope Douglas Fir Forests | 8 | | 38% | 38% | 88% | 38% | G5 | S4 |
| Natural Communities | *Quercus gambelii/stipa comata* | *Quercus gambelii/Stipa comata* | 1 | | | | 100% | 100% | GU | SU |
| Natural Communities | *Quercus gambelii-amelanchier utahensis* | Mixed Mountain Shrubland | 4 | | | | 75% | 75% | G4 | SU |
| Natural Communities | *Quercus gambelii-cercocarpus montanus/muhlenbergia montana* | Mesic Oak Thickets | 4 | | | 25% | 75% | 50% | GU | SU |
| Natural Communities | *Redfieldia flexuosa* | *Redfieldia flexuosa* | 1 | 100% | 100% | 100% | 100% | 100% | G1 | S1 |
| Natural Communities | *Rhus trilobata* | Skunkbrush Riparian Shrubland | 9 | | 11% | 11% | 89% | 89% | G2 | S2 |
| Natural Communities | *Salicornia rubra* | Western Slope Salt Meadows | 4 | | | | 100% | 100% | G2 | S1? |
| Natural Communities | *Salix boothii/calamagrostis canadensis* | Booth Willow/Canadian Reed Grass | 1 | | | 100% | 100% | 100% | G3 | S2 |
| Natural Communities | *Salix boothii/carex utriculata* | Booth's Willow/Beaked Sedge | 4 | | 25% | 25% | 100% | 100% | G4 | S3 |
| Natural Communities | *Salix boothii/mesic forb* | Booth's Willow/Mesic Forb | 18 | 22% | 50% | 61% | 100% | 89% | G3 | S3 |
| Natural Communities | *Salix boothii/mesic graminoid* | Riparian Willow Carr | 26 | | 23% | 31% | 73% | 77% | G3 | S3 |
| Natural Communities | *Salix brachycarpa/carex aqualtils* | Subalpine Riparian/Wetland Carr | 4 | 50% | 100% | 100% | 100% | 100% | G2 | S2S3 |
| Natural Communities | *Salix brachycarpa/deschampsia cespitosa-geum rossii* | Alpine Willow Scrub | 16 | 19% | 75% | 75% | 100% | 100% | G4 | S3S4 |
| Natural Communities | *Salix drummondiana/calamagrostis canadensis* | Lower Montane Willow Carrs | 23 | 52% | 83% | 83% | 91% | 87% | G3 | S3 |
| Natural Communities | *Salix drummondiana/carex aquatilis* | Drummond Willow/Aquatic Sedge | 1 | | | | 100% | | G2 | S2 |
| Natural Communities | *Salix drummondiana/mesic forb* | Drummonds Willow/Mesic Forb | 34 | 41% | 56% | 59% | 94% | 71% | G4 | S4 |
| Natural Communities | *Salix eriocephala var ligulifolia-salix exigua* | Strapleaf Willow-Coyote Willow | 3 | | | | 67% | 67% | G2 | S2S3 |
| Natural Communities | *Salix exigua/bare ground* | Coyote Willow/Bare Ground | 15 | 20% | 40% | 40% | 93% | 80% | G5 | S5 |

| Group | Scientific Name | Common Name | Total Available | Current | CW | CA | VISION | TNC | G-rank | S-rank |
|---|---|---|---|---|---|---|---|---|---|---|
| Natural Communities | *Salix exigua/mesic graminoid* | Coyote Willow/Mesic Graminoid | 13 | 8% | 31% | 38% | 85% | 77% | G5 | S5 |
| Natural Communities | *Salix geyeriana/carex aquatilis* | Montane Willow Carr | 54 | 4% | 52% | 69% | 83% | | G3 | S3 |
| Natural Communities | *Salix geyeriana/carex utriculata* | Geyer's Willow/Beaked Sedge | 10 | 10% | 20% | 40% | 80% | 100% | G5 | S3 |
| Natural Communities | *Salix geyeriana/mesic graminoid* | Geyer's Willow/Mesic Graminoid | 1 | | | | | | G3 | S3 |
| Natural Communities | *Salix geyeriana-salix monticola/calamagrostis canadensis* | Montane Willow Carrs | 28 | 21% | 36% | 46% | 82% | 82% | G3 | S3 |
| Natural Communities | *Salix geyeriana-salix monticola/mesic forb* | Geyer's Willow-Rocky Mountain Willow/Mesic Forb | 14 | 7% | 36% | 50% | 57% | 86% | G3 | S3 |
| Natural Communities | *Salix geyeriana-salix monticola/mesic graminoid* | Montane Riparian Willow Carr | 93 | 1% | | | | 77% | GU | S3 |
| Natural Communities | *Salix planifolia/carex aquatilis* | Subalpine Riparian Willow Carr | 113 | 12% | 71% | 74% | 97% | 92% | G5 | S4 |
| Natural Communities | *Salix planifolia/mesic forbs* | Planeleaf Willow/Mesic Forbs | 2 | 100% | 100% | 100% | 100% | 100% | G4 | S4 |
| Natural Communities | *Scirpus maritimus* | Emergent Wetland (Marsh) | 3 | | | | 67% | 100% | G4 | S2 |
| Natural Communities | *Scirpus pungens* | Bulrush | 8 | 38% | 38% | 38% | 50% | 88% | G3 | S3 |
| Natural Communities | *Scirpus tabernaemontani-scirpus acutus* | Great Plains Marshes | 6 | | 17% | 17% | 50% | 100% | G3 | S2S3 |
| Natural Communities | *Sparganium angustifolium* | Montane Floating/Submergent Palustrine Wetlands | 5 | 20% | | 40% | 80% | 100% | G4 | SU |
| Natural Communities | *Sparganium eurycarpum* | Foothills/Plains Floating/Submergent Palustrine Wetlands | 3 | | | 33% | 67% | 100% | GU | S2 |
| Natural Communities | *Sporobolus airoides* | Great Plains Salt Meadows | 2 | 50% | | 50% | 50% | 100% | G3 | S3 |
| Natural Communities | *Stipa comata - east* | Great Plains Mixed Grass Prairies | 7 | | | | | 100% | G2 | S2 |
| Natural Communities | *Stipa comata-oryzopsis hymenoides* | *Stipa comata-Oryzopsis hymenoides* | 3 | 100% | 100% | 100% | 100% | 100% | G2 | S1 |
| Natural Communities | *Typha angustifolia-typha latifolia* | Narrow-leaf Cattail Marsh | 5 | | | | 40% | 20% | G5 | S3 |
| Natural Communities | *Utricularia vulgaris* | Montane Floating/Submergent Wetland | 1 | | | 100% | 100% | 100% | G3 | S1 |
| Natural Communities | *Vaccinium cespitosum/vaccinium scoparium* | Alpine Scrub | 1 | 100% | 100% | 100% | 100% | 100% | G4 | S1? |

| Group | *Scientific Name* | Common Name | Total Available | Current | CW | CA | VISION | TNC | G-rank | S-rank |
|---|---|---|---|---|---|---|---|---|---|---|
| Plants | *Adiantum aleuticum* | Aleutian Maidenhair Fern | 1 | | | | | 100% | G5 | S1 |
| Plants | *Adiantum capillus-veneris* | Southern Maidenhair | 1 | | 100% | 100% | 100% | 100% | G5 | S2 |
| Plants | *Agastache foeniculum* | Lavender Hyssop | 4 | 25% | 25% | 25% | 75% | 75% | G4 | S1 |
| Plants | *Aletes humilis* | Larimer Aletes | 39 | | | 44% | 67% | 97% | G2 | S2S3 |
| Plants | *Aletes lithophilus* | Rock-Loving Neoparrya | 28 | | 11% | 25% | 68% | 93% | G3 | S3 |
| Plants | *Aletes nuttallii* | Dog Parsley | 2 | | | | | 100% | G3 | S1 |
| Plants | *Allium schoenoprasum var sibiricum* | Wild Chives | 7 | | | 14% | 57% | 100% | T5 | S1 |
| Plants | *Amorpha nana* | Dwarf Wild Indigo | 2 | | | | | 50% | G5 | S2S3 |
| Plants | *Aquilegia chrysantha var rydbergii* | Golden Columbine | 5 | | | | 60% | 100% | T1 | S1 |
| Plants | *Aquilegia saximontana* | Rocky Mountain Columbine | 36 | 53% | 72% | 75% | 94% | 94% | G3 | S3 |
| Plants | *Aralia racemosa* | American Spikenard | 1 | | | | 100% | 100% | G4 | S1 |
| Plants | *Argillochloa dasyclada* | Utah Fescue | 13 | | 31% | 38% | 54% | 62% | G3 | S3 |
| Plants | *Aristida basiramea* | Forktip Three-Awn | 1 | | | | | 100% | G5 | S1 |
| Plants | *Armeria scabra ssp sibirica* | Sea Pink | 2 | | 50% | 100% | 100% | 100% | T5 | S1 |
| Plants | *Arnica alpina var tomentosa* | Alpine Arnica | 1 | | | 100% | 100% | 100% | T5 | S1 |
| Plants | *Asclepias uncialis* | Dwarf Milkweed | 1 | | | | 200% | | G3 | S1S2 |
| Plants | *Askellia nana* | Dwarf Hawksbeard | 29 | 45% | 48% | 66% | 100% | 97% | G5 | S2 |
| Plants | *Asplenium trichomanes-ramosum* | Green Spleenwort | 6 | 67% | 83% | 83% | 100% | 83% | G4 | S1S2 |
| Plants | *Aster alpinus var vierhapperi* | Alpine Aster | 2 | 50% | | 50% | 50% | 100% | T5 | S1 |
| Plants | *Astragalus anisus* | Gunnison Milkvetch | 27 | 4% | 19% | 33% | 56% | 96% | G2 | S2 |
| Plants | *Astragalus argophyllus var martinii* | Meadow Milkvetch | 1 | | | | 100% | 100% | T4 | S1 |
| Plants | *Astragalus bodinii* | Bodin Milkvetch | 6 | | | | 17% | 100% | G4 | S2 |
| Plants | *Astragalus brandegeei* | Brandegee Milkvetch | 5 | | 20% | 20% | 60% | 80% | G3 | S1S2 |
| Plants | *Astragalus cibarius* | Browse Milkvetch | 1 | | | | | 100% | G4 | S1 |
| Plants | *Astragalus debequaeus* | Debeque Milkvetch | 18 | | 67% | 67% | 94% | 100% | G2 | S2 |
| Plants | *Astragalus detritalis* | Debris Milkvetch | 1 | | | | | | G3 | S2 |
| Plants | *Astragalus iodopetalus* | A milkvetch | 1 | | | 100% | 100% | 100% | G2 | S1 |
| Plants | *Astragalus linifolius* | Grand Junction Milkvetch | 22 | 5% | 77% | 95% | 100% | 95% | G3 | S3 |
| Plants | *Astragalus microcymbus* | Skiff Milkvetch | 38 | | | 3% | 97% | 100% | G1 | S1 |
| Plants | *Astragalus missouriensis var humistratus* | Missouri Milkvetch | 1 | | 100% | 100% | 100% | | T2 | S1? |
| Plants | *Astragalus molybdenus* | Leadville Milkvetch | 17 | 18% | 41% | 53% | 100% | 100% | G3 | S2 |
| Plants | *Astragalus naturitensis* | Naturita Milkvetch | 12 | | | | 83% | 83% | G2 | S2S3 |
| Plants | *Astragalus osterhoutii* | Osterhout Milkvetch | 11 | | | | | 100% | G1 | S1 |
| Plants | *Astragalus proximus* | Aztec Milkvetch | 7 | | 14% | 29% | 114% | 43% | G4 | S2 |

| Group | Scientific Name | Common Name | Total Available | Current | CW | CA | VISION | TNC | G-rank | S-rank |
|---|---|---|---|---|---|---|---|---|---|---|
| Plants | *Astragalus ripleyi* | Ripley Milkvetch | 51 | | | 4% | 100% | 98% | G3 | S2 |
| Plants | *Astragalus wetherillii* | Wetherill Milkvetch | 33 | 9% | 24% | 27% | 58% | 85% | G3 | S3 |
| Plants | *Astragalus wootonii var wootonii* | Wooton Milkvetch | 1 | | | | 100% | | T3 | S1 |
| Plants | *Azaleastrum albiflorum* | White-flowered Azalea | 16 | 100% | 100% | 100% | 100% | 100% | G4 | S2 |
| Plants | *Betula papyrifera* | Paper Birch | 1 | | | | | 100% | G5 | S1 |
| Plants | *Boechera crandallii* | Crandall's Rock-Cress | 3 | | | 67% | 67% | 100% | G2 | S2 |
| Plants | *Bolophyta tetraneuris* | Arkansas River Feverfew | 1 | | | | 200% | | G3 | S3 |
| Plants | *Botrychium campestre* | Prairie Moonwort | 1 | 100% | 100% | 100% | 100% | 100% | G3 | S1 |
| Plants | *Botrychium echo* | Reflected Moonwort | 40 | 23% | 58% | 58% | 93% | 93% | G2 | S2 |
| Plants | *Botrychium hesperium* | Western Moonwort | 26 | 35% | 58% | 62% | 96% | 88% | G3 | S2 |
| Plants | *Botrychium lineare* | Narrowleaf Grapefern | 5 | | | 20% | 80% | 80% | G1 | S1 |
| Plants | *Botrychium minganense* | Mingan Moonwort | 24 | 25% | 46% | 50% | 83% | 79% | G4 | S1 |
| Plants | *Botrychium multifidum ssp coulteri* | Leathery Grape Fern | 4 | 50% | 50% | 50% | 75% | 100% | T? | S1 |
| Plants | *Botrychium pallidum* | Pale Moonwort | 12 | 8% | 75% | 75% | 83% | 92% | G3 | S2 |
| Plants | *Botrychium pinnatum* | Northern Moonwort | 8 | 13% | 50% | 50% | 88% | 100% | G4 | S1 |
| Plants | *Botrychium simplex* | Least Moonwort | 6 | 33% | 50% | 67% | 100% | 83% | G5 | S1 |
| Plants | *Botrypus virginianus ssp europaeus* | Rattlesnake Fern | 2 | | | | 50% | 100% | G5 | S1 |
| Plants | *Braya glabella var glabella* | Arctic Braya | 9 | 11% | 56% | 100% | 100% | 100% | T? | S1 |
| Plants | *Braya humilis* | Alpine Braya | 18 | 22% | 56% | 67% | 100% | 100% | G5 | S2 |
| Plants | *Bupleurum triradiatum ssp arcticum* | Thoroughwax | 1 | 100% | 100% | 100% | 100% | | G5 | S1 |
| Plants | *Calochortus flexuosus* | Weak-Stemmed Mariposa Lily | 1 | | | | | | G4 | S1 |
| Plants | *Camissonia eastwoodiae* | Eastwood Evening-Primrose | 1 | | | | 100% | | G2 | S1 |
| Plants | *Carex capitata ssp arctogena* | Round-headed Sedge | 1 | 100% | 100% | 100% | 100% | 100% | T4 | S1 |
| Plants | *Carex concinna* | Low Northern Sedge | 3 | 33% | 67% | 67% | 67% | 100% | G4 | S1 |
| Plants | *Carex diandra* | Lesser Panicled Sedge | 3 | 33% | 33% | 67% | 67% | 100% | G5 | S1 |
| Plants | *Carex lasiocarpa* | Slender Sedge | 4 | 75% | 75% | 100% | 100% | 100% | G5 | S1 |
| Plants | *Carex leptalea* | Bristle-Stalk Sedge | 3 | 33% | 67% | 67% | 100% | 100% | G5 | S1 |
| Plants | *Carex limosa* | Mud Sedge | 8 | 75% | 88% | 100% | 100% | 100% | G5 | S2 |
| Plants | *Carex livida* | Livid Sedge | 7 | 29% | 29% | 57% | 86% | 100% | G5 | S1 |
| Plants | *Carex oreocharis* | A sedge | 7 | 43% | 57% | 57% | 71% | 86% | G3 | S1 |
| Plants | *Carex peckii* | Peck Sedge | 2 | | | | 50% | 50% | G4 | SH |
| Plants | *Carex retrorsa* | Retrorse Sedge | 2 | 100% | 100% | 100% | 100% | 100% | G5 | S1 |
| Plants | *Carex saximontana* | Rocky Mountain Sedge | 1 | | | | | 100% | G5 | S1 |
| Plants | *Carex scirpoidea* | Canadian Single-Spike Sedge | 14 | 7% | 14% | 21% | 79% | 100% | G5 | S2 |

| Group | Scientific Name | Common Name | Total Available | Current | CW | CA | VISION | TNC | G-rank | S-rank |
|---|---|---|---|---|---|---|---|---|---|---|
| Plants | *Carex sprengelii* | Sprengel's Sedge | 1 | | | | | 100% | G5 | S2S3 |
| Plants | *Carex stenoptila* | Carex Stenoptila | 2 | 50% | 100% | 100% | 100% | 100% | G2 | S2? |
| Plants | *Carex sychnocephala* | Many-headed Sedge | 1 | | 100% | 100% | 100% | 100% | G4 | S1 |
| Plants | *Carex tenuiflora* | Slender-flower Sedge | 1 | 100% | 100% | 100% | 100% | 100% | G5 | S1 |
| Plants | *Carex torreyi* | Torrey Sedge | 2 | | | | | 100% | G4 | S1 |
| Plants | *Carex viridula* | Green Sedge | 4 | 25% | 25% | 75% | 100% | 100% | G5 | S1 |
| Plants | *Castilleja lineata* | Marsh-Meadow Indian-Paintbrush | 1 | | | | 100% | 100% | G4 | S1 |
| Plants | *Castilleja puberula* | Downy Indian-Paintbrush | 1 | | | | 100% | 100% | G2 | S? |
| Plants | *Ceanothus martinii* | Utah Mountain Lilac | 2 | | | | | 50% | G4 | S1 |
| Plants | *Centaurium arizonicum* | Arizona Centaury | 2 | | | | | | G5 | S1 |
| Plants | *Cheilanthes eatonii* | Eaton's Lip Fern | 1 | | | | 100% | 100% | G5 | S2 |
| Plants | *Cirsium perplexans* | Rocky Mountain Thistle | 22 | | 5% | 23% | 45% | 32% | G2 | S2 |
| Plants | *Cleome multicaulis* | Slender Spiderflower | 56 | 29% | 9% | 30% | 57% | 91% | G2 | S2S3 |
| Plants | *Collomia grandiflora* | Showy Collomia | 1 | | | | | | G5 | S1 |
| Plants | *Comarum palustre* | Marsh Cinquefoil | 7 | 29% | 43% | 43% | 100% | 57% | G5 | S1S2 |
| Plants | *Commelina dianthifolia* | Birdbill Day-Flower | 2 | | 50% | 100% | 100% | 100% | G5 | S1? |
| Plants | *Conimitella williamsii* | Williams Bishop's Cap | 1 | | | | 100% | 100% | G3 | SH |
| Plants | *Crataegus chrysocarpa* | Yellow Hawthorn | 1 | | | | | 100% | G5 | S1 |
| Plants | *Crataegus saligna* | Willow Hawthorn | 1 | | | | | | G2 | S2 |
| Plants | *Cryptantha pustulosa* | Catseye | 2 | 50% | 50% | 50% | 50% | 100% | T? | S1 |
| Plants | *Cryptogramma stelleri* | Slender Rock-Brake | 12 | 25% | 58% | 67% | 75% | 100% | G5 | S2 |
| Plants | *Cylactis arctica ssp acaulis* | Nagoon Berry | 4 | 50% | 100% | 100% | 100% | 100% | T5 | S1 |
| Plants | *Cypripedium calceolus ssp parviflorum* | Yellow Lady's-Slipper | 19 | 5% | 21% | 26% | 53% | 63% | G5 | S2 |
| Plants | *Cypripedium fasciculatum* | Purple Lady's-Slipper | 109 | 37% | 50% | 65% | 99% | 95% | G4 | S3 |
| Plants | *Cystopteris montana* | Mountain Bladder Fern | 10 | 30% | 70% | 90% | 100% | 90% | G5 | S1 |
| Plants | *Delphinium ramosum var alpestre* | Colorado Larkspur | 4 | 25% | 100% | 100% | 100% | 50% | G2 | S2 |
| Plants | *Draba borealis* | Northern Rockcress | 6 | | 33% | 50% | 83% | 83% | G4 | S2 |
| Plants | *Draba crassa* | Thick-Leaf Whitlow-Grass | 30 | 43% | 57% | 77% | 93% | 97% | G3 | S3 |
| Plants | *Draba exunguiculata* | Clawless Draba | 19 | 42% | 74% | 74% | 100% | 95% | G2 | S2 |
| Plants | *Draba fladnizensis* | Arctic Draba | 26 | 46% | 69% | 73% | 92% | 92% | G4 | S2S3 |
| Plants | *Draba globosa* | Rockcress Draba | 4 | 50% | 50% | 75% | 100% | 100% | G3 | S1 |
| Plants | *Draba graminea* | San Juan Whitlow-Grass | 16 | 44% | 75% | 75% | 75% | 75% | G2 | S2 |
| Plants | *Draba grayana* | Gray's Peak Whitlow-Grass | 21 | 48% | 71% | 81% | 100% | 95% | G2 | S2 |
| Plants | *Draba incerta* | Yellowstone Whitlow-Grass | 4 | | 25% | 75% | 75% | 100% | G5 | S1 |

| Group | Scientific Name | Common Name | Total Available | Current | CW | CA | VISION | TNC | G-rank | S-rank |
|---|---|---|---|---|---|---|---|---|---|---|
| Plants | *Draba lonchocarpa var lonchocarpa* | Lancepod Whitlowgrass | 9 | 9 | 78% | 100% | 100% | 100% | T4 | S2 |
| Plants | *Draba oligosperma* | Woods Draba | 15 | 15 | 40% | 60% | 100% | 100% | G5 | S2 |
| Plants | *Draba porsildii* | Porsild Draba | 10 | 10 | 50% | 60% | 100% | 90% | G3 | S1 |
| Plants | *Draba rectifructa* | Mountain Whitlow-Grass | 12 | 12 | 8% | 33% | 92% | 92% | G3 | S2 |
| Plants | *Draba smithii* | Smith Whitlow-Grass | 15 | 15 | 73% | 73% | 100% | 100% | G2 | S2 |
| Plants | *Draba spectabilis var oxyloba* | Draba | 26 | 26 | 46% | 50% | 100% | 88% | T3 | S3 |
| Plants | *Draba streptobrachia* | Colorado Divide Whitlow-Grass | 31 | 31 | 71% | 74% | 94% | 77% | G3 | S3 |
| Plants | *Draba ventosa* | Tundra Draba | 5 | 5 | 60% | 100% | 100% | 100% | G3 | S1 |
| Plants | *Draba weberi* | Weber's Draba | 1 | 1 |  | 100% | 100% | 100% | G1 | S1 |
| Plants | *Drosera rotundifolia* | Roundleaf Sundew | 7 | 7 | 57% | 71% | 100% | 86% | G5 | S2 |
| Plants | *Dryopteris expansa* | Spreading Wood Fern | 6 | 6 | 100% | 100% | 100% | 100% | G5 | S1 |
| Plants | *Epipactis gigantea* | Helleborine | 20 | 20 | 55% | 60% | 95% | 80% | G3 | S2 |
| Plants | *Erigeron humilis* | Low Fleabane | 5 | 5 | 40% | 80% | 100% | 100% | G4 | S1 |
| Plants | *Erigeron lanatus* | Woolly Fleabane | 6 | 6 | 67% | 100% | 100% | 67% | G3 | S1 |
| Plants | *Erigeron philadelphicus* | Philadelphia Fleabane | 2 | 2 | 50% | 50% | 100% | 50% | G5 | S1 |
| Plants | *Eriogonum brandegeei* | Brandegee Wild Buckwheat | 17 | 17 |  |  | 100% | 100% | G1 | S1S2 |
| Plants | *Eriogonum coloradense* | Colorado Wild Buckwheat | 15 | 15 | 60% | 73% | 93% | 93% | G2 | S2 |
| Plants | *Eriogonum pelinophilum* | Clay-Loving Wild Buckwheat | 35 | 35 |  | 29% | 46% | 14% | G2 | S2 |
| Plants | *Eriophorum altaicum var neogaeum* | Altai Cottongrass | 33 | 33 | 91% | 91% | 94% | 73% | T3 | S3 |
| Plants | *Eriophorum gracile* | Slender Cottongrass | 12 | 12 | 67% | 75% | 92% | 92% | G5 | S2 |
| Plants | *Eutrema edwardsii ssp penlandii* | Penland Alpine Fen Mustard | 31 | 31 | 13% | 19% | 100% | 100% | G1 | S1S2 |
| Plants | *Festuca campestris* | Big Rough Fescue | 1 | 1 | 100% | 100% | 100% | 100% | G4 | SH |
| Plants | *Festuca hallii* | Hall Fescue | 1 | 1 | 100% | 100% | 100% | 100% | G4 | SH |
| Plants | *Gaura neomexicana ssp coloradensis* | Colorado Butterfly Weed | 2 | 2 |  | 50% | 50% | 100% | T2 | S1 |
| Plants | *Gaura neomexicana ssp neomexicana* | New Mexico Butterfly Weed | 1 | 1 |  |  |  | 100% | T3 | S1 |
| Plants | *Gilia penstemonoides* | Black Canyon Gilia | 28 | 28 | 43% | 46% | 75% | 68% | G3 | S3 |
| Plants | *Gilia sedifolia* | Stonecrop Gilia | 1 | 1 | 100% | 100% | 100% | 100% | G1 | S1 |
| Plants | *Gymnocarpium dryopteris* | Oak Fern | 1 | 1 | 100% | 100% | 100% | 100% | G5 | S2S3 |
| Plants | *Herrickia horrida* | Canadian River Spiny Aster | 3 | 3 |  |  | 100% | 100% | G2 | S1 |
| Plants | *Heuchera richardsonii* | Richardson Alum-Root | 1 | 1 |  |  | 100% |  | G5 | S1 |
| Plants | *Heuchera rubescens* | Red Alum-Root | 2 | 2 |  |  | 100% |  | G5 | S1 |
| Plants | *Hippochaete variegata* | Variegated Scouringrush | 4 | 4 | 25% | 50% | 100% | 100% | G5 | S1 |
| Plants | *Hypoxis hirsuta* | Yellow Stargrass | 1 | 1 |  |  |  | 100% | G5 | SH |
| Plants | *Iliamna crandallii* | Crandall's Wild-Hollyhock | 1 | 1 | 100% | 100% | 100% | 100% | GH | SH |

| Group | *Scientific Name* | Common Name | Total Available | Current | CW | CA | VISION | TNC | G-rank | S-rank |
|---|---|---|---|---|---|---|---|---|---|---|
| Plants | *Iliamna grandiflora* | Large-Flower Globe-Mallow | 6 | 17% | 50% | 67% | 100% | 100% | G3 | S1 |
| Plants | *Ipomopsis aggregata ssp weberi* | Rabbit Ears Gilia | 42 | | 52% | 74% | 93% | 98% | T2 | S2 |
| Plants | *Ipomopsis globularis* | Globe Gilia | 15 | 7% | 67% | 73% | 100% | 100% | G2 | S2 |
| Plants | *Ipomopsis multiflora* | Many-flowered Gilia | 1 | | | | 100% | | G4 | S1 |
| Plants | *Ipomopsis polyantha* | Pagosa Gilia | 2 | | | | | 100% | G1 | S1 |
| Plants | *Isoetes echinospora ssp muricata* | Spiny-Spored Quillwort | 6 | 67% | 83% | 100% | 100% | 100% | T5 | S2 |
| Plants | *Juncus bryoides* | Minute Rush | 1 | | | | 100% | | G4 | S1 |
| Plants | *Juncus tweedyi* | Tweedy Rush | 1 | 100% | 100% | 100% | 100% | 100% | G3 | S1 |
| Plants | *Juncus vaseyi* | Vasey Bulrush | 7 | 86% | 86% | 86% | 86% | 100% | G5 | S1 |
| Plants | *Lesquerella calcicola* | Rocky Mountain Bladderpod | 2 | | | | 100% | 100% | G2 | S2 |
| Plants | *Lesquerella parviflora* | Piceance Bladderpod | 1 | | | 100% | 100% | | G2 | S2S3 |
| Plants | *Lesquerella pruinosa* | Pagosa Bladderpod | 11 | | | | 36% | 100% | G2 | S2 |
| Plants | *Lesquerella vicina* | Good-Neighbor Bladderpod | 24 | 8% | 17% | 21% | 38% | 46% | G2 | S2 |
| Plants | *Lewisia rediviva* | Bitteroot | 6 | | | | 67% | 83% | G5 | S2 |
| Plants | *Liatris ligulistylis* | Gay-Feather | 6 | 17% | 33% | 33% | 50% | 67% | G5 | S1S2 |
| Plants | *Limnorchis ensifolia* | Canyon Bog-Orchid | 31 | 45% | 68% | 71% | 90% | 97% | T4 | S3 |
| Plants | *Listera borealis* | Northern Twayblade | 21 | 57% | 62% | 76% | 95% | 86% | G4 | S2 |
| Plants | *Listera convallarioides* | Broad-Leaved Twayblade | 12 | 33% | 50% | 50% | 67% | 83% | G5 | S2 |
| Plants | *Lomatium bicolor var leptocarpum* | Oregon Biscuitroot | 3 | 33% | 67% | 67% | 67% | 100% | T? | S2 |
| Plants | *Lomatium concinnum* | Colorado Desert-Parsley | 54 | 2% | 4% | 26% | 48% | 22% | G2 | S2 |
| Plants | *Lupinus crassus* | Payson Lupine | 8 | | | | 75% | 88% | G2 | S2 |
| Plants | *Lycopodium dubium* | Stiff Clubmoss | 3 | 67% | 67% | 100% | 100% | 100% | G? | SU |
| Plants | *Machaeranthera coloradoensis* | Colorado Tansy-Aster | 17 | 12% | 35% | 47% | 94% | 88% | G2 | S2 |
| Plants | *Malaxis monophyllos ssp brachypoda* | White Adder's-Mouth | 4 | | 25% | 25% | 50% | 75% | G4 | S1 |
| Plants | *Mertensia alpina* | Alpine Bluebells | 3 | | 67% | 67% | 100% | 100% | G4 | S1 |
| Plants | *Mimulus eastwoodiae* | Eastwood Monkey-Flower | 8 | | 75% | 88% | 100% | 100% | G3 | S1 |
| Plants | *Mimulus gemmiparus* | Weber Monkey-Flower | 8 | 88% | 88% | 88% | 100% | 100% | G2 | S2 |
| Plants | *Monardella odoratissima* | Mountain Wild Mint | 5 | | 20% | 20% | 40% | 100% | G4 | S2 |
| Plants | *Muscaria monticola* | Tundra Saxifrage | 3 | | 33% | 67% | 100% | 67% | T5 | S1 |
| Plants | *Myosurus cupulatus sensu lato* | Western Mouse-Tail | 1 | | | | 100% | 100% | G4 | S1? |
| Plants | *Nuttallia argillosa* | Arapien Stickleaf | 22 | | 41% | 41% | 55% | 77% | G3 | S2 |
| Plants | *Nuttallia chrysantha* | Golden Blazing Star | 4 | | | | 100% | 75% | G1 | S1S2 |
| Plants | *Nuttallia densa* | Arkansas Canyon Stickleaf | 20 | | 35% | 35% | 100% | 100% | G2 | S2 |
| Plants | *Nuttallia multicaulis* | Many-Stem Stickleaf | 14 | | 21% | 21% | 79% | 86% | G3 | S3 |

| Group | *Scientific Name* | Common Name | Total Available | Current | CW | CA | VISION | TNC | G-rank | S-rank |
|---|---|---|---|---|---|---|---|---|---|---|
| Plants | *Oenothera harringtonii* | Arkansas Valley Evening Primrose | 1 | | | | | | G2 | S2 |
| Plants | *Oenothera kleinii* | Wolf Creek Evening Primrose | 1 | | 100% | 100% | 100% | 100% | GU | SX |
| Plants | *Oonopsis foliosa* | Single-Head Goldenweed | 2 | | | | | | G2 | S2 |
| Plants | *Oreocarya cana* | Mountain Cat's-Eye | 1 | | | | | 100% | G5 | S2 |
| Plants | *Oreocarya longiflora* | Long-Flower Cat's-Eye | 15 | 20% | 27% | 47% | 87% | 27% | G3 | S2 |
| Plants | *Oreocarya weberi* | Weber's Catseye | 11 | | 18% | 18% | 45% | 100% | G3 | S3 |
| Plants | *Oreoxis humilis* | Pikes Peak Spring Parsley | 5 | | 40% | 40% | 100% | 100% | G1 | S1 |
| Plants | *Oxytropis parryi* | Parry Oxytrope | 2 | 50% | 50% | 50% | 50% | 100% | G5 | S1 |
| Plants | *Packera pauciflora* | Few-Flowered Ragwort | 15 | | | 7% | 60% | 100% | G4 | S1S2 |
| Plants | *Parnassia kotzebuei* | Kotzebue Grass-Of-Parnassus | 12 | 50% | 75% | 92% | 100% | 100% | G4 | S2 |
| Plants | *Pediocactus knowltonii* | Knowlton Cactus | 1 | | | | 100% | 100% | G1 | S1 |
| Plants | *Pellaea atropurpurea* | Purple Cliff-Brake | 4 | | 25% | 25% | 75% | 100% | G5 | S2S3 |
| Plants | *Pellaea breweri* | Brewer's Cliff-Brake | 1 | 100% | 100% | 100% | 100% | 100% | G5 | S2 |
| Plants | *Pellaea suksdorfiana* | Smooth Cliff-Brake | 2 | | | | 50% | 50% | T4 | S2 |
| Plants | *Penstemon breviculus* | Little Penstemon | 6 | | | | 100% | 100% | G3 | S2 |
| Plants | *Penstemon debilis* | Parachute Penstemon | 5 | | 40% | 40% | 80% | 80% | G1 | S1 |
| Plants | *Penstemon degeneri* | Degener Beardtongue | 8 | | 100% | 100% | 100% | 100% | G2 | S2 |
| Plants | *Penstemon harringtonii* | Harrington Beardtongue | 84 | | 23% | 23% | 52% | 100% | G3 | S3 |
| Plants | *Penstemon laricifolius* ssp *exilifolius* | Larch-Leaf Beardtongue | 8 | | | | 88% | 100% | T2 | S2 |
| Plants | *Penstemon lentus* | Abajo Penstemon | 3 | | 33% | 33% | 100% | 100% | G4 | S2 |
| Plants | *Penstemon mensarum* | Grand Mesa Penstemon | 33 | | 42% | 45% | 88% | 39% | G3 | S3 |
| Plants | *Penstemon penlandii* | Penland Beardtongue | 2 | | 50% | 50% | 50% | 100% | G1 | S1 |
| Plants | *Penstemon radicosus* | Matroot Penstemon | 2 | | | | | 100% | G5 | S1 |
| Plants | *Penstemon retrorsus* | Adobe Beardtongue | 37 | | 5% | 24% | 62% | 8% | G3 | S3 |
| Plants | *Phacelia formosula* | North Park Phacelia | 7 | | | | 86% | 100% | G1 | S1 |
| Plants | *Phacelia submutica* | Debeque Phacelia | 13 | | 8% | 46% | 100% | 100% | T2 | S2 |
| Plants | *Phippsia algida* | Snow Grass | 12 | 42% | 50% | 75% | 100% | 92% | G5 | S2 |
| Plants | *Phlox caryophylla* | Pagosa Phlox | 26 | | 15% | 15% | 42% | 88% | G4 | S2 |
| Plants | *Phlox kelseyi* ssp *salina* | Marsh Phlox | 1 | | | | 100% | 100% | T3 | S1 |
| Plants | *Physaria alpina* | Avery Peak Twinpod | 7 | | 43% | 43% | 100% | 100% | G2 | S2? |
| Plants | *Physaria bellii* | Bell's Twinpod | 6 | | | | | 83% | G2 | S2 |
| Plants | *Physaria rollinsii* | Rollin's Twinpod | 2 | 50% | 50% | 50% | 100% | 100% | G2 | S2 |
| Plants | *Picradenia helenioides* | Intermountain Bitterweed | 4 | 25% | 25% | 25% | 25% | 100% | G3 | S1 |
| Plants | *Polypodium hesperium* | Western Polypody | 7 | 14% | 29% | 29% | 29% | 86% | G5 | S1S2 |

| Group | *Scientific Name* | Common Name | Total Available | Current | CW | CA | VISION | TNC | G-rank | S-rank |
|---|---|---|---|---|---|---|---|---|---|---|
| Plants | *Potentilla ambigens* | Southern Rocky Mountain Cinquefoil | 8 | 13% | 75% | 75% | 75% | 88% | G3 | S1S2 |
| Plants | *Potentilla rupincola* | Rocky Mountain Cinquefoil | 20 | 30% | 30% | 35% | 65% | 100% | T2 | S2 |
| Plants | *Primula egaliksensis* | Greenland Primrose | 19 | | | 5% | 53% | 100% | G4 | S2 |
| Plants | *Ptilagrostis porteri* | Porter Feathergrass | 26 | 19% | 35% | 42% | 81% | 100% | T2 | S2 |
| Plants | *Puccinellia parishii* | Parish's Alkali Grass | 16 | | | | 6% | 100% | G2 | S1 |
| Plants | *Ranunculus gelidus* ssp *grayi* | Tundra Buttercup | 13 | 31% | 46% | 54% | 100% | 92% | G4 | S2 |
| Plants | *Ribes niveum* | Snow Gooseberry | 2 | | | | 100% | 100% | G3 | S1 |
| Plants | *Rorippa coloradensis* | Colorado Watercress | 1 | 100% | | 100% | 100% | 100% | GH | SH |
| Plants | *Sagittaria montevidensis ssp calycina* | Long-Lobe Arrowhead | 1 | 100% | 100% | 100% | 100% | 100% | T5 | S1 |
| Plants | *Salix candida* | Hoary or Silver Willow | 19 | 21% | 26% | 37% | 89% | 95% | G5 | S2 |
| Plants | *Salix lanata* ssp *calcicola* | Lime-loving Willow | 2 | 50% | 100% | 100% | 100% | 100% | T4 | S1 |
| Plants | *Salix myrtillifolia* | Low Blueberry Willow | 5 | | | 20% | 100% | 100% | G5 | S1 |
| Plants | *Salix serissima* | Autumn Willow | 6 | 67% | 67% | 83% | 83% | 83% | G4 | S1 |
| Plants | *Sarcostemma crispum* | Twinevine | 1 | | | | 100% | 100% | G4 | S1 |
| Plants | *Saussurea weberi* | Weber Saussurea | 20 | | 30% | 60% | 100% | 100% | G2 | S2 |
| Plants | *Sclerocactus glaucus* | Uinta Basin Hookless Cactus | 161 | | 81% | 90% | 94% | 85% | G3 | S3 |
| Plants | *Sisyrinchium demissum* | Blue-eyed Grass | 6 | | 17% | 17% | 33% | 67% | G5 | S2 |
| Plants | *Sisyrinchium pallidum* | Pale Blue-eyed Grass | 40 | 5% | 13% | 15% | 65% | 93% | G2 | S2 |
| Plants | *Sparganium eurycarpum* | Broodfruit Burreed | 1 | | | | 100% | 100% | G5 | S2? |
| Plants | *Spatularia foliolosa* | Leafy Saxifrage | 1 | 100% | 100% | 100% | 100% | 100% | G4 | S1 |
| Plants | *Spiranthes diluvialis* | Ute Ladies' Tresses | 8 | | | | | 100% | G2 | S2 |
| Plants | *Stellaria irrigua* | Altai Chickweed | 19 | 63% | 74% | 79% | 89% | 100% | G4 | S2 |
| Plants | *Subularia aquatica* | Water Awlwort | 1 | 100% | 100% | 100% | 100% | 100% | G5 | S1 |
| Plants | *Sullivantia hapemanii* var *purpusii* | Hanging Garden Sullivantia | 22 | 50% | 73% | 77% | 95% | 82% | T3 | S3 |
| Plants | *Telesonix jamesii* | James's Telesonix | 14 | | 29% | 29% | 100% | 100% | G2 | S2? |
| Plants | *Thalictrum heliophilum* | Sun-loving Meadowrue | 5 | | | 20% | 40% | 40% | G3 | S3 |
| Plants | *Thellungiella salsuginea* | Saltl-Lick Mustard | 4 | | | | 100% | 100% | G4 | S1 |
| Plants | *Thelypodium paniculatum* | Northwestern Thelypody | 1 | 100% | 100% | 100% | 100% | 100% | G2 | S1 |
| Plants | *Townsendia glabella* | Gray's Townsend-Daisy | 2 | | | | 50% | 50% | G2 | S2? |
| Plants | *Townsendia rothrockii* | Rothrock Townsend-Daisy | 1 | | 100% | 100% | 100% | 100% | G2 | S2? |
| Plants | *Townsendia strigosa* | Strigose Easter-Daisy | 1 | | 100% | 100% | 100% | 100% | G4 | S1 |
| Plants | *Trichophorum pumilum* | Little Bulrush | 16 | | 6% | 13% | 69% | 100% | G5 | S2 |
| Plants | *Trifolium kingii* | King Clover | 8 | | 13% | 13% | 63% | 100% | G5 | S1 |
| Plants | *Triteleia grandiflora* | Triteleia Grandiflora | 1 | | | 100% | 100% | | G4 | S1 |

| Group | Scientific Name | Common Name | Total Available | Current | CW | CA | VISION | TNC | G-rank | S-rank |
|---|---|---|---|---|---|---|---|---|---|---|
| Plants | *Unamia alba* | Prairie Goldenrod | 8 | 50% | 13% | 63% | 75% | 88% | G5 | S2S3 |
| Plants | *Utricularia ochroleuca* | Northern Bladderwort | 1 | | | 100% | 100% | 100% | G4 | S1? |
| Plants | *Viola pedatifida* | Prairie Violet | 10 | | 20% | 20% | 50% | 80% | G5 | S2 |
| Plants | *Viola selkirkii* | Selkirk Violet | 1 | | | | 100% | 100% | G5 | S1 |
| Plants | *Woodsia neomexicana* | New Mexico Cliff Fern | 16 | 50% | 50% | 63% | 63% | 69% | G4 | S2 |
| Reptiles | *Cnemidophorus neotesselatus* | Triploid Colorado Checkered Whiptail | 2 | | | | 50% | 50% | G2 | S2 |
| Reptiles | *Cnemidophorus velox* | Plateau Striped Whiptail | 5 | | 40% | 40% | 100% | 80% | G5 | S4 |
| Reptiles | *Coluber constrictor mormon* | Western Yellowbelly Racer | 11 | | 9% | 9% | 27% | 36% | T5 | S3 |
| Reptiles | *Crotalus viridis concolor* | Midget Faded Rattlesnake | 4 | | 75% | 75% | 75% | 75% | T3 | S3? |
| Reptiles | *Elaphe guttata* | Corn Snake | 6 | | | | 50% | 50% | G5 | S3 |
| Reptiles | *Eumeces multivirgatus epipleurotus* | Variable Skink | 1 | | | | 100% | 100% | T5 | S3 |
| Reptiles | *Liochlorophis vernalis* | Smooth Green Snake | 2 | | | | 100% | 50% | G5 | S4 |
| Reptiles | *Tantilla hobartsmithi* | Southwestern Blackhead Snake | 2 | | | | | | G5 | S2? |
| Reptiles | *Urosaurus ornatus* | Tree Lizard | 2 | | 50% | 50% | 100% | 100% | G5 | S4 |

Where: Current = Current Protected, CW = Core Wilderness, CA = Core Areas, VISION = the network design map, and TNC = The Nature Conservancy's conservation portfolio (Neely et al. 2002). G-rank and S-rank are the global and state rankings, respectively, see definitions below. Note that rankings may be adjusted over time to reflect changes and that these rankings may not be the most current available. The element occurrence data are products and property of the Colorado Natural Heritage Program (CNHP), a sponsored program at Colorado State University. Care should be taken in interpreting these data. The information provided should not replace field studies necessary for more localized planning efforts. Please note that the absence of any data does not mean that other resources of special concern do not occur. The data are provided on an as-is, as-available basis without warranties of any kind. CNHP, Colorado State University, and the State of Colorado further expressly disclaim any warranty that the data are error-free or current as of the date supplied.

G1 - Globally critically imperiled; typically 5 or fewer occurrences
G2 - Globally imperiled; typically 6 to 20 occurrences
G3 - Globally vulnerable; typically 21 to 100 occurrences
G4 - Globally apparently secure; usually > 100 occurrences
G5 - Globally demonstrably secure although it may be rare in parts of its range
G? - Unranked; element is not yet ranked globally
GU - Unrankable; not enough information is known
GH - Historically known with hopes of rediscovery
GX - Presumed extinct
T# - Rank applies to a subspecies or variety

C - Element is extant only in captivation or cultivation

S1 - State critically imperiled; typically 5 or fewer occurrences
S2 - State imperiled; typically 6 to 20 occurrences
S3 - State vulnerable; typically 21 to 100 occurrences
S4 - State apparently secure; usually > 100 occurrences
S5 - State demonstrably secure
S? - Unranked; element is not yet ranked in the state
SU - Unrankable; not enough information is known
SH - Historically known with hopes of rediscovery

SX - Extirpated in the state
SE - An exotic established in the state; native to a nearby region
SA - Accidental; includes species (usually birds or butterflies) recorded once or twice or only at very great intervals, hundreds or thousands of miles outside their usual range
SR - Reported in the state, but not confirmed
SZ - Zero occurrences; typically refers to nonbreeding bird populations
B - Rank refers to the breeding population of the element
N - Rank refers to the nonbreeding population of the element

**Table A3.3 Coverage of SITES inputs in the Wildlands Network Design.**

| Focal Species | Total in ecoregion (ha) | Goal (ha) | Included in Network Design (ha) | % of total | % of goal |
|---|---|---|---|---|---|
| Colorado River cutthroat | 56,190 | 56,190 | 53,178 | 95% | 95% |
| Greenback cutthroat | 55,723 | 55,723 | 54,241 | 97% | 97% |
| Rio Grande cutthroat | 456,896 | 228,493 | 341,349 | 75% | 149% |
| Wolf Core Reintroduction Area | 1,007,314 | 1,007,314 | 885,960 | 88% | 88% |
| Wolf Suitable Habitat | 5,651,101 | 2,825,550 | 4,437,852 | 79% | 157% |
| Bear Core Habitat | 3,519,784 | 2,639,840 | 2,431,496 | 69% | 92% |
| Pronghorn Habitat | 3,844,834 | 1,922,420 | 1,847,178 | 48% | 96% |
| **Special Elements** | | | | | |
| Roadless Areas | 2,978,613 | 2,233,960 | 2,819,355 | 95% | 126% |
| Wilderness & Park lands | 1,645,040 | 1,645,040 | 1,645,040 | 100% | 100% |
| **Vegetation Communities** | | | | | |
| Active sand dune & swale complex | 10,495 | 4,000 | 10,495 | 100% | 262% |
| Alpine dry tundra & moist meadow | 680,381 | 191,103 | 519,489 | 76% | 272% |
| Alpine substrate - ice field | 206,565 | 61,969 | 197,497 | 96% | 319% |
| Alpine tundra - dwarf shrub & fell field | 125,342 | 47,556 | 119,304 | 95% | 251% |
| Aspen forest | 1,336,483 | 399,827 | 995,935 | 75% | 249% |
| Bristlecone - limber pine forest & woodland | 77,710 | 23,312 | 42,351 | 54% | 182% |
| Douglas fir - ponderosa pine forest | 383,708 | 66,585 | 270,034 | 70% | 406% |
| Foothills riparian woodland & shrubland | 5,283 | 3,487 | 869 | 16% | 25% |
| Gambel's oak shrubland | 641,882 | 210,190 | 301,912 | 47% | 144% |
| Greasewood flat & ephemeral meadow complex | 180,650 | 58,457 | 36,752 | 20% | 63% |
| Intermontane - foothill grassland | 837,424 | 290,272 | 383,286 | 46% | 132% |
| Juniper savanna | 312,702 | 93,811 | 186,778 | 60% | 199% |
| Lodgepole pine forest | 1,108,412 | 332,450 | 894,145 | 81% | 269% |
| Lower montane - foothills shrubland | 759,922 | 246,402 | 408,029 | 54% | 166% |
| Marsh & wet meadow | 19,001 | 12,500 | 9,419 | 50% | 75% |
| Montane - foothill cliff & canyon | 21,055 | 6,142 | 15,022 | 71% | 245% |
| Montane grassland | 293,272 | 96,775 | 126,866 | 43% | 131% |
| Montane mixed conifer forest | 616,665 | 172,856 | 446,429 | 72% | 258% |
| Montane riparian shrubland | 13,254 | 8,747 | 10,467 | 79% | 120% |
| Mountain sagebrush shrubland | 1,339,987 | 398,787 | 499,412 | 37% | 125% |
| North park sand dunes | 342 | 103 | 342 | 100% | 332% |
| Piñon - juniper woodland | 1,726,696 | 518,009 | 1,035,405 | 60% | 200% |
| Ponderosa pine woodland | 1,985,827 | 663,227 | 1,238,637 | 62% | 187% |
| Sagebrush steppe | 272,677 | 81,803 | 160,133 | 59% | 196% |
| San Luis valley winterfat shrub steppe | 141,259 | 50,047 | 81,735 | 58% | 163% |
| South Park montane grasslands | 221,107 | 72,317 | 92,782 | 42% | 128% |
| Spruce-fir forest | 2,251,859 | 674,148 | 1,990,330 | 88% | 295% |
| Stabilized sand dune | 38,336 | 11,162 | 28,497 | 74% | 255% |
| Upper montane riparian forest & woodland | 19,381 | 12,791 | 10,533 | 54% | 82% |
| Winterfat shrub steppe | 131,048 | 39,314 | 75,292 | 57% | 192% |

**Table A3.4 Comparison of the Wildlands Network Design with current protected areas.**

| | Current Protected | Core Wilderness | Core Areas | Network Design |
|---|---|---|---|---|
| **Area (ha)** | **1,715,568** | **3,344,035** | **4,330,241** | **10,429,615** |
| **Focal Species (ha)** | | | | |
| Colorado River cutthroat subwater. | 11,061 | 34,771 | 38,039 | 53,178 |
| greenback cutthroat subwatershed | 31,715 | 38,954 | 42,182 | 54,241 |
| Rio Grande cutthroat subwatershed | 70,771 | 130,646 | 169,067 | 341,349 |
| Wolf Core Reintroduction Area | 304,502 | 475,777 | 666,127 | 885,960 |
| Wolf Suitable Habitat | 1,268,130 | 2,187,840 | 2,703,973 | 4,437,852 |
| Bear Core Habitat | 171,489 | 630,703 | 951,161 | 2,431,496 |
| Pronghorn Habitat | 59,936 | 214,217 | 317,841 | 1,847,178 |
| **Vegetation Communities (ha)** | | | | |
| Active sand dune & swale complex | 10,495 | 8,675 | 10,495 | 10,495 |
| Alpine dry tundra & moist meadow | 229,595 | 356,176 | 395,068 | 519,489 |
| Alpine substrate - ice field | 144,863 | 170,711 | 176,654 | 197,497 |
| Alpine tundra - dwarf shrub & fell field | 78,252 | 97,647 | 104,687 | 119,304 |
| Aspen forest | 113,484 | 378,454 | 464,347 | 995,935 |
| Bristlecone - limber pine forest & woodland | 4,356 | 6,819 | 11,450 | 42,351 |
| Douglas fir - ponderosa pine forest | 12,905 | 62,511 | 100,743 | 270,034 |
| Foothills riparian woodland & shrubland | 0 | 52 | 52 | 869 |
| Gambel's oak shrubland | 8,952 | 59,049 | 64,373 | 301,912 |
| Greasewood flat & ephemeral meadow complex | 5,219 | 175 | 5,749 | 36,752 |
| Intermontane - foothill grassland | 3,763 | 17,312 | 42,607 | 383,286 |
| Juniper savanna | 8,756 | 48,498 | 53,946 | 186,778 |
| Lodgepole pine forest | 141,680 | 352,630 | 440,138 | 894,145 |
| Lower montane - foothills shrubland | 67,843 | 114,419 | 140,295 | 408,029 |
| Marsh & wet meadow | 670 | 1,833 | 2,054 | 9,419 |
| Montane - foothill cliff & canyon | 3,567 | 6,940 | 10,539 | 15,022 |
| Montane grassland | 3,495 | 9,180 | 22,810 | 126,866 |
| Montane mixed conifer forest | 66,971 | 89,736 | 146,892 | 446,429 |
| Montane riparian shrubland | 2,284 | 2,615 | 2,650 | 10,467 |
| Mountain sagebrush shrubland | 17,555 | 80,384 | 114,591 | 499,412 |
| North Park sand dunes | 0 | 0 | 0 | 342 |
| Piñon - juniper woodland | 40,333 | 157,127 | 261,311 | 1,035,405 |
| Ponderosa pine woodland | 55,612 | 169,258 | 372,374 | 1,238,637 |
| Sagebrush steppe | 1,050 | 29,571 | 38,904 | 160,133 |
| San luis valley winterfat shrub steppe | 365 | 0 | 1,081 | 81,735 |
| South Park montane grasslands | 0 | 1,356 | 2,192 | 92,782 |
| Spruce-fir forest | 672,337 | 1,098,739 | 1,246,238 | 1,990,330 |
| Stabilized sand dune | 17,885 | 3,669 | 18,112 | 28,497 |
| Upper montane riparian forest & woodland | 145 | 956 | 2,567 | 10,533 |
| Winterfat shrub steppe | 1,285 | 8,499 | 19,742 | 75,292 |

# List of Figures



---

*There are some who can live without wild things, and some who cannot.*

-Aldo Leopold

---

# FIGURES

Figure 1.1 The Wildlands Project Spine of the Continent MegaLinkage.

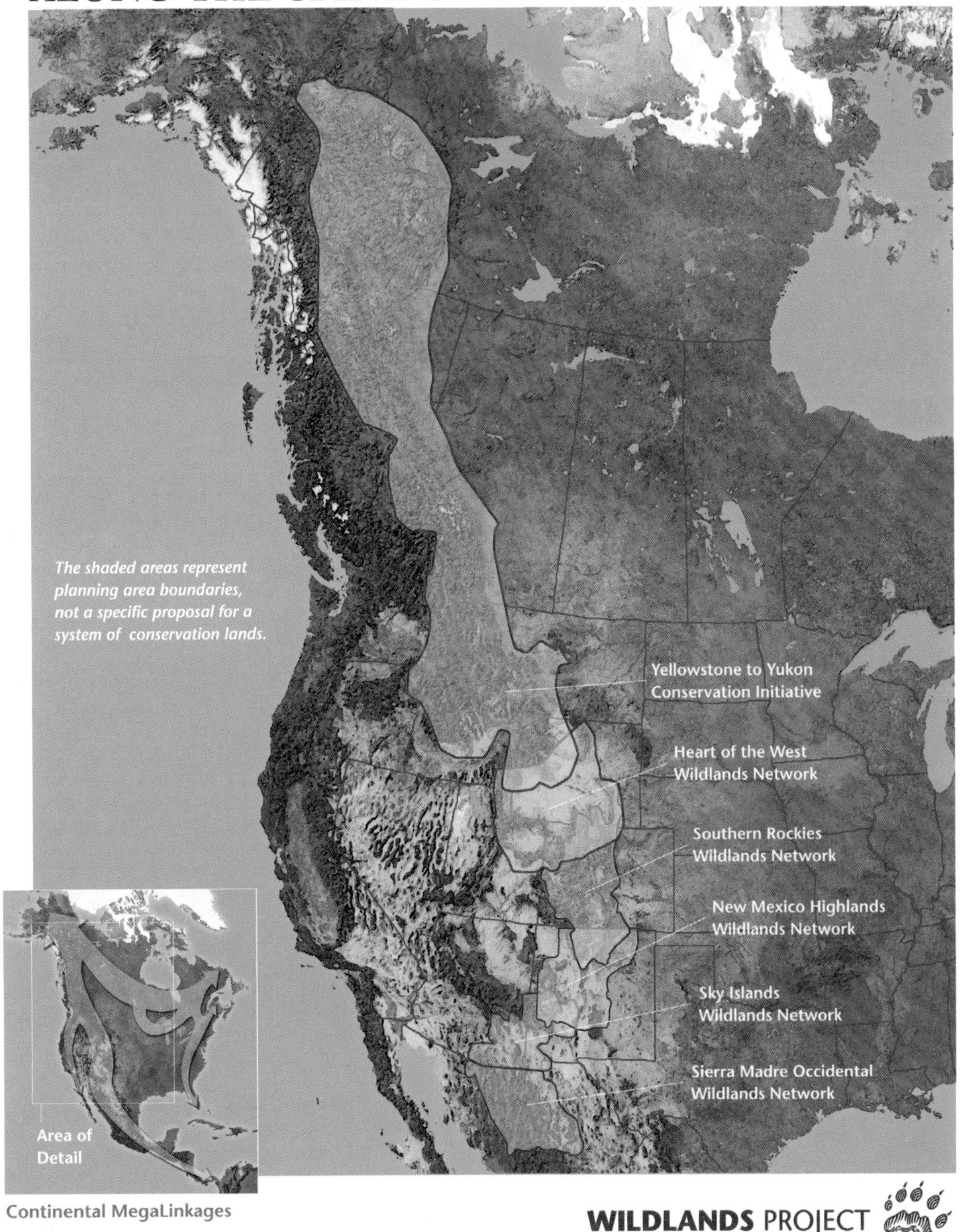

Figure 2.1 The Southern Rockies ecoregion.

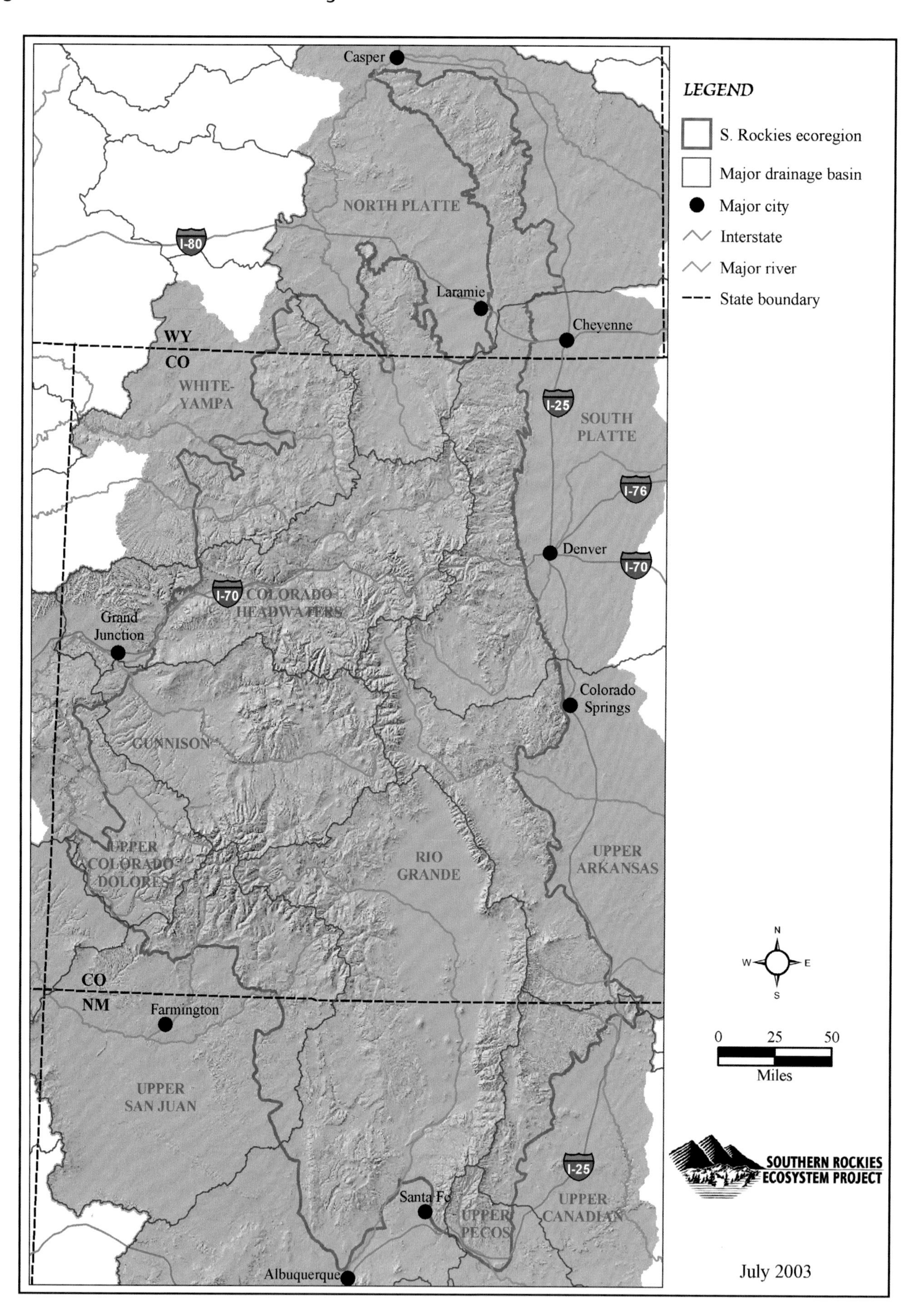

**Figure 2.2 Vegetation communities and other land cover in the Southern Rockies.**

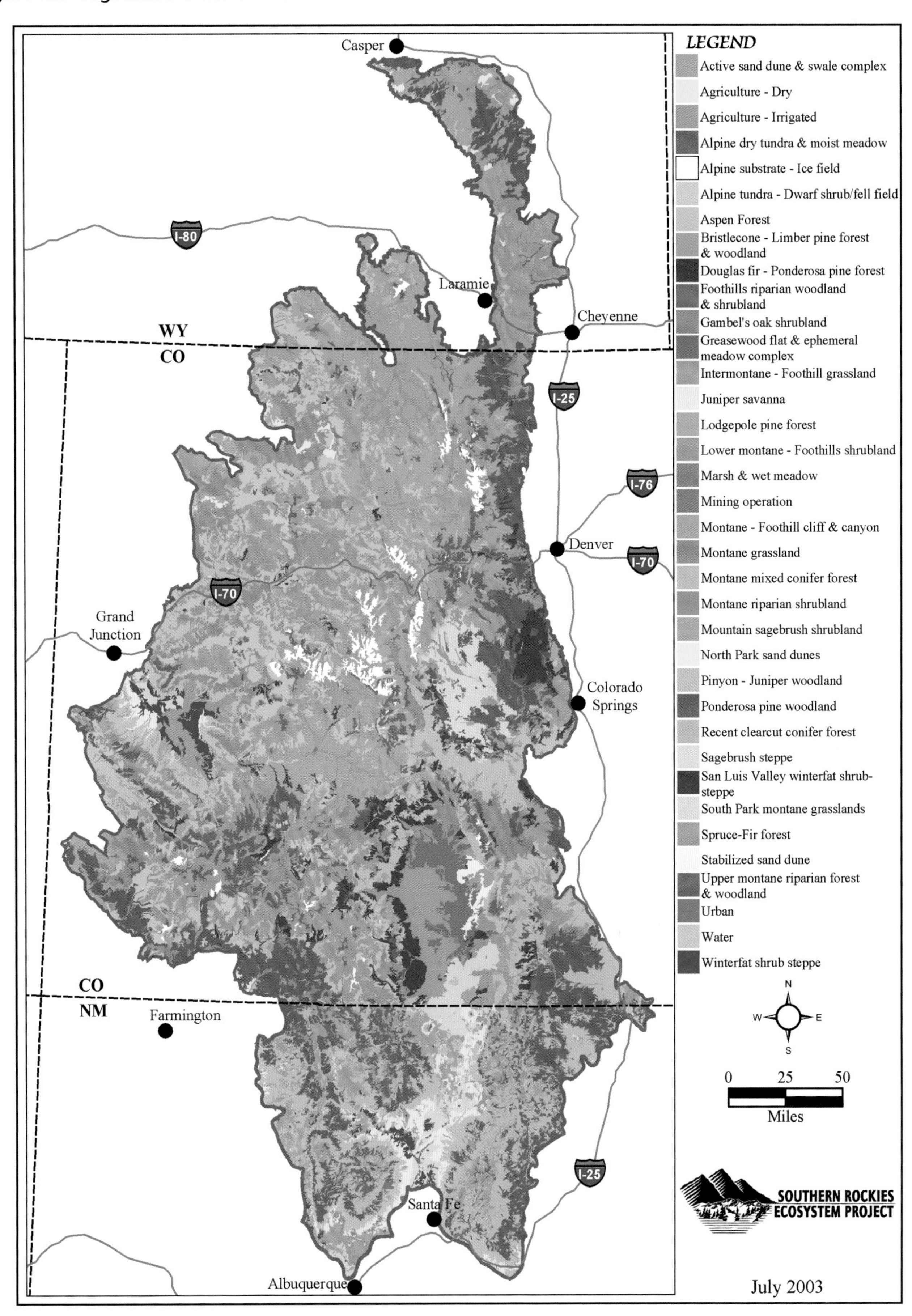

Figure 3.1 Land ownership around the Southern Rockies.

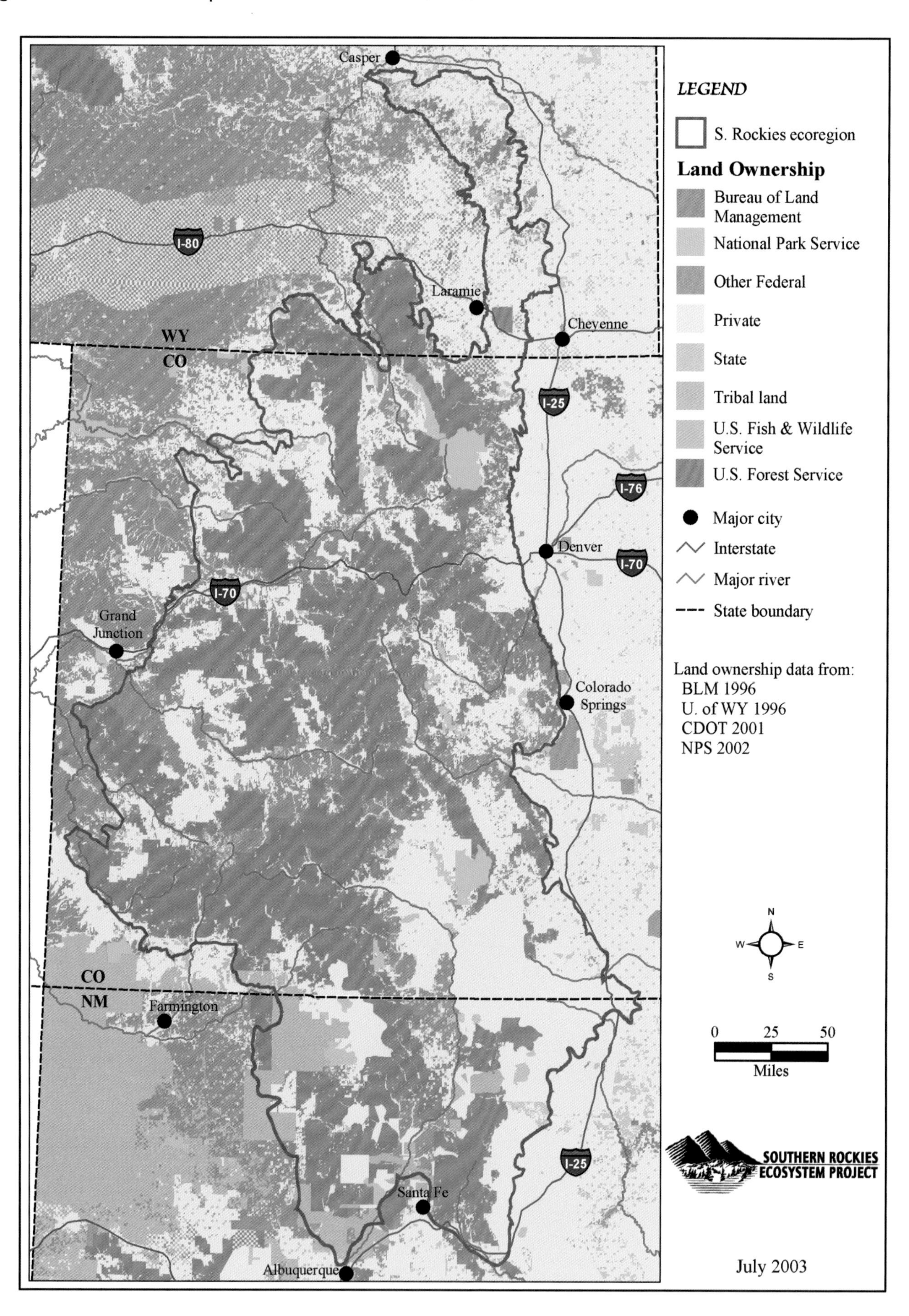

**Figure 3.2 Roads within the Southern Rockies.**

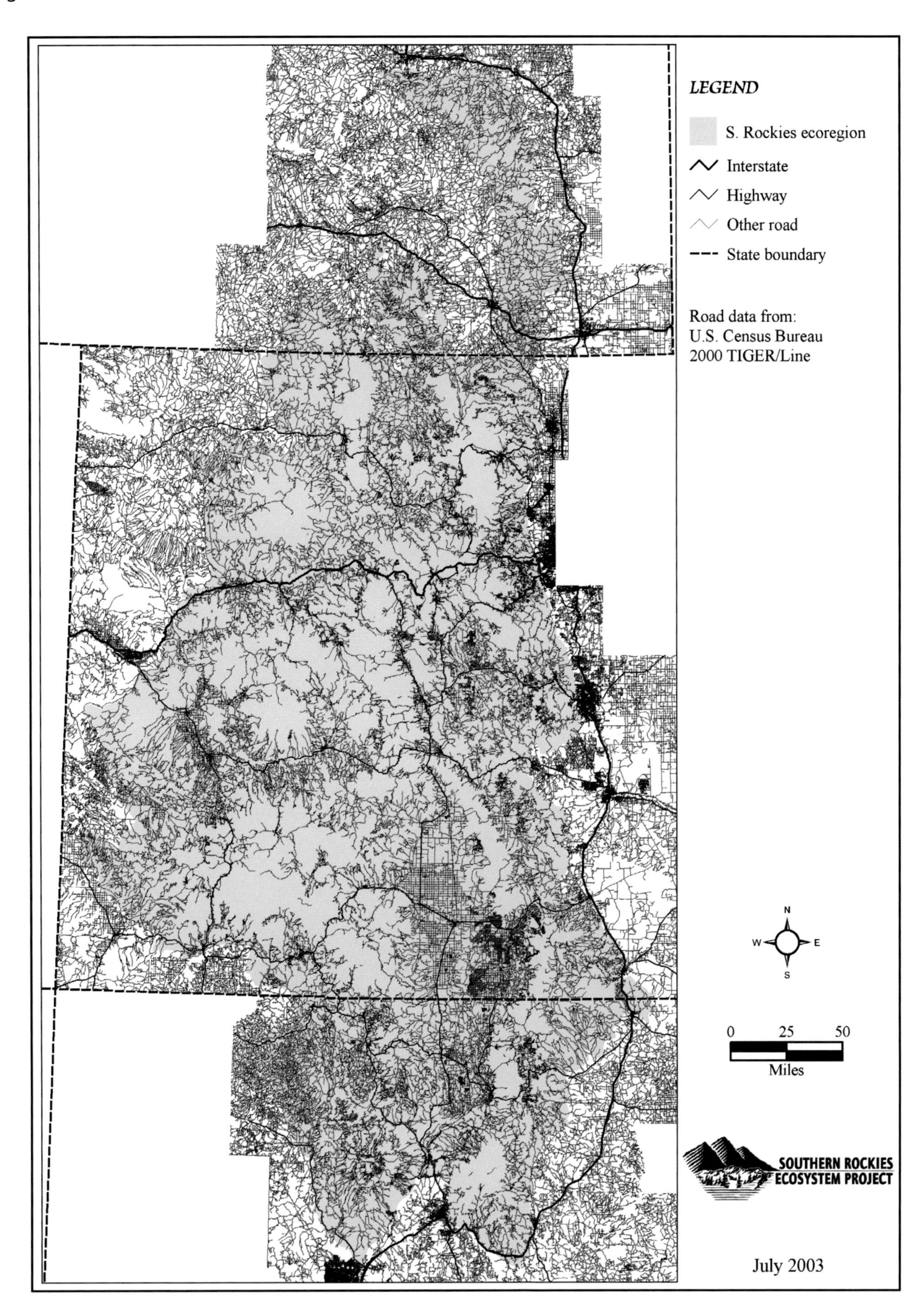

Figure 3.3 Livestock density by county in the Southern Rockies.

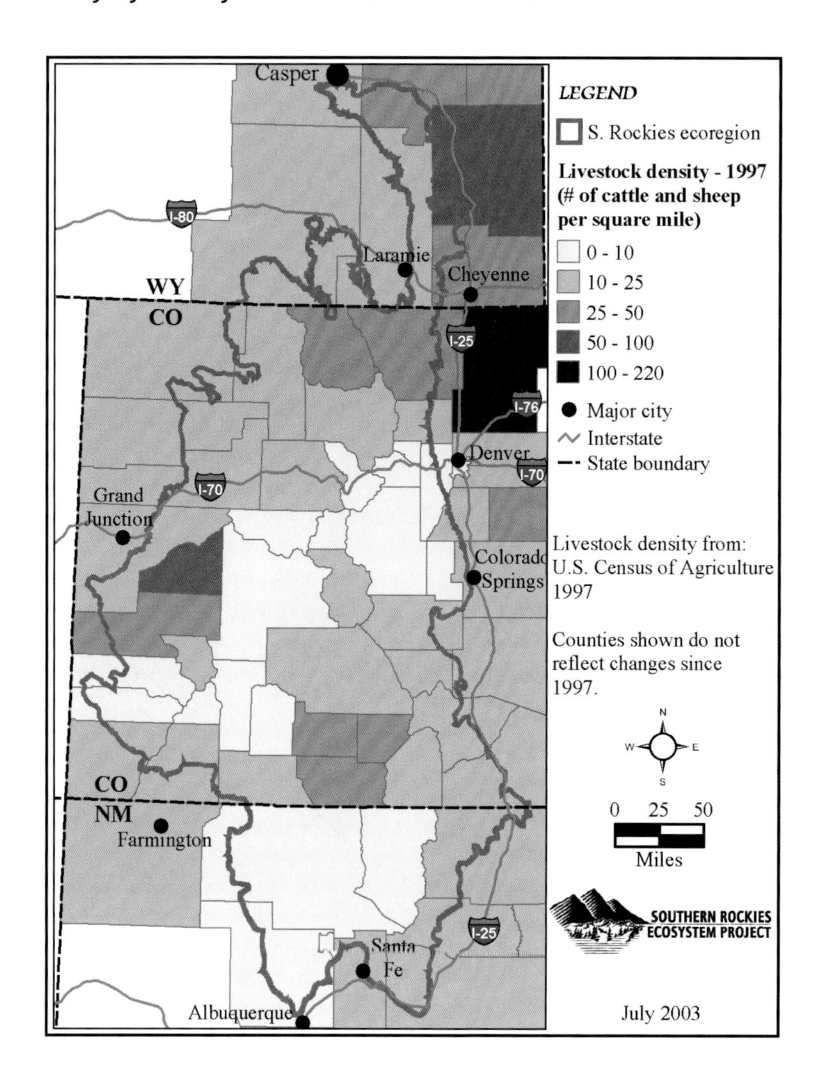


Figure 3.4 Projected number of recreationists in the Rocky Mountain region by activity and year.

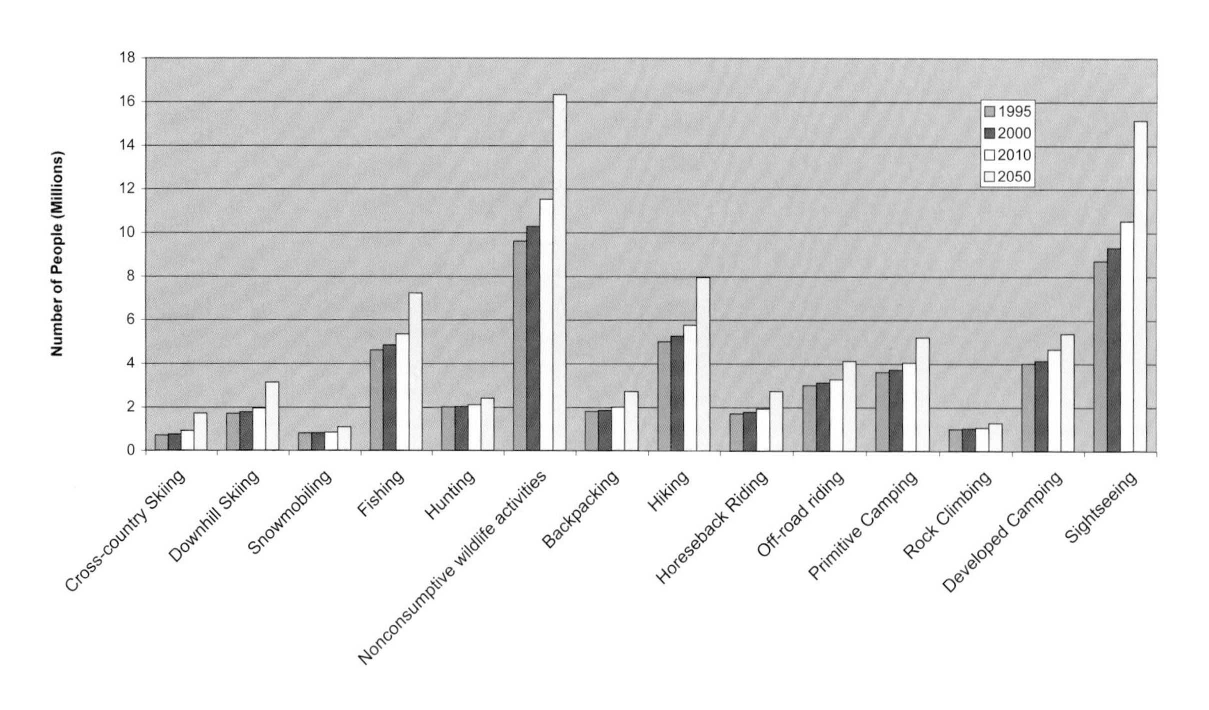

Figure 3.5 Population by county in the Southern Rockies.

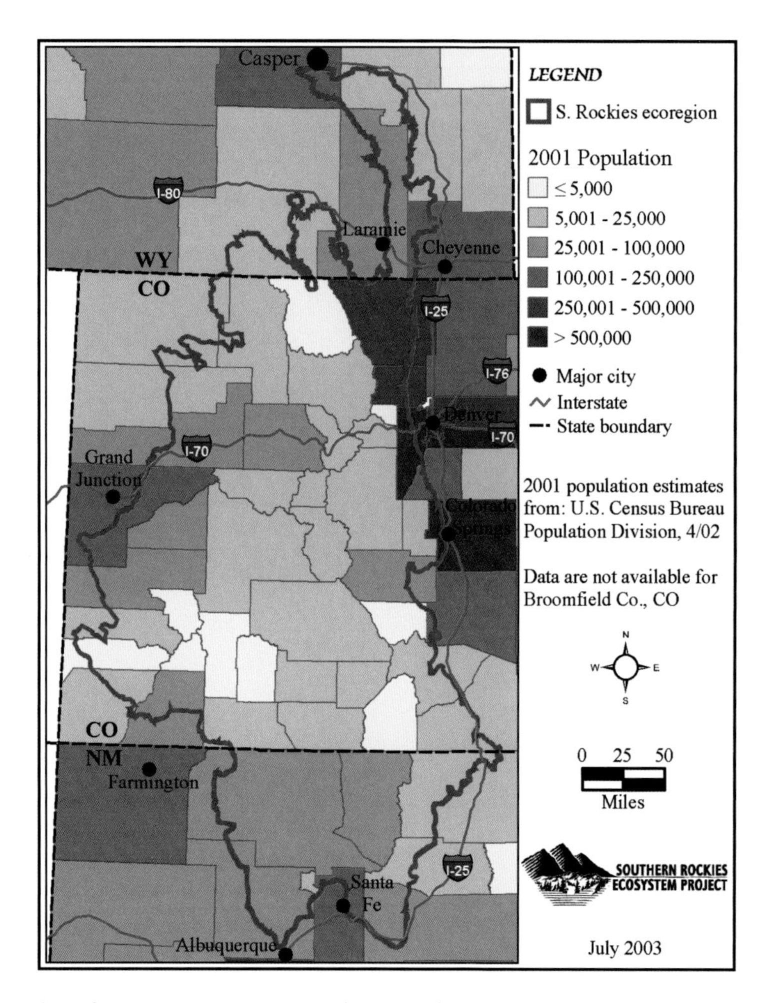


Figure 3.6 Housing density change over time in the Southern Rockies.

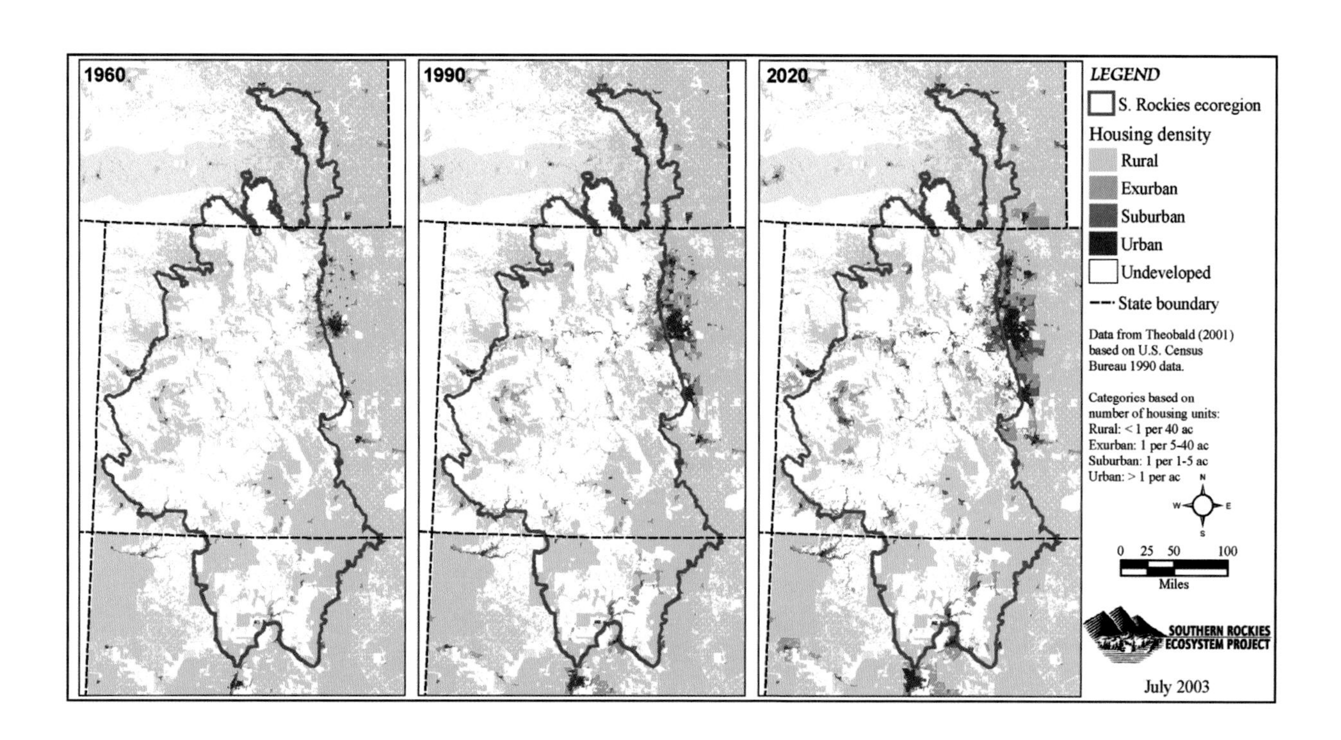

Figure 5.1 Relative level of threat and human impact in the Southern Rockies.

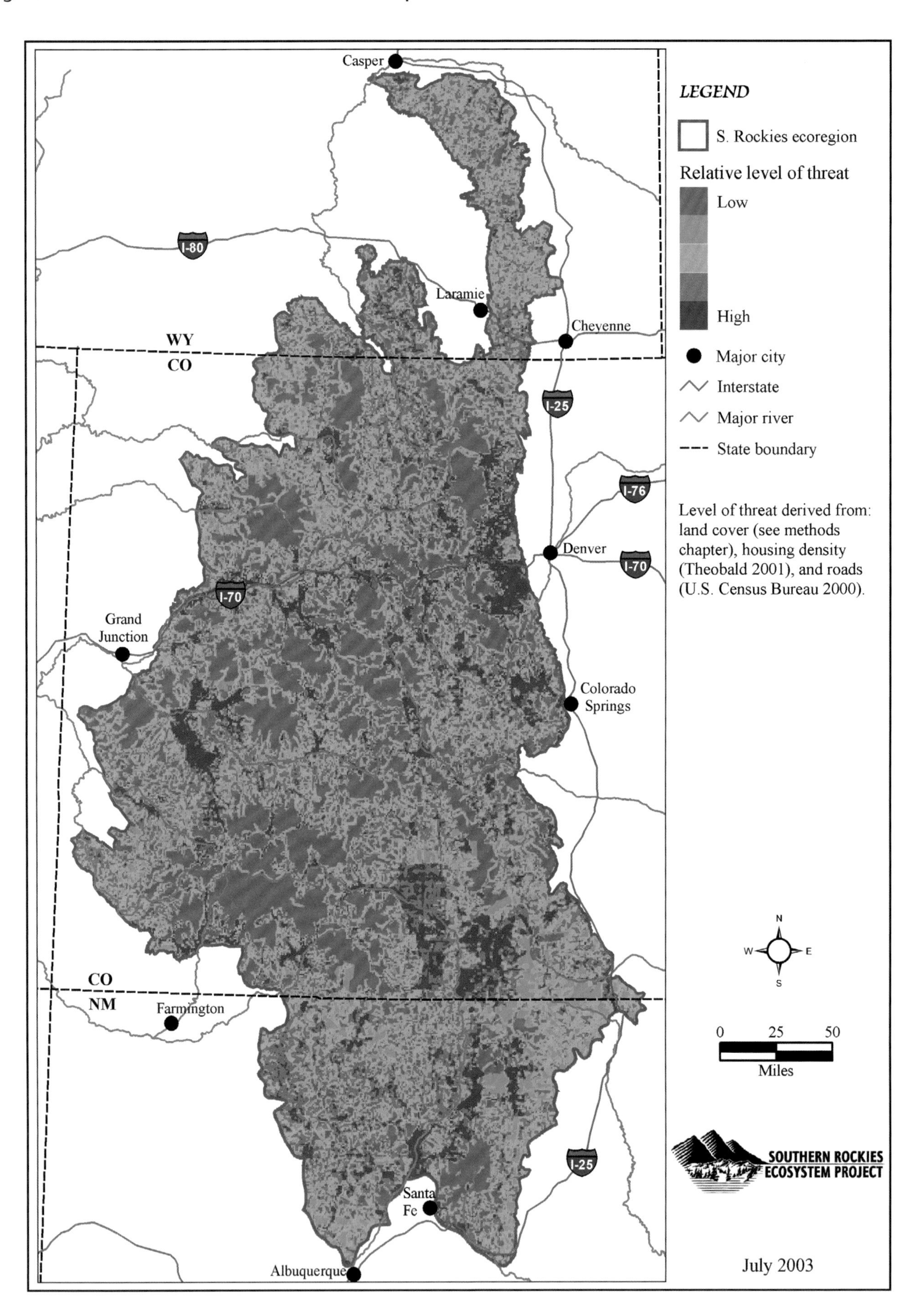

Figure 5.2 Density of dams by watershed.

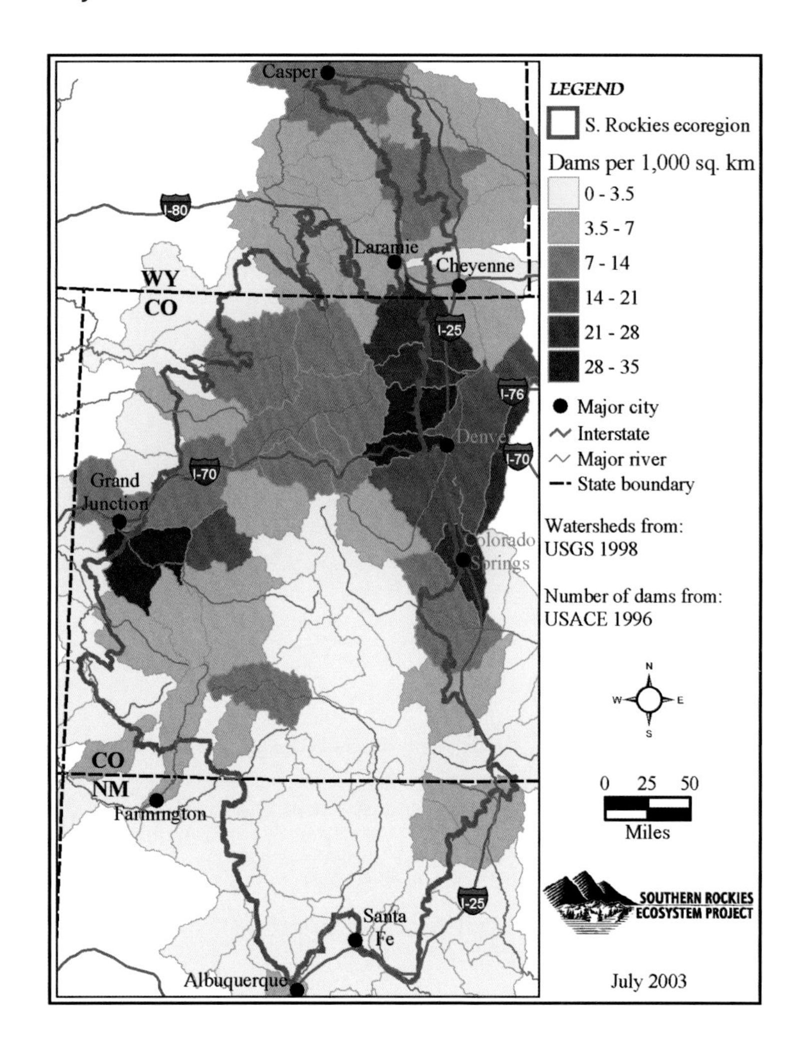


Figure 5.3 Total timber harvested from National Forests of the Southern Rockies from 1987-1997, measured in million board-feet (MMBF).

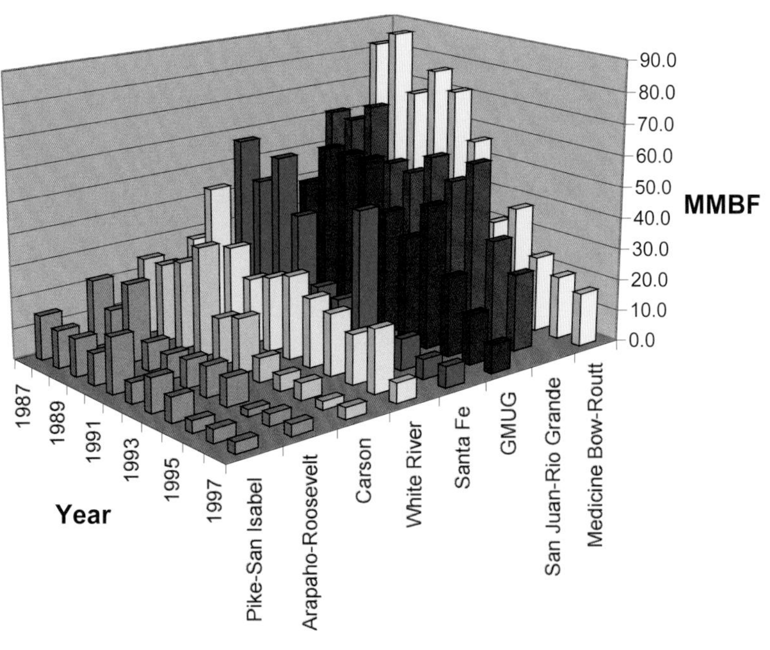

## Figure 5.4 Road edge effect on the Medicine Bow National Forest.

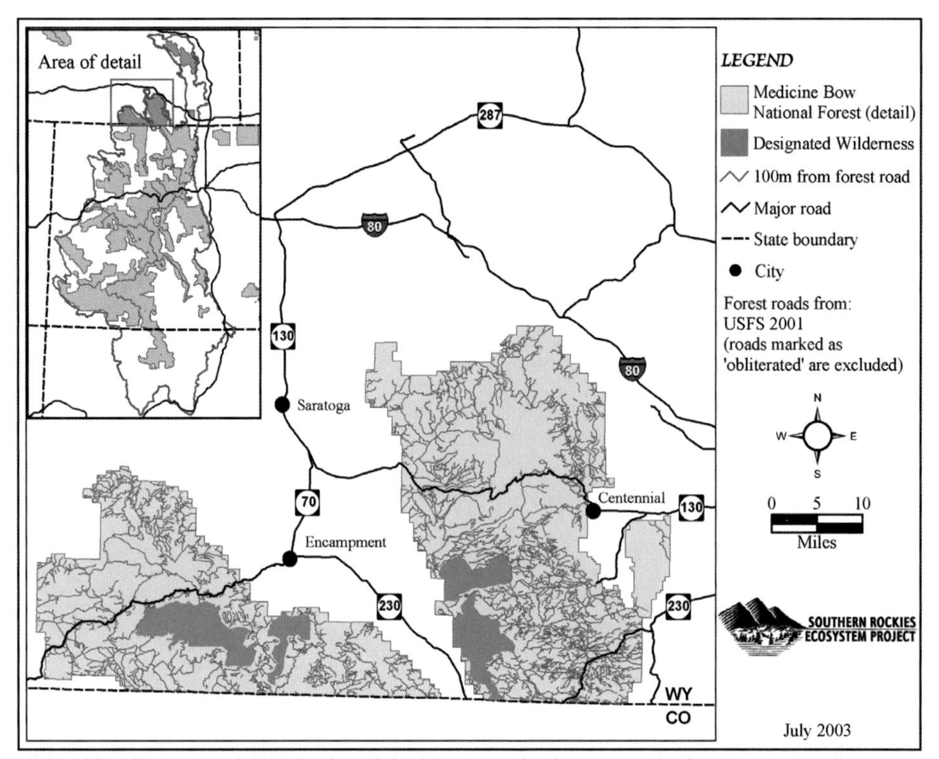


(Adapted from Shinneman et al. 2000) Roads on National Forests are often far more extensive than many people realize. Illustrated here are roads within two units of the Medicine Bow National Forest in Wyoming, buffered to 100 meters (328 feet) on either side to indicate those areas likely impacted by road edge-effect as well as how few large contiguous areas of forest remain. Not shown here is forest fragmentation from clearcut logging, which can be extensive.

**Figure 7.1 Cost layer input of SITES analysis.**

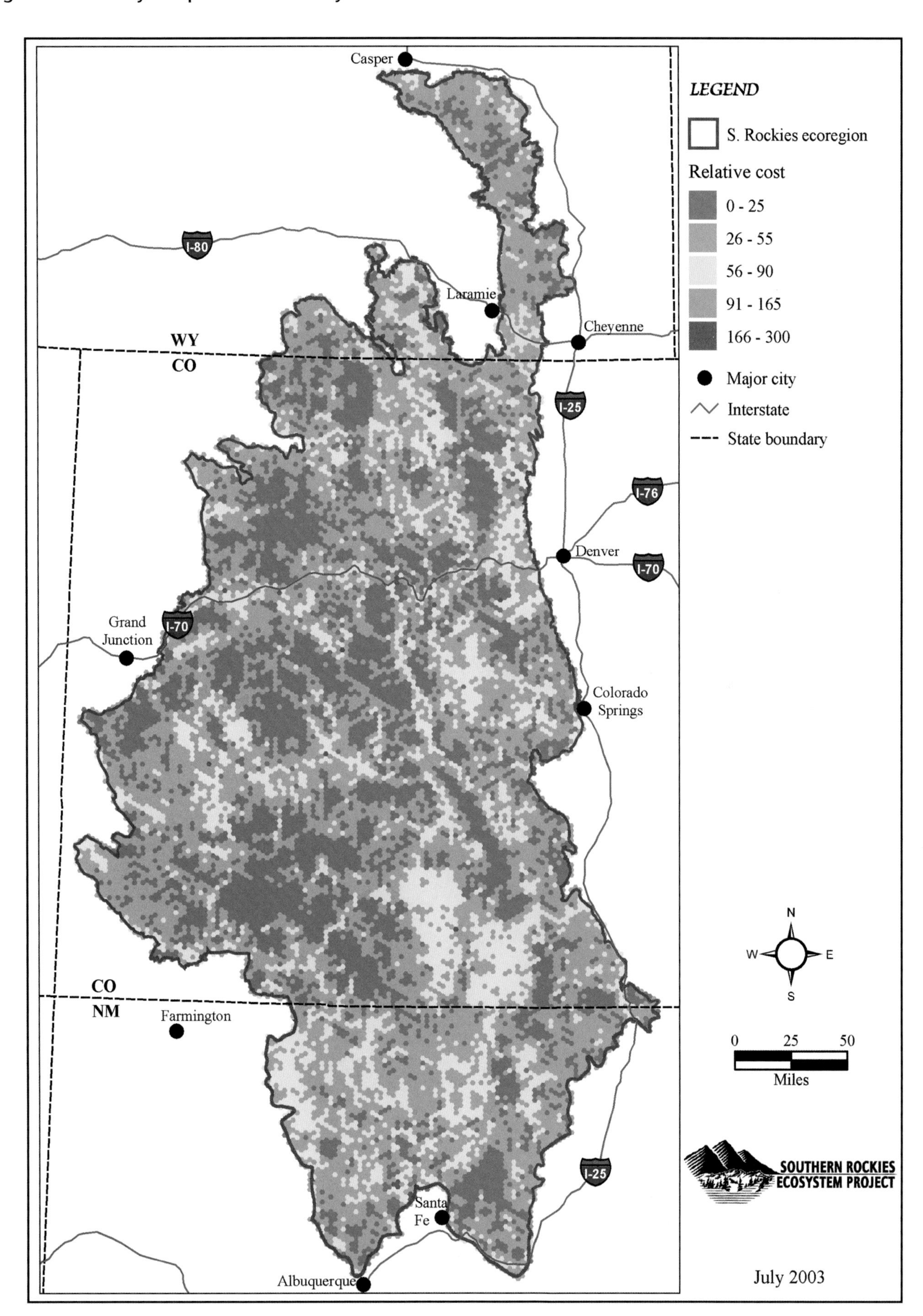

Figure 7.2 Designated Wilderness, Park Service lands, and other roadless areas in the Southern Rockies.

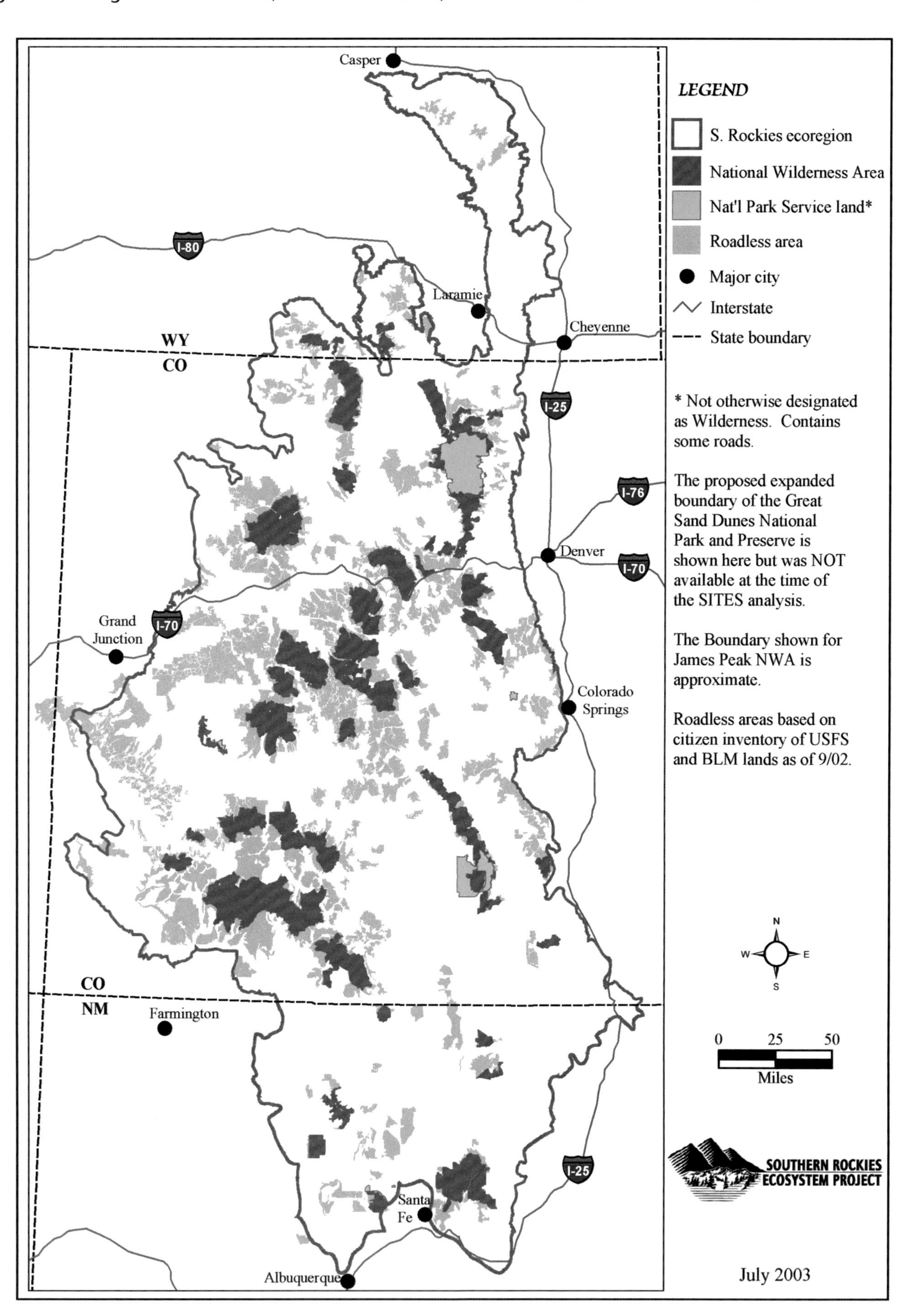

Figure 7.3 Native cutthroat trout subwatersheds in the Southern Rockies.

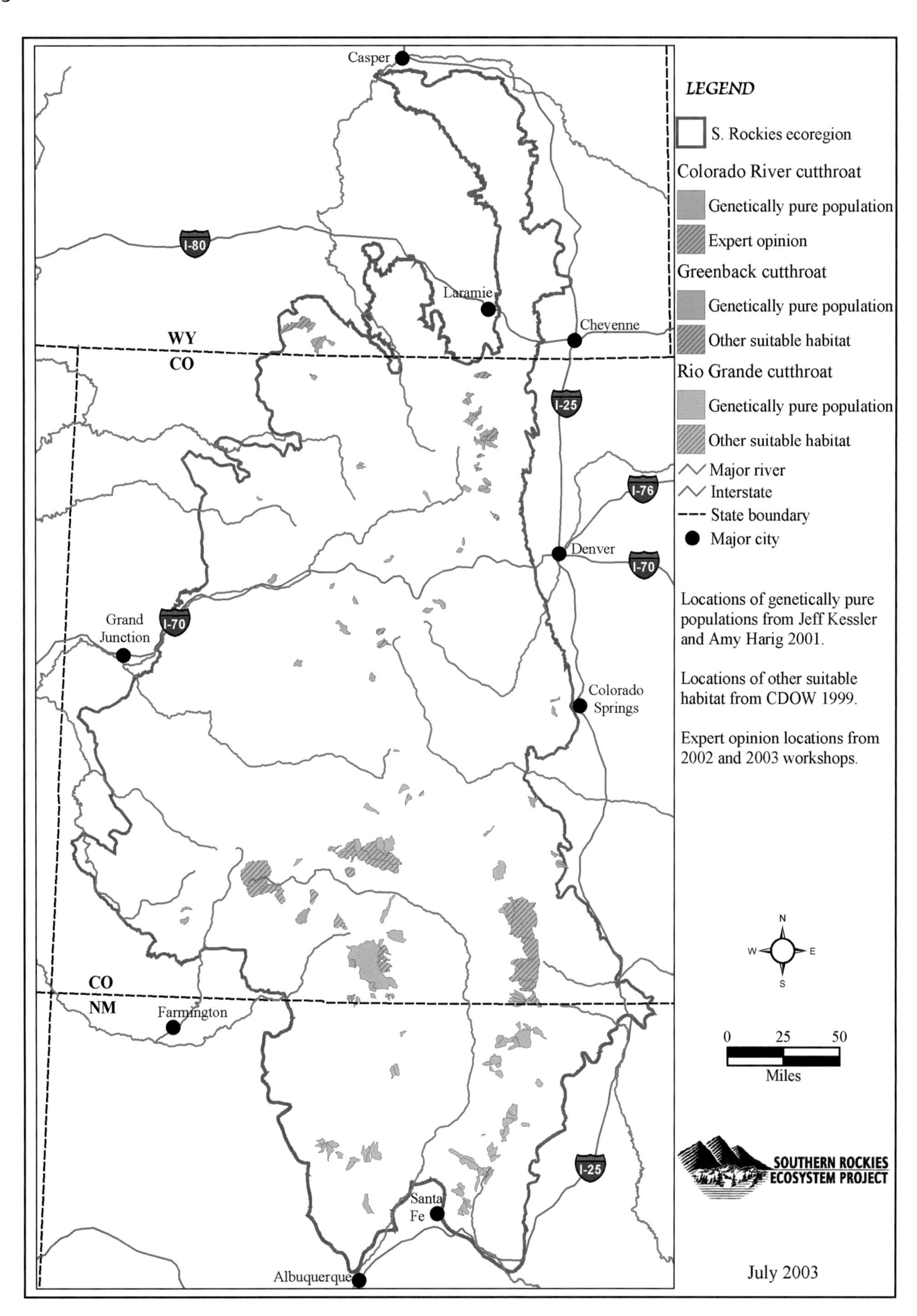

**Figure 7.4(a) Wolf primary and secondary suitable habitat in the Southern Rockies.**

**(b) Wolf suitability analysis final composite score for Colorado portion of the Greater Southern Rockies.**

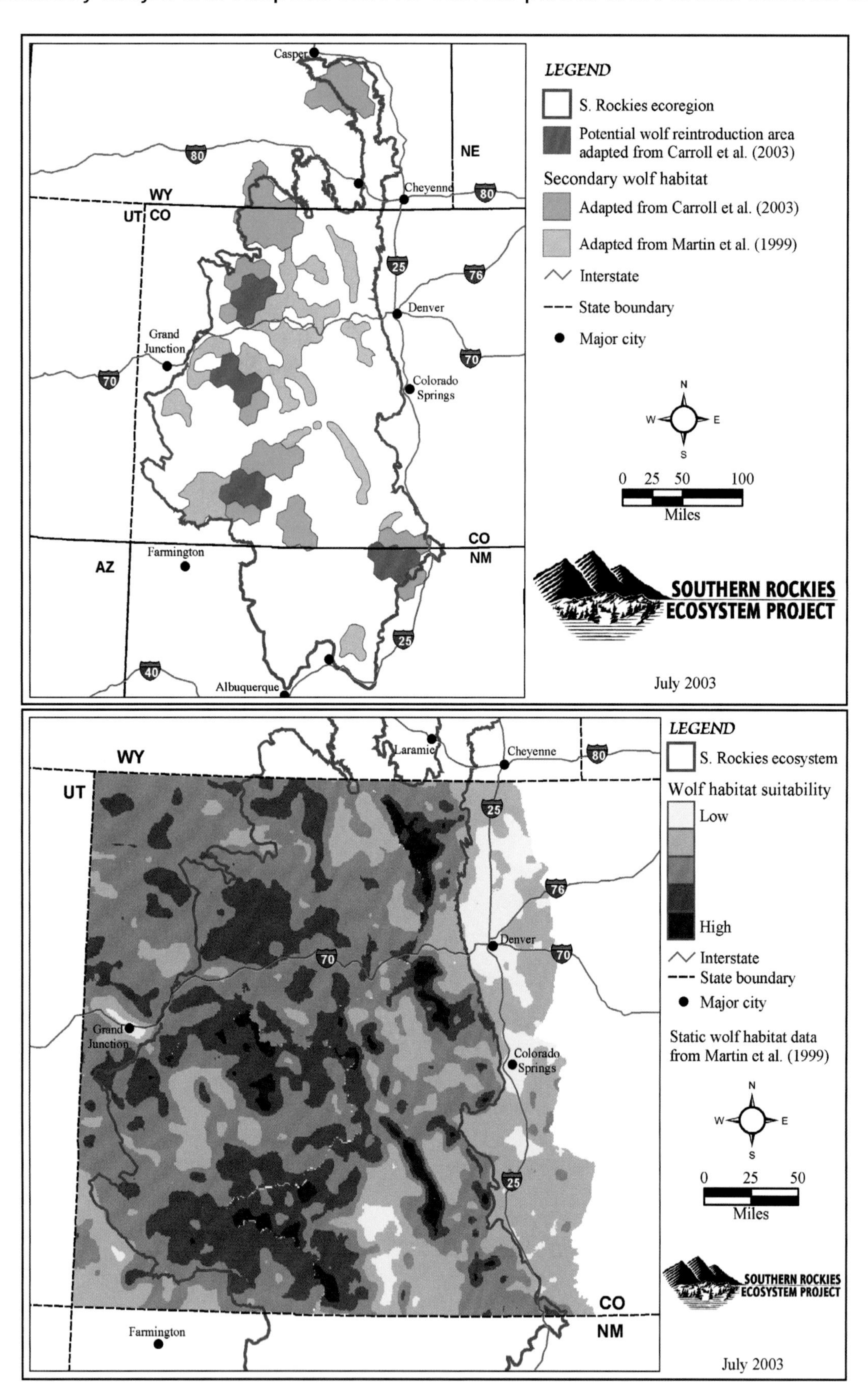

Figure 7.5 Pronghorn suitable habitat in the Southern Rockies.

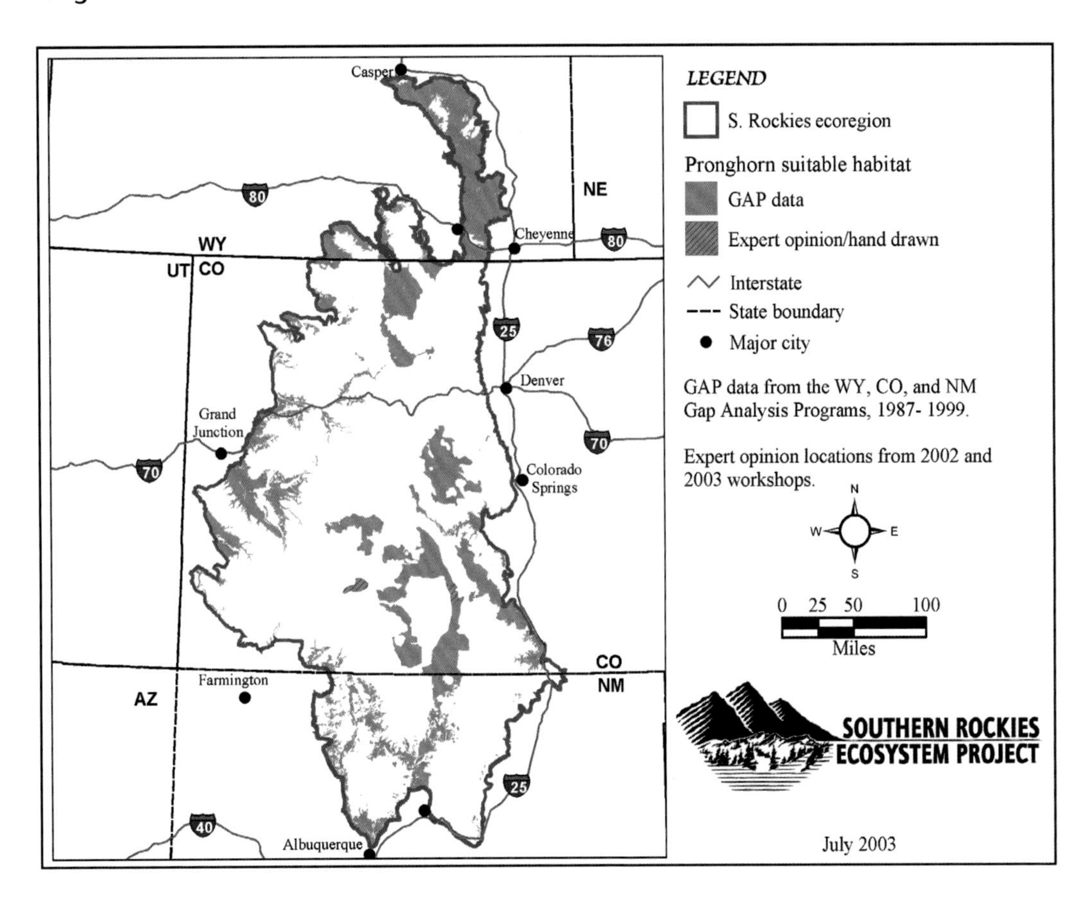


Figure 7.6 Bear core suitable habitat in the Southern Rockies.

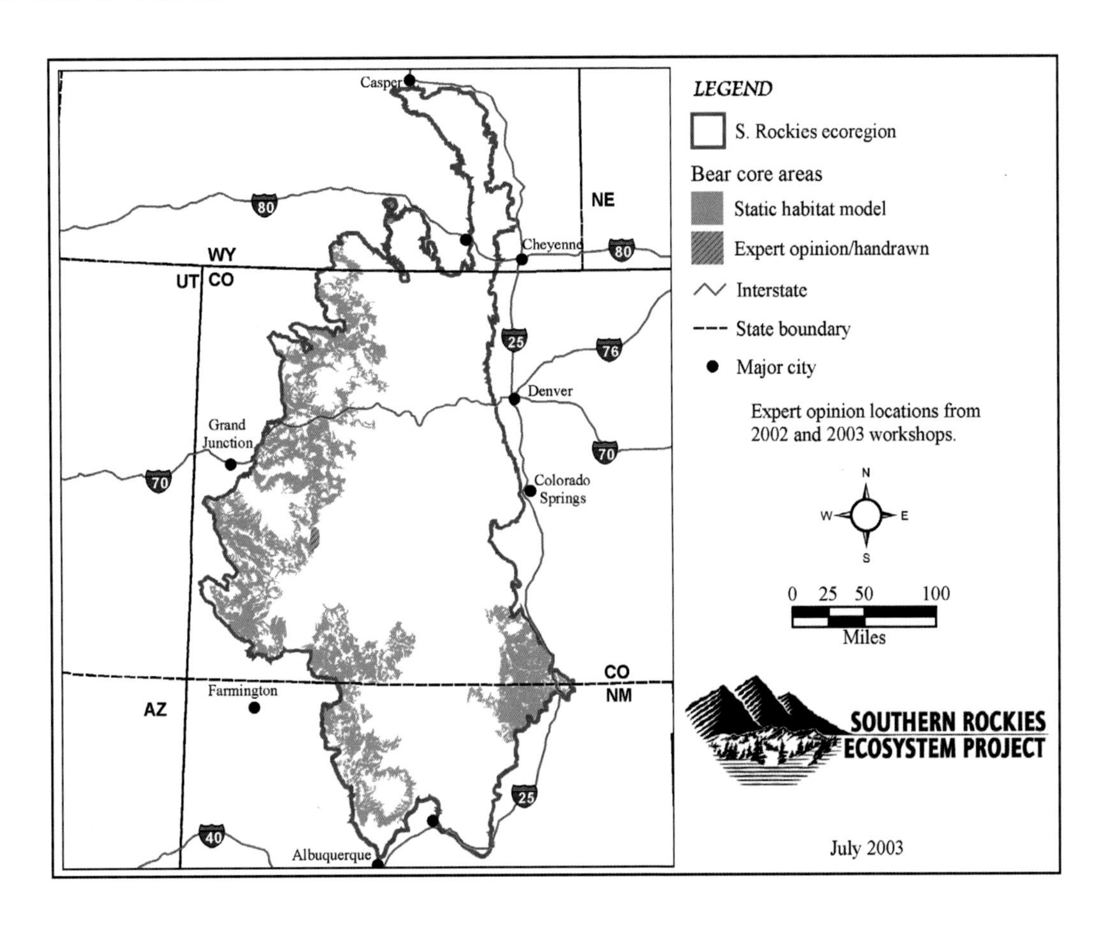

Figure 7.7 Comparison of probability function of permeability based on a weighted cost - distance function (in meters) used by SREP and Singleton et al. (2001).

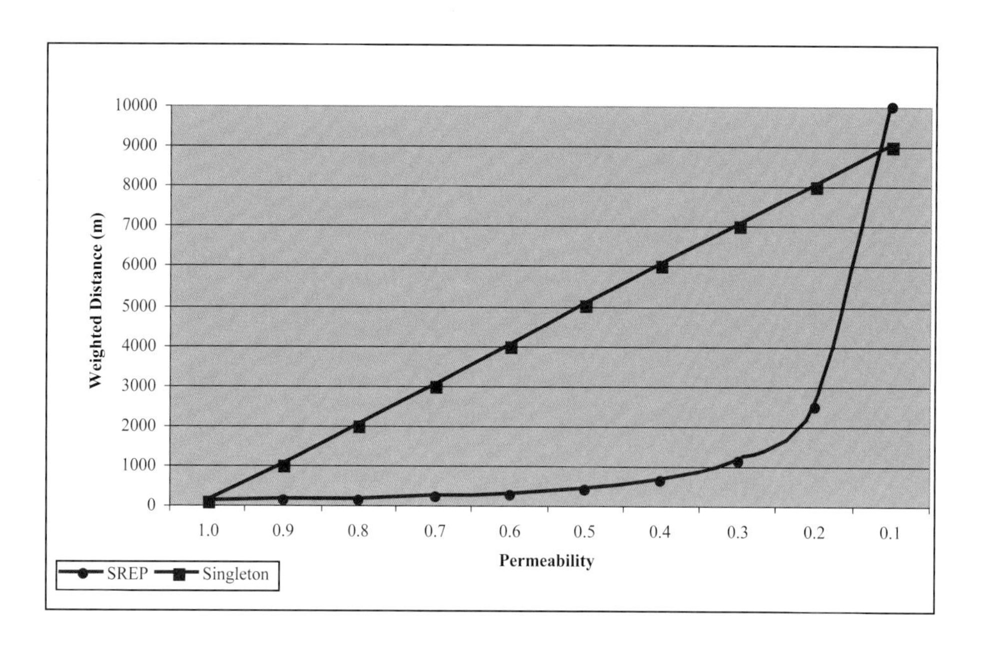


Figure 8.1 SITES final output.

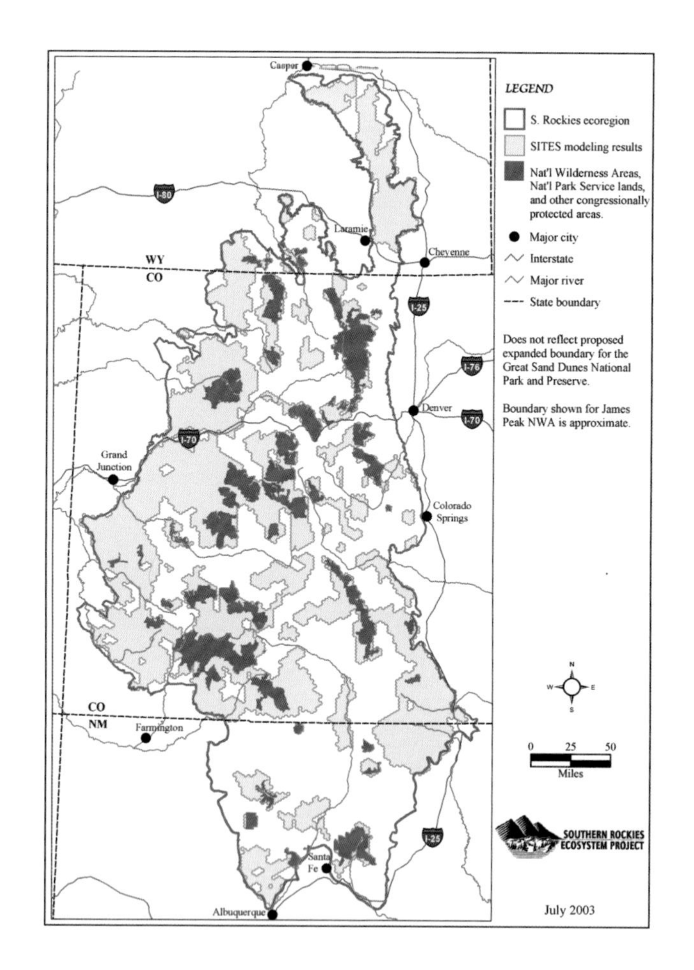

Figure 8.2 Least cost path analysis for wolves.

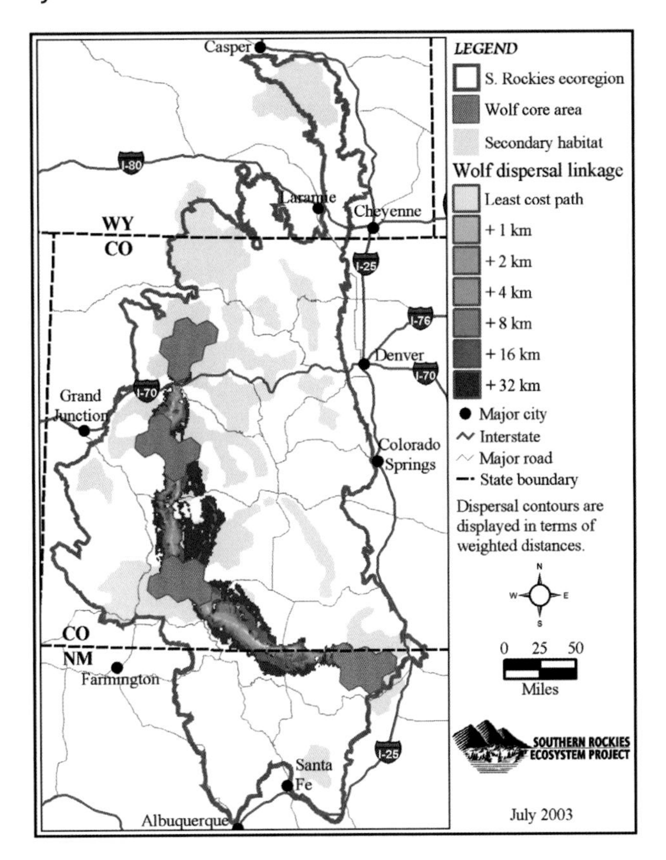


Figure 8.3 Least cost path analysis for bears.

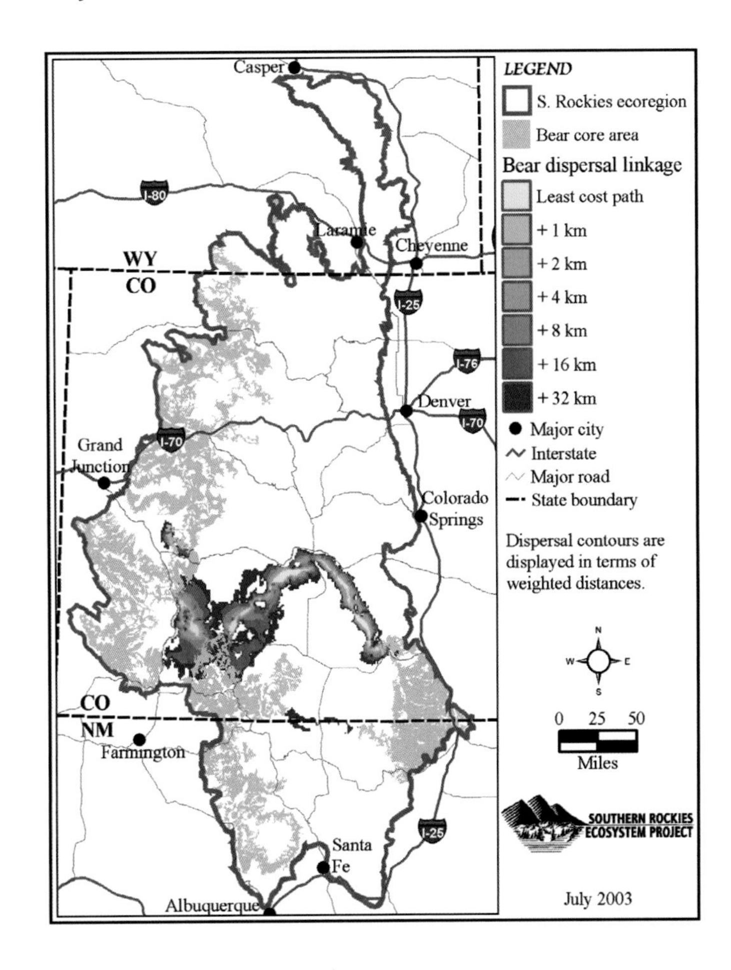

Figure 9.1 Wildlands Network Design for the Southern Rockies ecoregion.

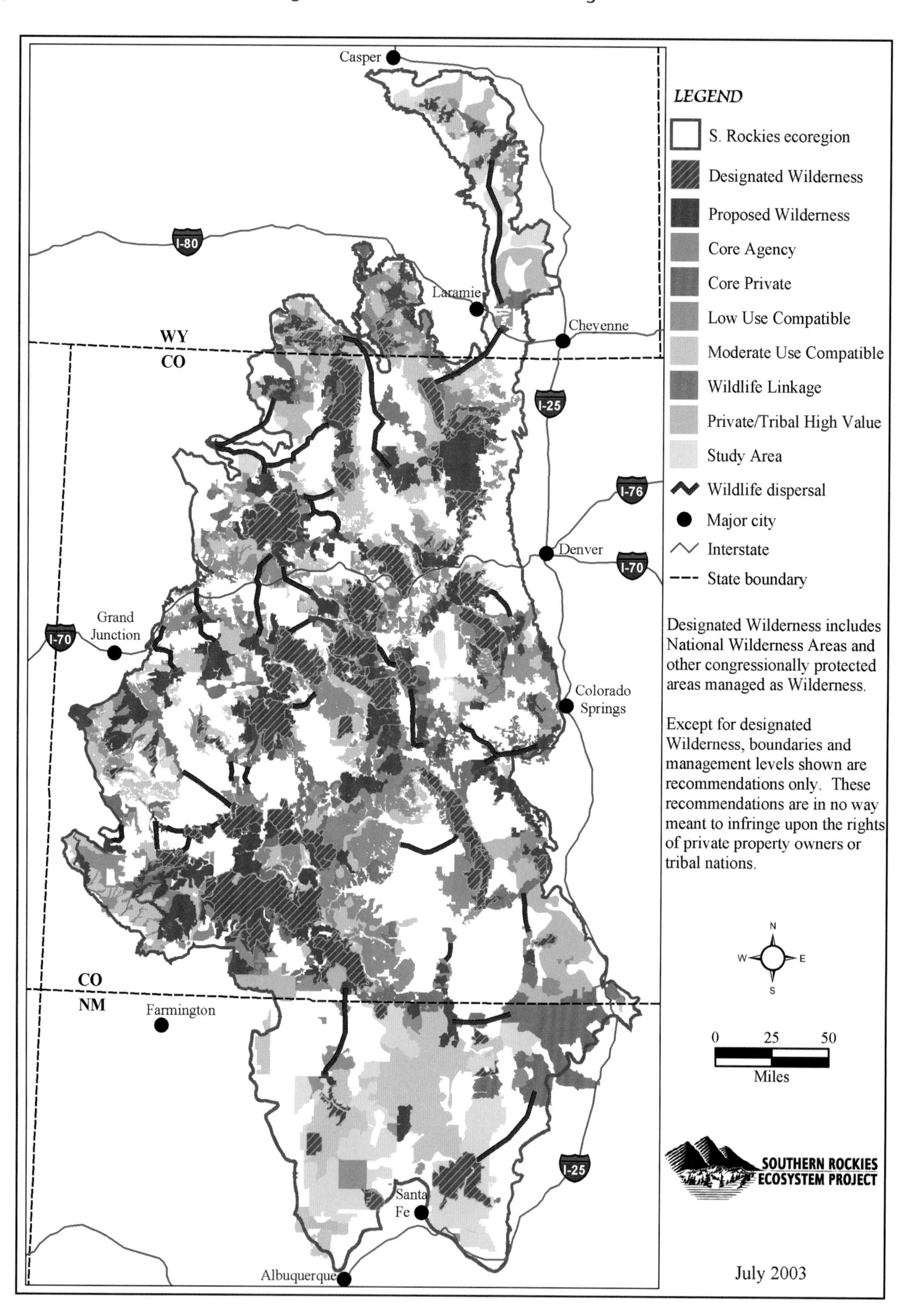

Figure 9.2 Major drainage basins of the Southern Rockies.

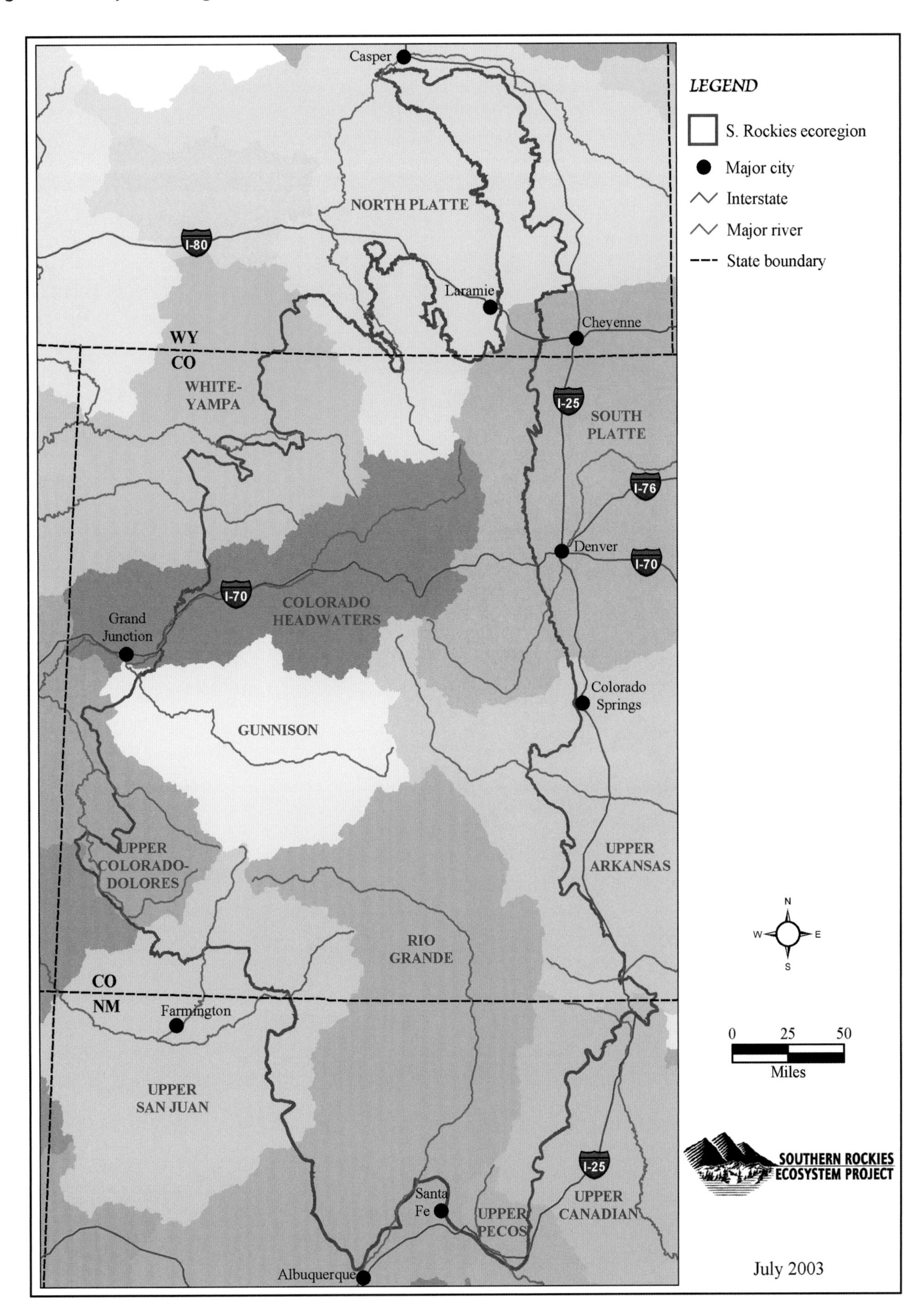

## Figure 9.3 Proportion of land ownership in the Southern Rockies.

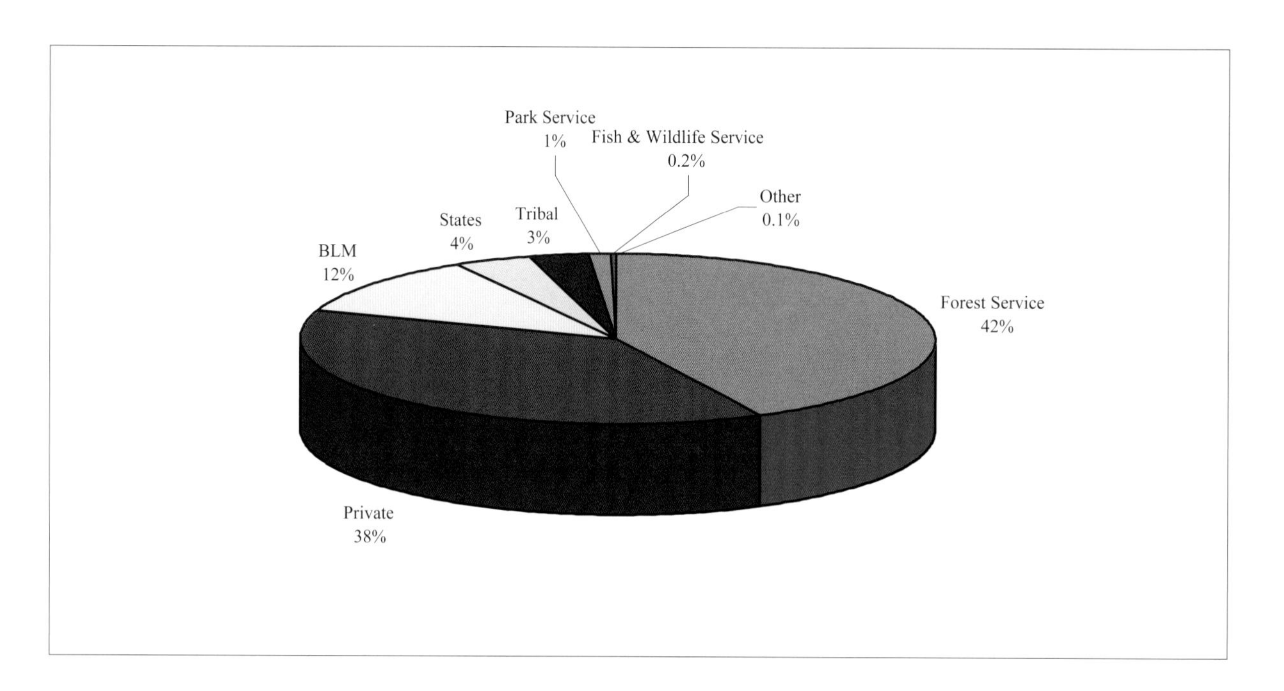


## Figure 9.4 Wilderness and other roadless areas on the Routt National Forest, Colorado.

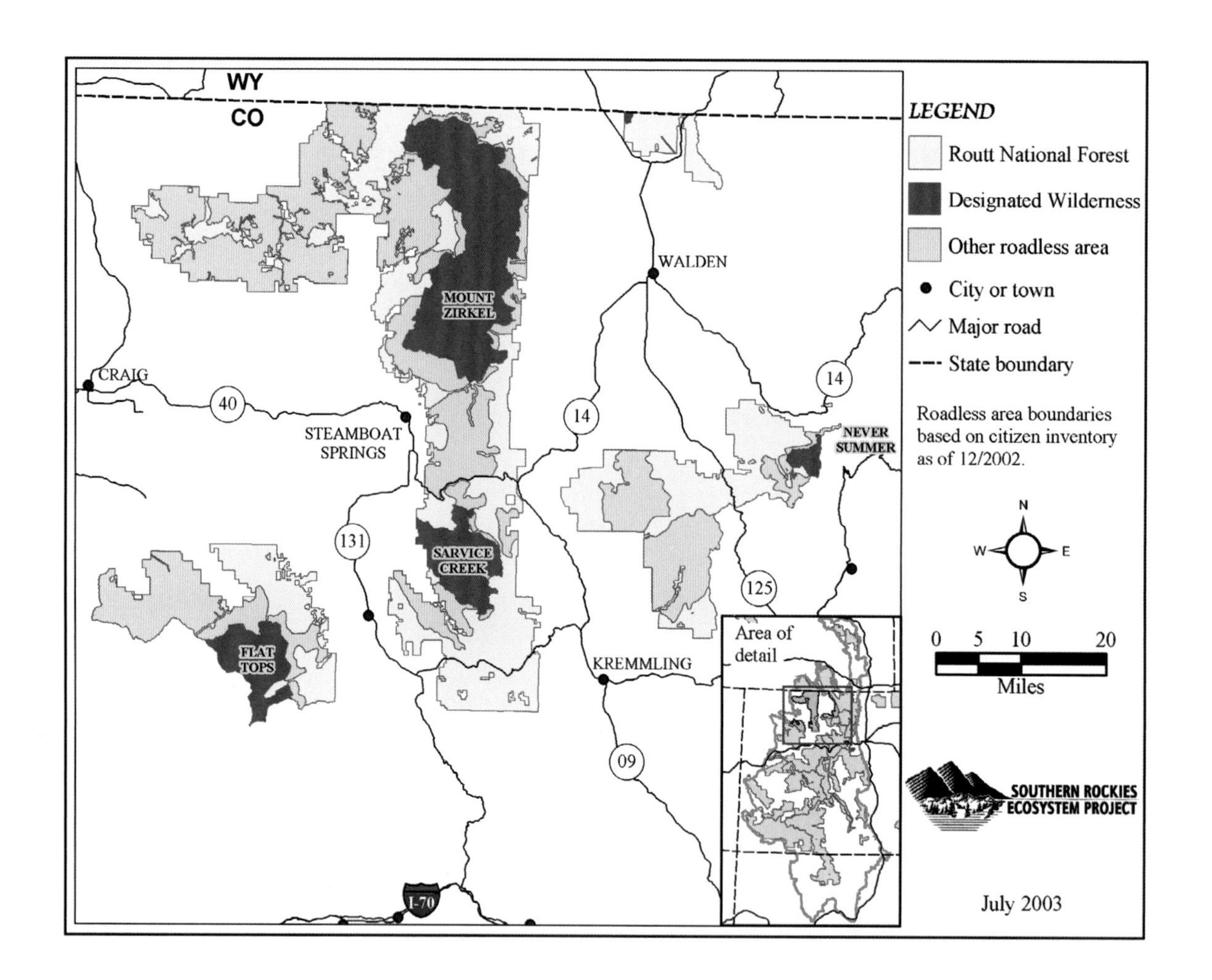

**Figure 10.1 Probable locations of wildlife crossings along the I-70 mountain corridor.**

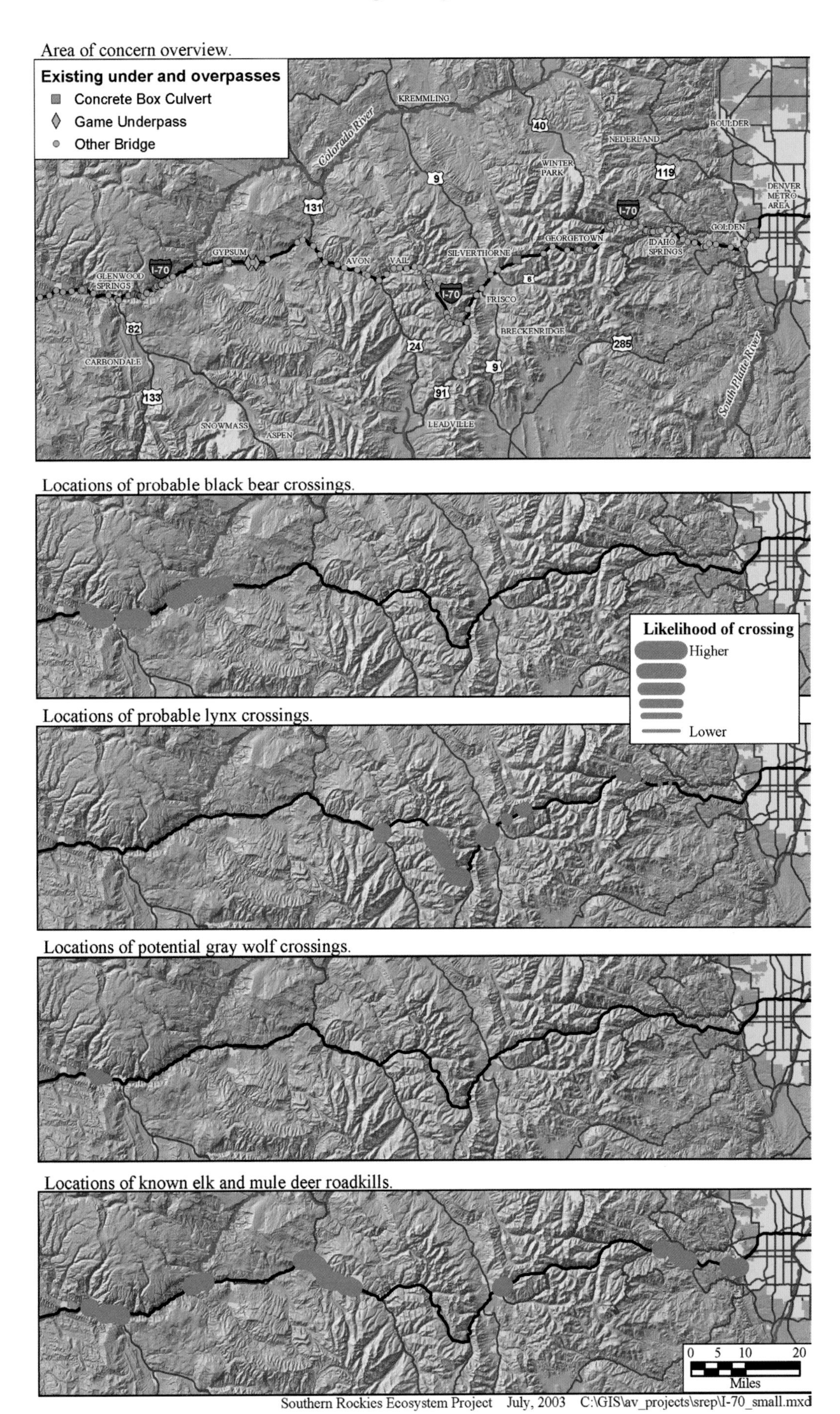

# NOTES